# BASIC ASPECTS OF THE QUANTUM THEORY OF SOLIDS

## Order and Elementary Excitations

Aimed at graduate students and researchers, this book covers the key aspects of the modern quantum theory of solids, including up-to-date ideas such as quantum fluctuations and strong electron correlations. It presents the main concepts of the modern quantum theory of solids, as well as a general description of the essential theoretical methods required when working with these systems.

Diverse topics such as the general theory of phase transitions, harmonic and anharmonic lattices, Bose condensation and superfluidity, modern aspects of magnetism including resonating valence bonds, electrons in metals, and strong electron correlations, are treated using the unifying concepts of order and elementary excitations. The main theoretical tools used to treat these problems are introduced and explained in a simple way, and their applications are demonstrated through concrete examples.

DANIEL I. KHOMSKII is a Professor in the II Physikalisches Institut, Cologne University. His main research interests are the theory of systems with strongly correlated electrons, metal–insulator transitions, magnetism, orbital ordering and superconductivity.

# BASIC ASPECTS OF THE QUANTUM THEORY OF SOLIDS

## Order and Elementary Excitations

DANIEL I. KHOMSKII

CAMBRIDGE
UNIVERSITY PRESS

**CAMBRIDGE**
**UNIVERSITY PRESS**

University Printing House, Cambridge CB2 8BS, United Kingdom

One Liberty Plaza, 20th Floor, New York, NY 10006, USA

477 Williamstown Road, Port Melbourne, VIC 3207, Australia

314-321, 3rd Floor, Plot 3, Splendor Forum, Jasola District Centre, New Delhi - 110025, India

79 Anson Road, #06-04/06, Singapore 079906

Cambridge University Press is part of the University of Cambridge.

It furthers the University's mission by disseminating knowledge in the pursuit of education, learning and research at the highest international levels of excellence.

www.cambridge.org
Information on this title: www.cambridge.org/9780521835213



First published 2010

*A catalogue record for this publication is available from the British Library*

*Library of Congress Cataloging in Publication data*
Khomskii, Daniel, 1938–
Basic aspects of the quantum theory of solids : order and elementary excitations / Daniel Khomskii.
p. cm.
Includes bibliographical references and index.
ISBN 978-0-521-83521-3
1. Solid-state physics. 2. Quantum theory. I. Title.
QC176.K47 2010
530.4′1 – dc22 2010018879

ISBN 978-0-521-83521-3 Hardback

# Contents

# Foreword and general introduction

## Foreword

There are many good books describing the foundations and basics of solid state physics, such as *Introduction to Solid State Physics* by C. Kittel (2004) or, on a somewhat higher level, *Solid State Physics* by N. W. Ashcroft and N. D. Mermin (1976). However there is a definite lack of books of a more advanced level which would describe the modern problems of solid state physics (including some theoretical methods) on a level accessible for an average graduate student or a young research worker, including experimentalists.

Usually there exists a rather wide gap between such books written for theoreticians and those for a wider audience. As a result many notions which are widely used nowadays and which determine 'the face' of modern solid state physics remain 'hidden' and are not even mentioned in the available literature for non-specialists.

The aim of the present book is to try to fill this gap by describing the basic notions of present-day condensed matter physics in a way understandable for an average physicist who is going to specialize in both experimental and theoretical solid state physics, and more generally for everyone who is going to be introduced to the exciting world of modern condensed matter physics – a subject very much alive and constantly producing new surprises.

In writing this book I tried to follow a unifying concept throughout. This concept, which is explained in more detail below, may be formulated as the connection between an *order* in a system and *elementary excitations* in it. These are the notions which play a crucial role in condensed matter physics in general and in solid state physics in particular. I hope that this general line will help the reader to see different parts of condensed matter physics as different sides of a unified picture and not as a collection of separate unrelated topics.

The plan of the book is the following. After discussing the general theory of phase transitions (Chapter 2) which forms the basis for describing *order* in solids,

I go step by step through different specific situations: systems of bosons (phonons in crystals – Chapter 4, and general Bose systems, including Bose condensation and superfluidity – Chapter 5). Then follows the important chapter on magnetism, Chapter 6 (strictly speaking dealing neither with bosons, nor with fermions), and after that we switch to the discussions of fermions – electrons in solids, Chapters 7–13. In each topic I have tried to follow the general line which I have already described above: to discuss first the type of order we have in one or the other situation, then introduce different types of elementary excitations in them, first independent excitations, but then paying most attention to the *interaction* between them and to their *quantum nature*. Thus altogether the material presented in the book is supposed to cover the main situations met in solids.

The theoretical methods used to describe these phenomena are introduced not so much separately, as such, but in the appropriate places where they are needed, and in a way which immediately shows how they work in specific problems. Thus, in studying Bose systems I introduce the widely used Bogolyubov canonical transformation, which later on is also used for treating magnons in antiferromagnets and for certain problems for electrons. Discussing spin waves, I introduce the method of equations of motion with corresponding decoupling, later on also used, e.g. for studying correlated electrons (the Hubbard model). When going to electron systems, I describe the Green function method and the Feynman diagram technique – without complete and rigorous derivations, but with the aim of demonstrating how these methods really *work* in different situations.

I hope the material covered in this book will give the reader a relatively complete picture of the main phenomena in modern solid state physics and of the main theoretical methods used to study them. But of course it is impossible to cover in one book of modest size this whole field. The most important and evident omissions are:

I do not practically touch on the broad and important field of *transport phenomena* (resistivity, thermal conductivity, thermopower, the Hall effect, etc.) This is a very big topic in itself, but it lies somewhat outside the main scope of this book. I also do not discuss specific features of such important, but well-known materials as semiconductors, ferroelectrics, etc. Also the wide field of superconductivity is touched upon only to the extent it is required to illustrate the general treatment.

Yet another relatively recent and very beautiful topic is missing – the phenomenon of the quantum Hall effect. Hopefully I can 'repair' this omission later.

On the theoretical side probably two important methods are not sufficiently discussed in the book. One is the renormalization group method used to treat complicated situations with strong interaction. I only briefly mention this method, but do not describe it in detail. Interested readers may find its description, e.g. in the books by Chaikin and Lubensky (2000) or Stanley (1987).

Another theoretical technique widely used nowadays is the use of different types of numerical calculations. This is a very broad and rapidly developing field which proved its efficiency for studying real materials and for theoretical investigations of many situations not accessible to analytical calculations. This is quite a special field, and it requires special treatment – although when appropriate I present some of the results obtained in this way.

With all these omissions, I still hope that the material which *is* included will be useful for a broad audience and will give the reader a relatively complete picture of the main phenomena and main problems in modern solid state physics.

A few words about the style of the presentation. This book has grown out of a lecture course which I gave for several years at Groningen University and at Cologne University. Therefore it still has some features of the lecture notes. I present in it all the main ideas, but often not the full derivations of corresponding results. This is also caused by the fact that the material touched upon in this book in fact covers a huge field, and it is impossible to present all the details in one book. There are many monographs and textbooks discussing in detail the separate subfields presented below. However I have tried to choose the topics and present them in such a way that the general ideas underlying modern solid state physics and the internal logic of this field become clear. For more detailed discussions of particular problems and/or corresponding methods of their theoretical treatment the readers should go to the specialized literature.

In accordance with this general concept of the book, I did not include in it a special 'problems' section. In some places, however, especially in the first part of the book, I formulate parts of the material, as Problems. Those who want to get a deeper understanding of the subject are recommended to stop reading the text at these places and try to find the answers themselves; the material presented before usually makes this task not too difficult. The answers, however, are usually given right after the problems, so that readers can also go on along the text if they do not have a desire, or time, to do these exercises themselves. Actually most of the problems, with their answers, form an integral part of the text.

In several places in the text I have also put some more special parts of the text in smaller type. These parts usually relate to more specialized (although useful) material.

In addition to the main material I have also included three very short chapters (Chapters 1, 3 and 7) with a short summary of some of the basic facts from statistical mechanics. I think it would be useful for readers to have this information at hand.

Some important notions are mentioned several times in different parts of the text. I did this intentionally, so that different chapters would become somewhat more independent – although of course there are a lot of cross-references in the text.

I hope that the book gives a coherent presentation of the main ideas and methods of the quantum theory of solids.

There are many good books which cover parts of the material contained in the present book (and actually much more!). I can recommend the following, already classical books:

1. J. M. Ziman (1979), *Principles of the Theory of Solids*. A very good and clear book covering the main topics in solid state physics. Highly recommended. However, it does not contain more modern methods.
2. N. W. Ashcroft and N. D. Mermin (1976), *Solid State Physics*. Also a very good and widely used book, covering the topics in more detail, on a somewhat more elementary level. Very transparent and useful.
3. M. P. Marder (2000), *Condensed Matter Physics*. A rather complete book describing the main phenomena in solid state physics, but not going into much theoretical detail.
4. C. Kittel (1987), *Quantum Theory of Solids*. Contains detailed discussion of many problems in quantum theory, using more modern methods such as diagram techniques. Somewhat more theoretical.
5. G. D. Mahan (2000), *Many-Particle Physics*. Gives a very complete treatment of the topics discussed; it is a kind of 'encyclopedia'. It uses the Green function method all the way through. Very useful for theoreticians, and contains all the necessary details and derivations, etc. However not all topics are discussed there.
6. L. D. Landau and I. M. Lifshits, *Course of Theoretical Physics*, especially *Statistical Physics* (old one-volume edition 1969, or new edition v. I 1980), and *Quantum Mechanics* (1977). These classical books contain virtually all the basic material necessary, and many particular topics important for our problems. If one can call the book by Mahan an encyclopedia, then the course of Landau and Lifshits is a 'bible' for all theoreticians, especially those working in condensed matter physics. But these books are very useful not just for theoreticians, but for everyone looking for clear and precise description of all the basic ideas of theoretical physics.
7. J. R. Schrieffer (1999), *Theory of Superconductivity*. A very clear book; contains in particular a very good and condensed treatment of the Green function method and diagram technique, in a form used now by most theoreticians.
8. A. A. Abrikosov, L. P. Gor'kov and E. Dzyaloshinsky (1975), *Methods of the Quantum Field Theory in Statistical Physics*. One of the first (and still the best) books on the modern methods applied to condensed matter physics. It gives a very detailed treatment of the theoretical methods and a good discussion of specific problems (Fermi and Bose liquids; plasma; electron–phonon interaction and the basics of the theoretical treatment of superconductivity).

9. P. M. Chaikin and T. C. Lubensky (2000), *Principles of Condensed Matter Physics*. A very good book containing in particular detailed discussion of different questions connected with phase transitions. The accent is on general statistical mechanics; specifically it contains a lot of material on soft condensed matter physics, but does not discuss such topics as electrons in metals, magnetism, etc.

Some other references will be given later, in the main body of the book. But, keeping in mind the character of the book (which is practically expanded lecture notes and which still retain that character), I deliberately refrained from including too many references – it would be simply impossible to cite all the relevant works. Therefore I mostly refer not to original publications but rather to monographs and review papers. Interested readers may find more detailed information on particular subjects in these references.

## General introduction

The unifying concept in this book is the concept of *order* and *elementary excitations*; these are the key words.

One can argue as follows. In general in macroscopic systems with many degrees of freedom the internal state, or internal motion on the microscopic scale, is random. However as $T \to 0$ the entropy of the system should go to zero, $S \to 0$; this is the well-known Nernst theorem, or the third law of thermodynamics. Accordingly, at $T = 0$ there should exist *perfect order* of some kind.

Such ordering sets in at a certain characteristic temperature $T^*$, often with a phase transition, but not necessarily.

At $T \ll T^*$ we can describe the state of the system as predominantly ordered, or maybe in a weakly excited state. Such relatively weakly excited states will be thermally excited, but can appear also due to small external perturbations. Usually in such a state we can speak of a small number of *elementary excitations*, or *quasiparticles*. Examples are, e.g. *phonons* in crystals, *magnons* or *spin waves* in ferromagnets, *excitons* in semiconductors, etc.

Sometimes such elementary excitations are rather strange: instead of electrons they may be excitations with spin, but no charge (*spinons*), or vice versa (*holons*). There exist also *topological* excitations (solitons, vortices, etc.).

In a first approximation we can consider these excitations as *noninteracting*. Such are, e.g. phonons in a harmonic crystal, etc. However in the next step we have to include in general an *interaction* between quasiparticles.

There may exist interactions between the same quasiparticles. They lead, e.g. to anharmonic effects in crystals (phonon–phonon interactions); they are included in the Landau Fermi-liquid theory, and give rise to screening for electrons in metals; the magnon–magnon interaction can lead, e.g. to the formation of bound states of magnons, etc. There also exist interactions between different quasiparticles: electron–phonon interactions, pairing in conventional superconductors, and interactions between many other elementary excitations. Often due to these interactions the properties of quasiparticles are strongly changed, or renormalized: an example is the formation of polarons (electron + strong lattice distortion). Also new quasiparticles may be formed (plasmons due to the Coulomb interaction of electrons; excitons – bound states of electrons and holes in semiconductors). Even the ground state itself, the very type or ordering, may change because of such interactions. An example is the superconducting state instead of the normal state of a metal.

It is important that these quasiparticles, or elementary excitations, are *quantum* objects. Consequently, one should not visualize the order as completely classical: there are *quantum fluctuations* (zero-point motion, or zero-point oscillations) even at $T = 0$. Sometimes they lead only to minor numerical changes, but there are cases, especially in low-dimensional or frustrated systems, when they can completely modify the properties of a system, e.g. destroying the long-range order totally. They can also modify the properties of the phase transitions themselves, e.g. leading to *quantum phase transitions.* Thus the classical picture is always very useful, but one should be cautious and aware of its possible failures in some cases – but very often these cases are the most interesting!

In treating these problems a lot of different approaches were used, and different methods developed. These methods are often used not only in solid state physics or condensed matter physics in general; many of them are also widely used (and often have been developed!) in other parts of physics: in elementary particle physics, in field theory, and in nuclear physics. Methods such as the Green function method and Feynman diagrams were introduced in field theory, but are now widely used in condensed matter physics. On the other hand, some methods and concepts which first appeared in solid state physics (the mean field, or self-consistent field method, the concept of a phase transition) are now used in nuclear physics, in elementary particle physics (e.g. quark–gluon plasma), and even in cosmology.

In this book I will try to describe the main concepts and ideas used in modern many-particle physics, and the methods used to study these problems, such as the equation of motion method, canonical transformations, diagram techniques, and the Green function method. Once again, the key words will be *elementary excitations*, in connection with *order*. The illustrations will be predominantly given for the examples of solid state systems, although, as I said before, many concepts, notions and methods are also applicable in different fields, and even not only in physics! Thus, some of the ideas of many-particle physics are now widely used for treating problems in biology, and even economics and sociology. I hope that such a general view will help readers to form a unified concept of the main phenomena in condensed matter physics and related fields.

# 1

# Some basic notions of classical and quantum statistical physics

## 1.1 Gibbs distribution function and partition function

In this short chapter some of the basic notions from thermodynamics and statistical physics are summarized.

The probability to observe a state $|n\rangle$ with energy $E_n$ is

$$w_n = Ae^{-E_n/T} \; ; \tag{1.1}$$

this is called the Gibbs distribution. (Here and below we put the Boltzmann constant $k_{\mathrm{B}} = 1$, i.e. the temperature is measured in units of energy, and vice versa.) The normalization constant $A$ is determined by the condition that the total sum of probabilities of all states is 1:

$$\sum_n w_n = 1 \; , \tag{1.2}$$

from which we find

$$\frac{1}{A} = \sum_n e^{-E_n/T} \equiv Z \; . \tag{1.3}$$

Here $Z$ is the partition function

$$Z = \sum_n e^{-E_n/T} = \mathrm{Tr}\left(e^{-\hat{\mathcal{H}}/T}\right) \; , \tag{1.4}$$

where $\hat{\mathcal{H}}$ is the Hamiltonian of the system. Thus

$$w_n = \frac{e^{-E_n/T}}{Z} \; . \tag{1.5}$$

The entropy is defined as

$$S = -\langle \ln w_n \rangle = -\frac{\sum_n \ln w_n \, e^{-E_n/T}}{Z} \tag{1.6}$$

($\langle \ldots \rangle$ is the symbol for the average). When we put (1.5) into (1.6), we obtain

$$S = \ln Z + \frac{E}{T} , \tag{1.7}$$

where $E$ is the average energy of the system, $E = \frac{1}{Z}\sum_n E_n e^{-E_n/T}$. We can introduce the quantity

$$F = E - TS = -T \ln Z , \tag{1.8}$$

which is called the (Helmholtz) free energy:

$$F = -T \ln Z = -T \ln \sum_n e^{-E_n/T} . \tag{1.9}$$

## 1.2 Thermodynamic functions

The Helmholtz free energy, $F$, is a function of the temperature $T$ and of the density $n = N/V$, or of the volume: $F = F(V, T)$. One can also introduce other so-called thermodynamic potentials, expressed as functions of different variables. These are:

At fixed pressure and temperature – the Gibbs free energy

$$\Phi(P, T) = E - TS + PV = F + PV . \tag{1.10}$$

If instead of the temperature $T$ we chose as free variable its conjugate, the entropy, then we obtain the enthalpy

$$W(P, S) = E + PV . \tag{1.11}$$

Enthalpy is often used in discussions of chemical reactions, thermodynamics of formation of different phases, etc.

The energy itself is also one of the thermodynamic potentials; it is a function of volume and entropy, $E(V, S)$.

Similar to mechanics, where the system at equilibrium tends to a state with mimimum energy, many-particle systems at finite temperature tend to minimize the free energy, i.e. the corresponding thermodynamic potential $F$ or $\Phi$.

From these definitions it is clear that, e.g.

$$dF = -S\,dT - P\,dV , \tag{1.12}$$

from which we obtain

$$S = -\left(\frac{\partial F}{\partial T}\right)_V , \tag{1.13}$$

$$P = -\left(\frac{\partial F}{\partial V}\right)_T . \tag{1.14}$$

Similarly

$$d\Phi = -S\,dT + V\,dP , \tag{1.15}$$

$$S = -\left(\frac{\partial \Phi}{\partial T}\right)_P , \tag{1.16}$$

$$V = \left(\frac{\partial \Phi}{\partial P}\right)_T . \tag{1.17}$$

Other useful thermodynamic quantities are, e.g. the specific heat at constant volume, $c_V$, and at constant pressure, $c_P$:

$$c_V = \left(\frac{\partial E}{\partial T}\right)_V = T\left(\frac{\partial S}{\partial T}\right)_V , \tag{1.18}$$

$$c_P = \left(\frac{\partial W}{\partial T}\right)_P = T\left(\frac{\partial S}{\partial T}\right)_P . \tag{1.19}$$

One can express $c_P$, $c_V$ through $F$, $\Phi$, using (1.12), (1.15).

Using the expressions given above, one can obtain useful relations between different thermodynamic quantities, e.g. between the specific heat, the thermal expansion coefficient (the volume coefficient of the thermal expansion $\beta = 3\alpha$, where $\alpha$ is the linear thermal expansion)

$$\beta = +\frac{1}{V}\frac{\partial V}{\partial T} , \tag{1.20}$$

and the compressibility

$$\kappa = -\frac{1}{V}\frac{\partial V}{\partial P} . \tag{1.21}$$

The resulting connection has the form (see, e.g. Landau and Lifshits 1980, Section 16):

$$c_P - c_V = -T\,\frac{(\partial V/\partial T)_P^2}{(\partial V/\partial P)_T} = VT\,\frac{\beta^2}{\kappa} . \tag{1.22}$$

Similarly one can also find relations between other thermodynamic quantities; some examples will be given below, especially in Chapter 2.

## 1.3 Systems with variable number of particles; grand partition function

One can also introduce thermodynamic quantities for systems with variable number of particles $N$. The thermodynamic potentials introduced above depend on the particle density $N/V$, i.e.

$$\begin{aligned} F &= N\, f_1\left(\frac{V}{N}, T\right) \\ \Phi &= N\, f_2(P, T) \\ E &= N\, f_3\left(\frac{S}{N}, \frac{V}{N}\right) . \end{aligned} \tag{1.23}$$

From these equations we get:

$$\begin{aligned} dF &= -S\,dT - P\,dV + \mu\,dN \\ d\Phi &= -S\,dT + V\,dP + \mu\,dN \\ dE &= \quad T\,dS - P\,dV + \mu\,dN . \end{aligned} \tag{1.24}$$

Here we have introduced the chemical potential $\mu$ which is defined by

$$\mu = \left(\frac{\partial E}{\partial N}\right)_{S,V} = \left(\frac{\partial F}{\partial N}\right)_{T,N} = \left(\frac{\partial \Phi}{\partial N}\right)_{P,T} . \tag{1.25}$$

From (1.25) and (1.23) we obtain

$$\mu = \frac{\Phi}{N} , \tag{1.26}$$

i.e. the chemical potential is the Gibbs free energy per particle.

One important remark is relevant here. If the number of (quasi)particles $N$ is not conserved, such as for example the number of phonons in a crystal, then the value of $N$ is determined by the condition of minimization of the free energy in $N$, e.g. $\partial F/\partial N = 0$, etc. One sees then that in such cases the chemical potential is $\mu = 0$. This fact will be used in several places later on.

The chemical potential $\mu$ and the number of particles $N$ are conjugate variables (like $T$ and $S$; $P$ and $V$). One can introduce a new thermodynamic potential with $\mu$ as a variable; it is usually denoted $\Omega(V, T, \mu)$. Using equations (1.3), (1.8) we can write down the distribution function (1.1) as

$$w_n = \frac{e^{-E_n/T}}{Z} = \exp\left(\frac{F - E_n}{T}\right) . \tag{1.27}$$

For a variable particle number $N$, it takes the form

$$w_{nN} = \exp\left(\frac{\Omega + \mu N - E_{nN}}{T}\right) , \tag{1.28}$$

where we have used this new thermodynamic potential $\Omega$, instead of the free energy:

$$\Omega(V, T, \mu) = F - \mu N \, . \tag{1.29}$$

Thus $\Omega$ is a generalization of the free energy to the case of variable number of particles. Similar to (1.24), we have:

$$d\Omega = -S\,dT - P\,dV - N\,d\mu \, , \tag{1.30}$$

i.e. the total number of particles is connected to the chemical potential by the relation

$$N = -\left(\frac{\partial \Omega}{\partial \mu}\right)_{T,V} \, . \tag{1.31}$$

**Problem:** One can show that $\Omega = -PV$; try to prove this.

**Solution:** From (1.29) $\Omega = F - \mu N$. But, by (1.26), $\mu N = \Phi$, and, by (1.10), $\Phi = F + PV$. Thus $\Omega = F - \mu N = F - \Phi = -PV$.

Analogously to (1.5), (1.9), we can write down

$$\Omega = -T \ln Z_{\mathrm{Gr}} \, , \tag{1.32}$$

where $Z_{\mathrm{Gr}}$ is called *grand partition function*:

$$Z_{\mathrm{Gr}} = \sum_N \left( e^{\mu N/T} \sum_n e^{-E_{nN}/T} \right) . \tag{1.33}$$

# 2

# General theory of phase transitions

The state of different condensed matter systems is characrerized by different quantities: density, symmetry of a crystal, magnetization, electric polarization, etc. Many such states can have a certain ordering. Different types of ordering can be characterized by *order parameters*.

Examples of order parameters are, for instance: for ferromagnets – the magnetization $M$; for ferroelectrics – the polarization $P$; for structural phase transitions – the distortion $u_{\alpha\beta}$, etc. Typically the system is disordered at high temperatures, and certain types of ordering may appear with decreasing temperature. This is clear already from the general expressions for thermodynamic functions, see Chapter 1: at finite temperatures the state of the system is chosen by the condition of the minimum of the corresponding thermodynamic potential, the Helmholtz free energy (1.8) or the Gibbs free energy (1.10), and from those expressions it is clear that with increasing temperature it is favourable to have the highest entropy possible, i.e. a disordered state. But some types of ordering are usually established at lower temperatures, where the entropy does not play such an important role, and the minimum of the energy is reached by establishing that ordering.

The general order parameter $\eta$ depends on temperature, and in principle also on other external parameters – pressure, magnetic field, etc. Typical cases of the dependence of the order parameter on temperature are shown in Fig. 2.1. The situation shown in Fig. 2.1($a$), where the order parameter changes continuously, is called a *second-order phase transition*, and that shown in Fig. 2.1($b$), where $\eta$ changes in a jump-like fashion, is a *first-order phase transition*. The temperature $T_c$ below which there exists order in a system ($\eta \neq 0$) is called the *critical temperature* (sometimes the Curie temperature, the notion coming from the field of magnetism).

## 2.1 Second-order phase transitions (Landau theory)

For the second-order phase transitions close to $T_c$ the order parameter $\eta$ is small, and we can expand the (Gibbs) free energy $\Phi(P, T, \eta)$ in a Taylor series. This

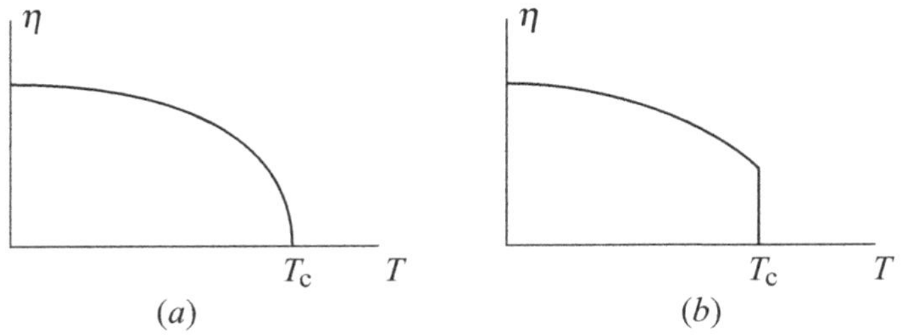

(a) (b)

Fig. 2.1

approach was first developed by Landau, and in this section we largely follow the classical presentation of Landau and Lifshits (1980).

The expansion of the free energy in small $\eta$ is, in general,

$$\Phi = \Phi_0 + \alpha\eta + A\eta^2 + C\eta^3 + B\eta^4 + \cdots \ . \tag{2.1}$$

(It will be clear below why we have chosen such an 'unnatural' notation with the sequence of coefficients $A$, $C$, $B$.) As mentioned above, the state of the system, in particular the value of the order parameter $\eta$ (magnetization, or spontaneous polarization, or distortion, etc.) is determined by the condition that the free energy, in this case $\Phi$, has a minimum. The coefficients $\alpha$, $A$, $C$, $B$ are functions of $P$, $T$ such that the minimum of $\Phi(P, T, \eta)$ as a function of $\eta$ should correspond to $\eta = 0$ above $T_c$ (disordered state), and to $\eta \neq 0$ (and small) below $T_c$. From this requirement it is clear that the coefficient $\alpha$ in a system without external fields should be $\alpha = 0$, otherwise $\eta \neq 0$ at all temperatures: in the presence of the linear term in (2.1) the free energy would never have a minimum at $\eta = 0$, which should be the case in a disordered system at $T > T_c$.

The same requirement that $\eta = 0$ above $T_c$, but $\eta \neq 0$ for $T < T_c$, leads to the requirement that the first nonzero term $A\eta^2$ in the expansion (2.1) should obey the condition

$$\begin{aligned} A(P, T) &> 0 \quad \text{for } T > T_c \\ A(P, T) &< 0 \quad \text{for } T < T_c \ . \end{aligned} \tag{2.2}$$

As a result the dependence of $\Phi(\eta)$ would have the form shown in Fig. 2.2.

Thus at the critical temperature $T_c$ the coefficient $A(P, T)$ should pass through zero and change sign. (We assume that it changes continuously with temperature. We also assume that the other coefficients in equation (2.1) are such that $C = 0$, which is often the case, see Section 2.2 below, and $B > 0$.) Again, making the simplest assumption, we can write close to $T_c$:

$$A(P, T) = a\,(T - T_c)\ , \tag{2.3}$$

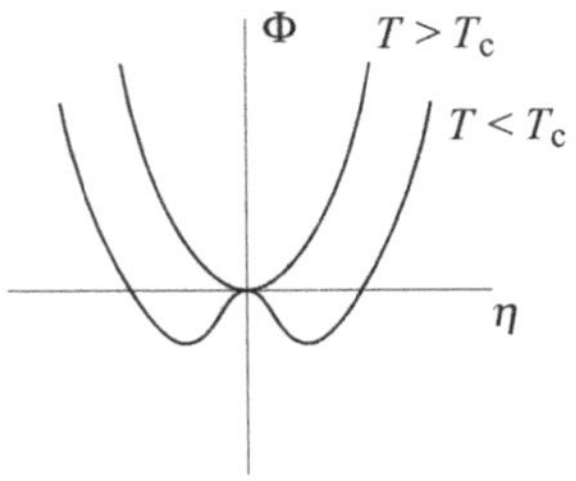


Fig. 2.2

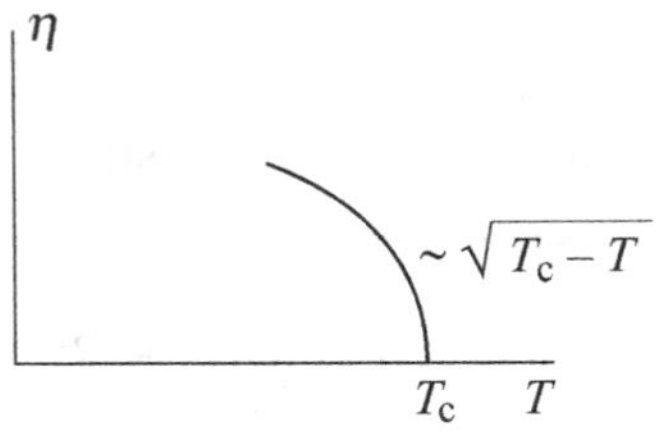


Fig. 2.3

with the coefficient $a > 0$. Then

$$\boxed{\Phi = \Phi_0 + A\eta^2 + B\eta^4 = \Phi_0 + a\,(T - T_c)\,\eta^2 + B\eta^4} \,. \tag{2.4}$$

The behaviour of $\eta(T)$ can be easily found from (2.4) by minimizing the free energy with respect to $\eta$:

$$\frac{\partial \Phi}{\partial \eta} = 0 \implies 2A\eta + 4B\eta^3 = 2a(T - T_c)\eta + 4B\eta^3 = 0\,, \tag{2.5}$$

$$\boxed{\eta^2 = -\frac{A}{2B} = \frac{a}{2B}(T_c - T)} \,. \tag{2.6}$$

This behaviour is shown in Fig. 2.3.

Here in principle all coefficients may be functions of pressure (or other external variables), $a = a(P)$, $B = B(P)$, $T_c = T_c(P)$. But in practice the dependence of $T_c(P)$ is the most important one; the coefficients $a$ and $B$ can usually be taken as constants.

The equilibrium free energy itself at $T < T_c$ is obtained by putting the equilibrium value of the order parameter (2.6) back into the free energy (2.4):

$$\Phi_{\min} = \Phi_0 - \frac{A^2}{4B} = \Phi_0 - \frac{a^2}{4B}(T_c - T)^2 \tag{2.7}$$

(and $\Phi = \Phi_0$ for $T > T_c$). Thus $\Phi$ (and other thermodynamic potentials – e.g. the Helmholtz free energy $F$ if we work at fixed volume $V$ and have a second-order phase transition) are continuous, see Fig. 2.4($a$). However the derivatives $(\partial\Phi/\partial T)$,

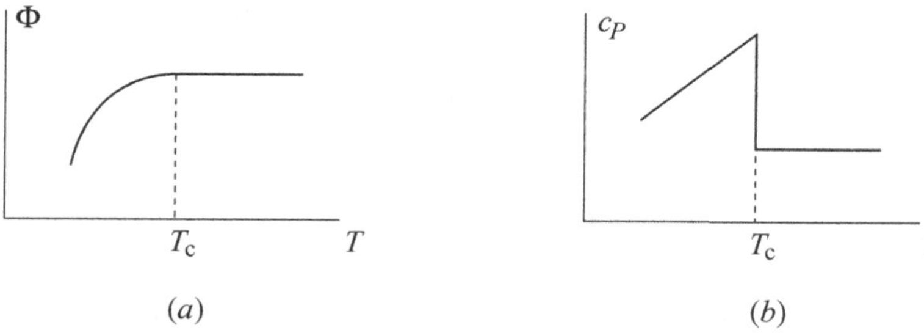


Fig. 2.4

etc. have *kinks* at $T_c$, and the second derivative would have *jumps*; this is typical behaviour of thermodynamic functions at the second-order phase transitions in the Landau theory.

**Problem:** Using the definition of specific heat $c_P$ (1.19), equations (1.16), (2.6), (2.7), find the behaviour of specific heat at the second-order phase transition.

**Solution:** The entropy $S$, by (1.16) and (2.7), is

$$S = -\frac{\partial \Phi}{\partial T} = \begin{cases} S_0 & (T > T_c) \\ S_0 - \frac{a^2}{2B}(T_c - T) & (T < T_c) \end{cases} \tag{2.8}$$

and, by (1.19), the specific heat is

$$c_P = T\left(\frac{\partial S}{\partial T}\right)_P = \begin{cases} 0 & (T > T_c) \\ a^2 T/2B & (T < T_c)\,. \end{cases} \tag{2.9}$$

Note that this expression is valid only close to $T_c$; at lower temperatures the specific heat may and will deviate from this simple linear behaviour.

At $T_c$ the entropy has a kink, see (2.8), and there exists at $T_c$ a jump in the specific heat at the second-order phase transition:

$$\Delta c_P = \frac{a^2 T_c}{2B}\,. \tag{2.10}$$

This behaviour is shown in Fig. 2.4(*b*).

The total entropy connected with the ordering is

$$S_{\text{ord}} = \int_0^{T_c} \frac{1}{T}\, c_P(T)\, dT\,. \tag{2.11}$$

The experimental measurements of specific heat and of the total entropy of the transition give very important information: the observation of the behaviour of $c_P$ of the type shown in Fig. 2.4 proves that we are dealing with a second-order phase transition (see however Section 2.5 later), and the measurement of the total entropy of the transition (part of the total entropy, connected with the ordering) tells us which degrees of freedom participate in ordering. Thus, e.g. if we have a magnetic ordering of spins $\frac{1}{2}$, the total entropy of the transition in the ideal case should be $S_{\text{tot}} = k_B \ln 2$ (or $k_B \ln(2S+1)$ for spin $S$, where $2S+1$ is the number of possible states of spin $S$ in a fully disordered state, and this entropy has to be removed in the ordered state at $T = 0$). If experimentally one finds $S_{\text{tot}}$ smaller than this value, then this means that there is still a certain degree of ordering (or short-range correlations) above $T_c$. If, however, one finds the value of $S_{\text{tot}}$ *larger* than the expected one, one can conclude that some other degrees of freedom also order at $T_c$, not only the ones initially assumed. This is an important test, often used experimentally.

**Problem:** Find the connection between the specific heat jump $\Delta c_P$ and other properties of the solid (compressibility, thermal expansion).

**Solution:** By definition, second-order phase transitions are continuous, so that along the transition line there is no jump in volume and in entropy, $\Delta V = 0, \Delta S = 0$ (these would be nonzero at the first-order phase transition). Let us differentiate these relations along the curve $T_c(P)$: we thus obtain, e.g.

$$\Delta\left(\frac{\partial V}{\partial P}\right)_T + \left.\frac{\partial T}{\partial P}\right|_{T_c} \Delta\left(\frac{\partial V}{\partial T}\right)_P = 0\,. \tag{2.12}$$

Remembering that the thermal expansion coefficient is $\beta = 3\alpha = \frac{1}{V}\frac{dV}{dT}$, and the compressibility $\kappa = -\frac{1}{V}\frac{\partial V}{\partial P}$, we can rewrite equation (2.12) as

$$\boxed{\Delta\kappa = 3\frac{dT_c}{dP}\Delta\alpha = \frac{dT_c}{dP}\Delta\beta}\,. \tag{2.13}$$

Similarly, from the condition $\Delta S = 0$, we obtain:

$$\Delta\left(\frac{\partial S}{\partial P}\right)_T + \left.\frac{\partial T}{\partial P}\right|_{T_c} \Delta\left(\frac{\partial S}{\partial T}\right)_P = 0\,. \tag{2.14}$$

As

$$S = -\left(\frac{\partial \Phi}{\partial T}\right)_P\,, \tag{2.15}$$

$$\left(\frac{\partial S}{\partial P}\right)_T = -\frac{\partial^2 \Phi}{\partial T\,\partial P} = -\frac{\partial}{\partial T}\left(\frac{\partial \Phi}{\partial P}\right) = -\left(\frac{\partial V}{\partial T}\right)_P\,, \tag{2.16}$$

this gives the relation

$$\Delta\left(\frac{\partial V}{\partial T}\right)_P = \frac{dT_c}{dP}\frac{\Delta c_P}{T} \tag{2.17}$$

(here we have used the expression (1.19) for the specific heat $c_P$). In effect we obtain

$$\boxed{3\Delta\alpha = \Delta\beta = \frac{dT_c}{dP}\frac{1}{VT_c}\Delta c_P} . \tag{2.18}$$

The relations (2.13), (2.18) are known as the Ehrenfest relations. They are the analogues for the second-order phase transition of the well-known Clausius–Clapeyron relations valid for first-order transitions, e.g. the relation between the jump in volume and the latent heat of transition $\Delta Q = T\Delta S$:

$$\Delta V = \frac{\partial T_c}{\partial P}\Delta S = \frac{1}{T_c}\frac{dT_c}{dP}\Delta Q . \tag{2.19}$$

One can easily check that in the limit in which all the jumps at the first-order transition go to zero, i.e. when the first-order phase transition goes over to the second-order one, this expression gives equation (2.18): one can obtain this by applying the operation $\frac{1}{V}\frac{\partial}{\partial T}$ to (2.19) and using $\frac{\partial(\Delta Q)}{\partial T} = \Delta c_p$, and $\beta = 3\alpha = \frac{1}{V}\frac{\partial V}{\partial T}$.

## 2.2 (Weak) First-order phase transitions

Until now we have ignored the cubic term in the expansion (2.1). Very often it is indeed zero, just by symmetry. Thus, in an isotropic ferromagnet the states with positive and negative magnetization should be equivalent, which means that the free energy may contain only terms *even* in $M$ (which in this case is the order parameter), i.e. the term $C\eta^3$ in this and similar cases should be absent. But there may be other situations, in which such terms are allowed by symmetry and should be present in the expansion (2.1).

Suppose now that the term $C\eta^3$ (cubic invariant) in the free energy (2.1) is nonzero. One can easily analyse the resulting equation for $\eta$, analogous to (2.5), and find the properties of the solution. It is also very instructive just to look at the dependence $\Phi(\eta, T)$ in this case, which will immediately tell us what happens. The corresponding set of curves $\Phi(\eta)$ for different temperatures is shown in Fig. 2.5 for the case $C < 0$ (the case $C > 0$ can be studied similarly). For high enough $T$, when the coefficient $A$ in (2.1) is large, we have only one minimum of $\Phi(\eta)$, at $\eta = 0$ (curve 1 in Fig. 2.5.) With decreasing temperature and, consequently, decreasing coefficient $A$, the set of curves $\Phi(\eta)$ would look as shown in Fig. 2.5, curves 2–6. At a certain temperature $T^*$ there will appear a second minimum in $\Phi(\eta)$, curve 2.

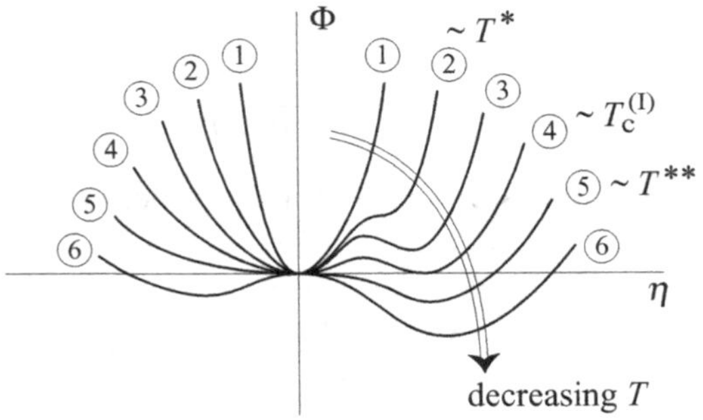


Fig. 2.5

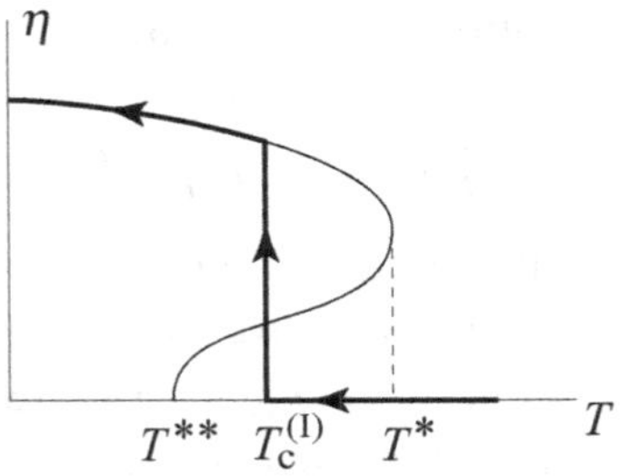


Fig. 2.6

Initially it is just a metastable state, the absolute minimum is still at $\eta = 0$. But with further decrease of temperature, at $T < T_c^{(I)}$, this new state will have free energy lower than the disordered state with $\eta = 0$, curve 4. Note that this will happen when the coefficient $A$ is still positive. If we wait long enough, at this temperature the system would jump from the state $\eta = 0$ to a new minimum with nonzero $\eta$, i.e. we will have a first-order transition. The disordered state $\eta = 0$ will still exist as a local minimum, i.e. as a metastable (overcooled) state. At a still lower temperature $T^{**}$ the coefficient $A$ in (2.1) will itself become negative, and the metastable state $\eta = 0$ will cease to exist. The temperatures $T^*$ and $T^{**}$ are the limits of overheating and overcooling in such a first-order transition; they determine the maximum width of hysteresis at such a transition. These points are called *spinodal* points, and if $T_c^{(I)}$ changes, e.g. under pressure, they will form spinodal lines, or simply spinodals.

Thus the behaviour of $\eta(T)$ has here the form shown in Fig. 2.6. We see that the presence of cubic invariants in the expansion of the free energy (2.1) always leads to first-order phase transitions. Note that here the real critical temperature, the point $T_c^{(I)}$ in Fig. 2.6, is not a singular point of the free energy – it is just the point at which two minima of the free energy in Fig. 2.5 have equal depths. On the other hand, the temperatures $T^*$ and $T^{**}$ (the limits of overheating and overcooling) *are*

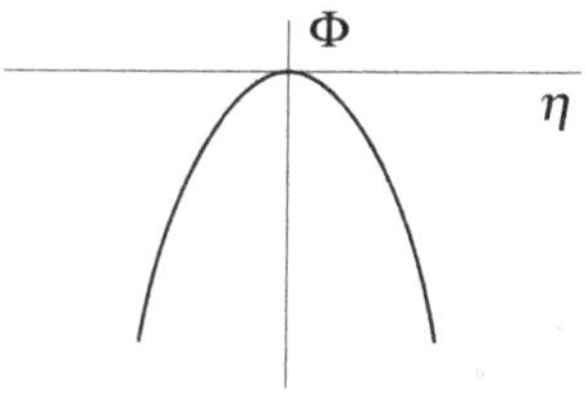


Fig. 2.7

singular points, so that, e.g. the susceptibility would not diverge at $T_c^{(1)}$, but it would at $T^*$ and $T^{**}$.

When do we meet such a situation? The general principle is the following: we should write down in the free energy functional (2.1) *all the terms* permitted by symmetry. Usually one has to check it using *group theory*. And if cubic invariants are allowed by symmetry, then generally speaking they should be present in the expansion (2.1), so that the corresponding transition should be first order.

There are many examples in which this is the case. One is the structural phase transition or charge (or spin) density wave transition in a triangular or honeycomb lattice with the formation of *three equivalent waves* $\eta_n = \eta e^{i\boldsymbol{Q}_n \cdot \boldsymbol{r}}$, with the wavevectors $\boldsymbol{Q}_1$, $\boldsymbol{Q}_2$, $\boldsymbol{Q}_3$ at 120° to one another, so that $\boldsymbol{Q}_1 + \boldsymbol{Q}_2 + \boldsymbol{Q}_3 = 0$. As a result there exists an invariant $C\eta_1\eta_2\eta_3 = C\eta^3 e^{i(\boldsymbol{Q}_1+\boldsymbol{Q}_2+\boldsymbol{Q}_3)\cdot\boldsymbol{r}} = C\eta^3$, so that such a phase transition should be *first order*. This is, for example, the situation in transition metal dichalcogenides such as $NbSe_2$, $TaS_2$, etc.

One can also show in general that at crystallization (formation of a periodic structure from a liquid) the situation is the same – and consequently crystallization, or melting, should be a first-order phase transition (Landau).

### 2.2.1 Another possibility of getting a first-order phase transition

Suppose the coefficient $C$ in (2.1) is zero, and we have

$$\Phi(P, T) = A(P, T)\eta^2 + B(P, T)\eta^4 \,. \tag{2.20}$$

We have assumed previously that the coefficients $A(P, T) = a(T - T_c)$ and $B(P, T) > 0$. But the coefficient $B$ is itself a function of pressure, and in principle it may become negative. Then at $T < T_c$, instead of the behaviour shown in Fig. 2.2, we would have the behaviour of the type shown in Fig. 2.7, i.e. the free energy (2.20) would have no minima at finite $\eta$. To stabilize the system we then have to include higher order terms in the expansion for the free energy. Thus, e.g. we should write

$$\Phi = A\eta^2 + B\eta^4 + D\eta^6 \tag{2.21}$$

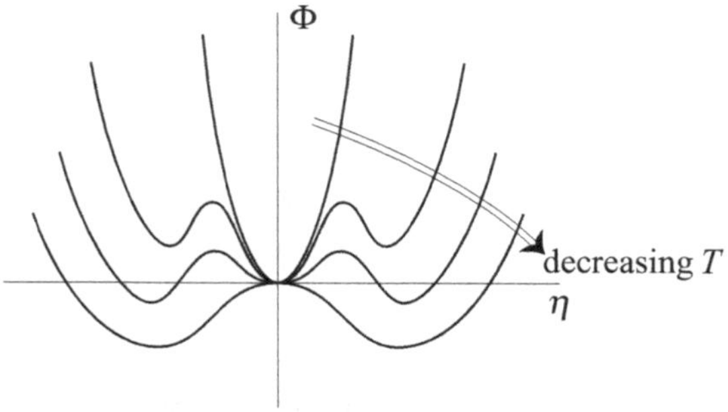


Fig. 2.8

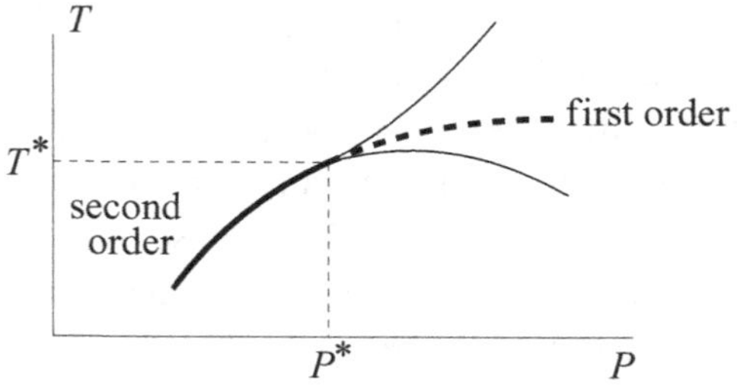


Fig. 2.9

with $D > 0$. Suppose that the coefficient $B < 0$, and that the temperature is slightly above the original $T_c$ (i.e. the coefficient $A$ is positive but small). Then the behaviour of $\Phi$ at different temperatures would have the form shown in Fig. 2.8. When we approach the 'old' $T_c$ from above, so that $A = a(T - T_c)$ becomes small but is still positive, the term $B\eta^4$ with $B < 0$ starts to be important at small $\eta$, leading to the situation with three minima of $\Phi$. That is, in this case the phase transition also becomes *first order*. Thus at the point where $B(P, T)$ changes sign, a second-order phase transition changes to a first-order one. (The point $(T^*, P^*)$ in the $(T, P)$ plane is indeed a *point*: it is the simultaneous solution of two equations $A(P, T) = 0$, $B(P, T) = 0$.) Such a point, at which a second-order phase transition changes to a first-order one, is called a *tricritical point*. Close to it the phase diagram of the system has the form shown in Fig. 2.9. Here we have marked the second-order phase transition by a bold solid line and the first-order transition by a bold dashed one; thin solid lines are spinodals – the lines of overheating and overcooling.

## 2.3 Interaction with other degrees of freedom

The Landau method of treating phase transitions is very simple, but extremely powerful. It permits us, e.g. to study the influence on the phase transition of the interaction with other degrees of freedom. Thus, for instance, we can study phase transitions (e.g. magnetic) in a compressible lattice. If $u$ is the deformation, we can write in general (this is a typical situation)

$$\Phi = A\eta^2 + B\eta^4 + \frac{bu^2}{2} + \lambda\eta^2 u \,. \tag{2.22}$$

Here the third term is the elastic energy ($b$ is the bulk modulus, i.e. the inverse compressibility), and the last term is the coupling of our order parameter with the deformation, for example the magnetoelastic coupling in the case of magnetic phase transitions.

Minimizing (2.22) with respect to $u$, we obtain

$$\frac{\partial\Phi}{\partial u} = bu + \lambda\eta^2 = 0 \,, \qquad u = -\frac{\lambda\eta^2}{b} \,. \tag{2.23}$$

Now, we put it back into $\Phi$ (2.22):

$$\Phi = A\eta^2 + B\eta^4 + \frac{\lambda^2\eta^4}{2b} - \frac{\lambda^2\eta^4}{b} = A\eta^2 + \left(B - \frac{\lambda^2}{2b}\right)\eta^4 \,. \tag{2.24}$$

Thus we see that if the coupling to the lattice $\lambda$ is sufficiently strong, or if the lattice compressibility is large (bulk modulus $b$ small), the renormalized coefficient of the $\eta^4$ term in (2.24) may become negative – and this, according to Figs. 2.8 and 2.9, makes the transition first order. This is a general rule: coupling to other degrees of freedom gives a tendency to make a phase transition first order (although it may remain second order if this coupling is not strong enough[1]).

Note that for the coupling included in equation (2.22) the resulting deformation is $u \sim \eta^2$, see (2.23). Thus, whereas below $T_c$ we have $\eta \sim \sqrt{T_c - T}$ (if the transition remains second order), the corresponding distortion changes linearly, $u \sim (T_c - T)$. In principle this effect can be measured directly. In particular, if the corresponding distortion breaks inversion symmetry and leads to ferroelectricity (i.e. if the polarization $P \sim u$), then the polarization will be proportional to the square of the primary order parameter $\eta$ and close to $T_c$ would also behave as $T_c - T$. Such systems are known as *improper ferroelectrics*, in contrast to the ordinary (proper) ferroelectrics in which the polarization itself is the main, primary order parameter, $\eta = P \sim \sqrt{T_c - T}$. A similar situation can exist also in other systems with coupled order parameters. The resulting properties depend on the specific type of coupling between such order parameters $\eta$ and $\zeta$: coupling of the type of equation (2.22), $\eta^2\zeta$, or of the type $\eta^2\zeta^2$, etc. (The detailed type of

[1] A more detailed treatment shows that the tendency to make a first-order transition due to coupling of the order parameter to the lattice is actually much stronger than that obtained above, especially when we take into account coupling to shear strain (Larkin and Pikin).

such coupling is determined by the symmetry properties of corresponding order parameters.)

## 2.4 Inhomogeneous situations (Ginzburg–Landau theory)

Up to now we have considered only homogeneous solutions, i.e. situations in which the order parameter $\eta$ – e.g. the magnetization – is taken as constant, the same throughout the whole sample, independent of the position $\boldsymbol{r}$. Often one has to consider inhomogeneous situations, for example in the presence of external fields, or close to the surface, etc. Then $\eta = \eta(\boldsymbol{r})$, and the total free energy should be written as an integral over the whole space, of the type

$$\Phi = \int d^3\boldsymbol{r}\,\{\text{free energy density }\Phi(\boldsymbol{r})\}\,. \tag{2.25}$$

Fig. 2.10

The free energy density $\Phi(\boldsymbol{r})$ contains terms of the type (2.1), with the order parameter $\eta(\boldsymbol{r})$ taken at a given point. But the variation of $\eta$ in space also costs some energy. Thus, e.g. in a ferromagnet all spins prefer to be parallel, $\uparrow\uparrow\uparrow\uparrow\uparrow\uparrow$. Variation in space means, e.g. the formation of structures of the type shown in Fig. 2.10 – spirals, or domain walls, etc. It is clear that the canting of neighbouring spins costs a certain energy (the usual exchange interaction in a ferromagnet preferring to keep neighbouring spins parallel). For *slow* variation the cost in energy should be proportional to the gradient of the order parameter, $\sim d\eta/d\boldsymbol{r} = \nabla\eta$. The only invariant containing $\nabla\eta$ (for scalar $\eta$) is $(\nabla\eta)^2$. Thus, in the same spirit as before, we can write the simplest generalization of equation (2.1) in the form

$$\Phi = \int d^3\boldsymbol{r}\left\{A\,\eta^2(\boldsymbol{r}) + B\,\eta^4(\boldsymbol{r}) + G\big(\nabla\eta(\boldsymbol{r})\big)^2\right\}. \tag{2.26}$$

This is called a Ginzburg–Landau (GL) functional (sometimes also Ginzburg–Landau–Wilson (GLW) functional) – the functional of the function $\eta(\boldsymbol{r})$. It is widely used, e.g. in the theory of superconductivity (where it was actually introduced), in discussion of domain walls in magnets, etc. Minimizing this functional in the order parameter $\eta(\boldsymbol{r})$ (which in the theory of superconductivity is usually a complex scalar denoted $\psi$, and which can be viewed as a wavefunction of the superconducting condensate) gives not the ordinary algebraic self-consistency equation (2.5), but, due to the presence of the gradient term in (2.26), a differential equation – the famous Ginzburg–Landau equation. In the theory of superconductivity it is usually

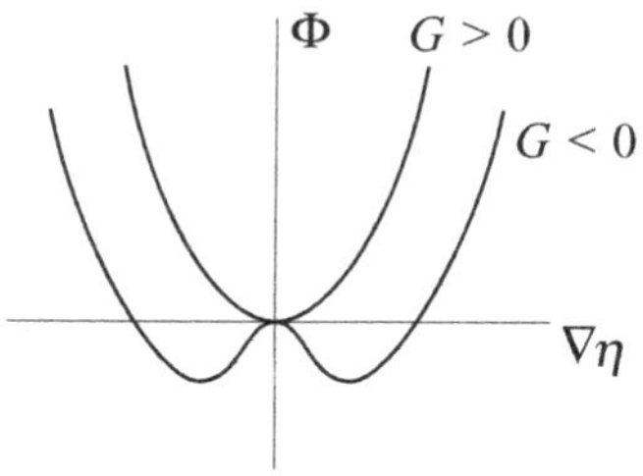


Fig. 2.11

written as

$$\frac{1}{2m}\left(-i\hbar\nabla - \frac{2e}{c}\boldsymbol{A}\right)^2 \psi + A\psi + 2B|\psi|^2\psi = 0\,. \qquad (2.27)$$

(Here $\boldsymbol{A}$ is the vector potential, introduced to describe the behaviour of superconductors in a magnetic field.) We see that the Ginzburg–Landau equation (2.27) has a form similar to the Schrödinger equation, but with a nonlinear term $\sim\psi^3$. Close to $T_c$, when the order parameter $\psi \to 0$, this equation can be linearized, and then it is indeed exactly equivalent to the corresponding Schrödinger equation. This analogy is very useful for the treatment of many problems in superconductivity, such as the upper critical magnetic field $H_{c2}$, the formation and properties of vortices, etc.

Similar equations can also be written down for other physical systems. Thus, the corresponding equation for ferromagnets (known as the Landau–Lifshits equation) is widely used for treating the domain structure of ferromagnets, the dynamics of domain walls, etc.

As we have discussed above, in the usual situation the homogeneous solutions correspond to the minimum energy, which means that the coefficient $G$ next to the gradient term in equation (2.26) is positive. But what would happen if the coefficient $G(P, T)$ becomes negative?

Let us suppose that $G < 0$. Then it is favourable to have $\nabla\eta \neq 0$, i.e. the homogeneous solution becomes unstable, and there should occur a transition to an inhomogeneous state. For instance, instead of a ferromagnet ↑ ↑ ↑ ↑ ↑ we may have a transition to a *spiral* state. To find such a solution and to describe its properties we have to minimize $\Phi$ now with respect to $\nabla\eta$. Again, if the term $G(\nabla\eta)^2$ becomes negative, we have to write down the next terms in the gradient (but still lowest order in $\eta$ itself), e.g. $E(\nabla^2\eta)^2$ with a positive coefficient $E$, to stabilize the system.

Thus we would have in the free energy the terms

$$G(\nabla\eta)^2 + E(\nabla^2\eta)^2\,. \qquad (2.28)$$

As a function of $\nabla\eta$, $\Phi$ has a form similar to the one shown in Fig. 2.2, see Fig. 2.11. It is convenient to go to the momentum representation: $(\nabla\eta)^2 \to q^2\eta^2$. Then $\Phi$

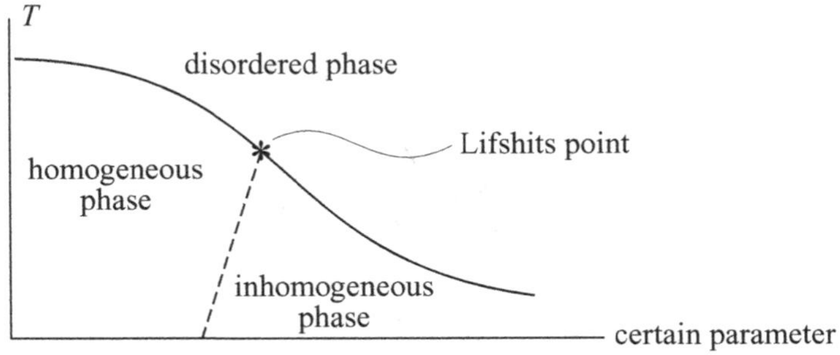


Fig. 2.12

would contain the terms

$$Gq^2\eta^2 + Eq^4\eta^2 \; . \tag{2.29}$$

Now we can find the value of $q$ which minimizes $\Phi$:

$$\frac{\partial \Phi}{\partial q^2} = G\eta^2 + 2Eq^2\eta^2 = 0 \; , \tag{2.30}$$

$$Q^2 = q^2_{\min} = -\frac{G}{2E} \qquad (G < 0) \; . \tag{2.31}$$

That is, in this case the structure with the wavevector $Q$, or with the period $l = 2\pi/Q$, will be formed. In general the period of this superstructure (lattice, or magnetic, etc.) is *incommensurate* with respect to the underlying lattice period.

The point where the coefficient $G$ changes sign and at which there is a change from homogeneous to inhomogeneous ordering is called the *Lifshits point*. The typical phase diagram in this case looks like Fig. 2.12. Here the solid lines are second-order phase transitions, and the dashed line is a first-order transition. Such is, for example, the phase diagram of $CuGeO_3$ (spin-Peierls system, see Section 11.2 below) in a magnetic field.

In general, for certain symmetry classes, there may exist in the free energy also invariants *linear* in the derivatives (the so-called *Lifshits invariants*), for example the term $\boldsymbol{M} \cdot \operatorname{curl} \boldsymbol{M}$ in magnetic systems. These terms will give helicoidal structures even for the case when the coefficient $G > 0$ (instead of (2.29) we will have an equation for the wavevector of a new periodic structure of the type $\text{const} \cdot q + Gq^2\eta^2 = 0$). This is actually the situation in many rare earth metals having different types of such incommensurate magnetic structures. These terms also play a crucial role in magnetically driven ferroelectricity in the so-called *multiferroics*. The microscopic origin of such linear terms is a special problem which we do not discuss here.

## 2.5 Fluctuations at the second-order phase transitions

The theory of second-order phase transitions described above is essentially a mean field theory. However, close to $T_c$, fluctuations become important. They modify the behaviour at $T_c$ of all thermodynamic functions, e.g. the specific heat $c$, the thermal expansion $\alpha$, the compressibility $\kappa$, etc. For example, instead of a jump in $c$, there may appear a real singularity, e.g. $c(T) \sim \ln(|T - T_c|/T_c)$ or $\sim |T - T_c|^{-\alpha}$. One can estimate the region $|T - T_c|/T_c$ in which fluctuations become very strong and the behaviour of many quantities, such as the specific heat, will deviate from that described in Section 2.1. This region is determined by the condition that the average fluctuations of the order parameter become comparable to the order parameter itself, $\langle(\Delta\eta)^2\rangle \sim \eta^2$. The corresponding criterion for the width of this region, due to Ginzburg and Levanyuk, has the form

$$\tau \sim \frac{|T - T_c|}{T_c} \simeq \frac{B^2}{8\pi^2 a^4 T_c^2 \xi_0^6} = \frac{B^2 T_c}{8\pi^2 a G^3} \,. \tag{2.32}$$

Here

$$\xi_0 = \sqrt{\frac{G}{aT_c}} \tag{2.33}$$

is the so-called *correlation length* at zero temperature; the parameters $a$, $B$, $G$ are the parameters of the free energy expansion (2.1), (2.26). The correlation length (2.33) determines, e.g. the typical length-scale at which, in the ordered state at $T = 0$, the order parameter recovers its equilibrium value when 'disturbed' at some point, or shows at which length-scale the order parameter changes close to the surface of the sample.

Inside the temperature interval (2.32) fluctuations are important and modify the behaviour of all physical characteristics of the system, but outside it (but still close to $T_c$) we can use the Landau theory. The parameter $\mathrm{Gi} = B^2 T_c/8\pi^2 a G^3$ is called the Ginzburg number; if $\mathrm{Gi} \ll 1$, one can use mean field results practically everywhere without restrictions.

One sees from (2.32) that $\mathrm{Gi} \ll 1$ if the correlation length $\xi_0$ is large enough. This is, for example, the case in the usual superconductors, in which $\xi_0$ (also called there the coherence length) is $\xi_0 \sim 10^4$ Å – much larger than the lattice parameter $a$. However, for example, in liquid $^4$He, $\xi_0$ is of atomic order, and the fluctuation region is very large, so that the specific heat at the transition to the superfluid phase has the shape shown in Fig. 2.13, which is sometimes called the λ-anomaly. The qualitative explanation of this behaviour is that due to fluctuations, e.g. above $T_c$, the ordering does not disappear immediately, but there exists a certain *short-range order*, so that the whole entropy is not released at $T_c$, but part of it remains also above $T_c$.

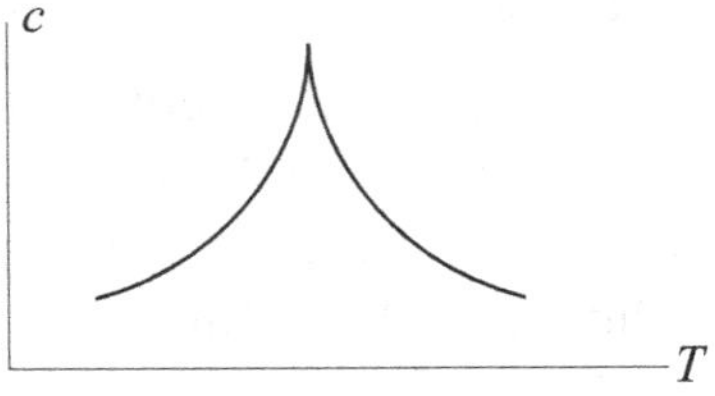


Fig. 2.13

This explains the tail of the specific heat at high temperatures. On the other hand, very close to $T_c$ the anomalies of thermodynamic functions may differ from those assumed in the Landau expansion (2.1), see below.

One can show that the size of the region in which fluctuations are important is determined microscopically by the spatial extension of the interaction leading to the ordering, such as the exchange interaction in a ferromagnet: if the interaction is long-range, the Ginzburg parameter is small, and consequently the fluctuation region is small as well, and the mean field description, which the Landau theory in fact is, is applicable. If, however, the corresponding interactions are short-range (as, for example, the interactions between helium atoms in liquid He), then fluctuations are important in a broad temperature interval; this is actually the microscopic reason for the λ-anomaly in helium.

The quantity $\xi$ has a definite physical meaning, which gave rise to its name 'correlation length'. If one considers fluctuations of the order parameter at different points, they are not completely independent but are correlated, very often of the form

$$\langle \Delta\eta(0)\,\Delta\eta(r)\rangle \sim \frac{T}{r}\, e^{-r/\xi}\,, \tag{2.34}$$

where $\xi$ is the correlation length. In general it depends on temperature:

$$\xi(T) = \sqrt{\frac{G}{a\,|T - T_c|}}\,. \tag{2.35}$$

The quantity $\xi_0$ in (2.33) is equal to $\xi(T = 0)$. Close to $T_c$, $\xi$ becomes very large both when approaching $T_c$ from below and from above. For $T \to T_c + 0$ the average equilibrium value is $\eta = 0$, but there exist fluctuation regions in which order is already established; and $\xi(T)$ is the typical size of such regions, which tends to infinity for $T \to T_c$. Below $T_c$ there exists long-range order, $\eta \neq 0$, but on approaching $T_c$ from below there appear disordered regions, whose size is again of the order $\xi(T)$.

It is often convenient to describe fluctuations not in real, but in momentum space. Equation (2.34) then looks as follows:

$$\langle \Delta\eta_{\boldsymbol{q}}\, \Delta\eta_{-\boldsymbol{q}} \rangle = \left\langle |\eta_{\boldsymbol{q}}|^2 \right\rangle = \frac{T}{2a|T-T_{\rm c}|}\frac{1}{(1+q^2\xi^2)} = \frac{T}{2G}\frac{\xi^2}{(1+q^2\xi^2)}\,. \qquad (2.36)$$

This is the famous Ornstein–Zernike theory of fluctuations. One can also obtain the expression for the susceptibility close to the second-order phase transition. It is given by similar expressions:

$$\chi(T) = \left.\frac{1}{2a(T-T_{\rm c})}\right|_{T>T_{\rm c}}, \qquad \chi(T) = \left.\frac{1}{4a(T_{\rm c}-T)}\right|_{T<T_{\rm c}}, \qquad (2.37)$$

i.e. the susceptibility also diverges when $T \to T_{\rm c}$ (see also later, Chapter 6, (6.40), (6.44)).

In principle the generalized susceptibility is a function of $\boldsymbol{q}$ and $\omega$, $\chi(\boldsymbol{q}, \omega)$ (e.g. the well-known dielectric function $\varepsilon(\boldsymbol{q}, \omega)$, which is a response function to the external electric field $\boldsymbol{E}(\boldsymbol{q}, \omega)$, just as the usual susceptibility of magnetics is a response function with respect to the magnetic field). There exists a very important general connection between $\chi(\boldsymbol{q}, \omega)$ and the corresponding correlation functions $\left\langle \eta(\boldsymbol{r}, 0)\, \eta(\boldsymbol{r}', t) \right\rangle$ or their Fourier transforms. For static susceptibility this relation has the form

$$\chi(\boldsymbol{q}) = \frac{1}{T}\int \frac{d^3\boldsymbol{r}}{(2\pi)^3} e^{i\boldsymbol{q}\cdot\boldsymbol{r}} \left\langle \eta(\boldsymbol{0}, t)\, \eta(\boldsymbol{r}, t) \right\rangle . \qquad (2.38)$$

For the usual susceptibility, e.g. in magnetic systems for which the order parameter $\eta$ is the magnetization, or the average spin $\boldsymbol{S}$, it gives a convenient relation:

$$\chi = \frac{1}{T}\sum_n \langle \boldsymbol{S}_0 \cdot \boldsymbol{S}_n \rangle\,. \qquad (2.39)$$

In general close to $T_{\rm c}$, e.g. for $T > T_{\rm c}$, one obtains from (2.36) and (2.38), that

$$\left.\chi(\boldsymbol{q})\right|_{T>T_{\rm c}} = \frac{1}{2a(T-T_{\rm c})}\frac{1}{\left(1+q^2\xi^2(T)\right)} = \frac{1}{2G}\frac{\xi^2(T)}{\left(1+q^2\xi^2(T)\right)}, \qquad (2.40)$$

which for $\boldsymbol{q} = 0$ gives (2.37).

### 2.5.1 Critical indices and scaling relations

All the considerations presented above are valid when we proceed from the expansion (2.26) for the free energy and treat a second-order phase transition in essentially a mean field way. However, as we have already seen above, close to $T_{\rm c}$ the fluctuations are always strong (and the width of the region in which they are important, given by the Ginzburg–Levanyuk criterion, may be sufficiently broad). In this region the description of second-order phase transitions should be modified, and all the anomalies of thermodynamic functions such as specific heat, compressibility, thermal expansion, susceptibility, etc. are very different from those predicted by the Landau theory. This is really the field of the theory of second-order phase transitions, which was especially active in the 1960s and 1970s and which resulted in a rather deep understanding of the phenomena close to $T_c$ for such transitions. This is a very big field in itself, which we cannot cover here; one can find corresponding results and references in many books and review articles, e.g. Chaikin and Lubensky (2000) and Stanley (1987). The basic conclusion of these very elaborate studies is that close to $T_c$ the properties of the system usually have singularities of the type $|T - T_c|^{-\lambda_i}$, where the exponents $\lambda_i$ (different for different quantities) are called *critical indices*. Thus, e.g. the specific heat behaves as $\tau^{-\alpha}$, where $\tau = |T - T_c|/T_c$; the order parameter itself changes close to $T_c$ as $\tau^{\beta}$, the susceptibility $\chi$ as $\tau^{-\gamma}$, the correlation length $\xi$ diverges as $\tau^{-\nu}$, etc. (This is the standard notation for these critical exponents.)

The exact values of the critical indices are known only in very few cases; a notable example is the exactly soluble two-dimensional Ising model, see Section 6.4.3 below. However, despite the absence of exact solutions in most cases, there exist very powerful general results in this field known as *scaling relations* (Kadanoff; Patashinskii and Pokrovskii). The underlying idea is that when the system approaches $T_c$, the correlation length diverges and becomes infinitely large. In this case all microscopic details, important at short distances, become irrelevant, and the properties of the system become universal. Moreover, as $T \to T_c$ all length-scales become equivalent, and, simply from dimensional arguments, one can show that for instance when one changes all distances by a certain factor, $L \to kL$, e.g. doubling the size of the system, different quantities would change accordingly. From these arguments one can find the relations between different quantities in the vicinity of the critical temperature. Some examples of these relations are:

$$\alpha + 2\beta + \gamma = 2\,, \tag{2.41}$$

$$d\,\nu + \alpha = 2 \tag{2.42}$$

(here $d$ is the space dimensionality), etc. The critical indices may be different from system to system, but relations of the type (2.41), (2.42) are universal. The values of the indices depend only on the dimensionality of the system $d$ and on the symmetry of the order parameter (real or complex scalar; vector; isotropic or anisotropic system, etc.), and for each particular case the critical indices should be the same, independent of the specific physical situation. These are called *universality classes*.

As mentioned above, the values of the critical indices for different universality classes are known exactly only in very few cases. But numerical calculations using powerful computer algorithms have given pretty accurate values of these indices in many cases. Thus, for example, we have:

1. For the 2d Ising model: $\alpha = 0$ (logarithmic divergence, which is weaker than any power law divergence); $\beta = \frac{1}{8}$; $\gamma = \frac{7}{4}$; $\nu = 1$.
2. For the 3d Ising model: $\alpha = 0.10$; $\beta = 0.33$; $\gamma = 1.24$; $\nu = 0.63$.
3. For the 3d Heisenberg model: $\alpha = -0.12$; $\beta = 0.36$; $\gamma = 1.39$; $\nu = 0.71$.
4. Compare this with the mean field behaviour following from the Landau theory: $\alpha = 0$ (the specific heat has a jump, equation (2.9), but no divergence at $T_c$; $\beta = \frac{1}{2}$, see (2.6); $\gamma = 1$ (the well-known Curie–Weiss law, see equation (6.40) below); $\nu = \frac{1}{2}$, see (2.35).

We see that indeed the specific behaviour (divergence) of different quantities as $T \to T_c$ is different for different cases (different universality classes), but the scaling relations (2.41), (2.42) and others are fulfilled for each of them.

Theoretical methods used to obtain these results are both analytical (those giving scaling relations) and numerical (giving specific values of critical indices for different quantities in different situations). One of the very powerful methods used to treat these problems, and also many others in which the interaction is strong and we do not have any small parameter which would allow us to use perturbation theory or expansion of the type of (2.1), is the *renormalization group* method. It originates from field theory, but is now widely used in many difficult problems in condensed matter physics, such as the theory of second-order phase transitions, the quantum Hall effect, etc., see, e.g. Chaikin and Lubensky (2000) and Stanley (1987).

## 2.6 Quantum phase transitions

Until now we have considered phase transitions occurring at finite temperatures. The description we used was actually that of classical statistical mechanics. Indeed, even if the physical nature of some phase transition is of essentially quantum origin, such as the superconducting transition in metals or the superfluid transition in liquid $^4$He, close to $T_c$ the correlation length $\xi$ becomes very large, see equation (2.35), i.e. it becomes much larger than the distance between particles or the radius of the interaction. In this case the behaviour of the system is essentially classical.

However, there may exist situations in which the critical temperature can be suppressed, e.g. by application of pressure, a magnetic field or some other control parameter $g$, see Fig. 2.14 (the grey region is here the region of classical fluctuations). In this case, if $T_c$ tends to zero, quantum effects start to play a more and more important role. The state of the system for $g > g_c$ (e.g. pressure $P > P_c$)

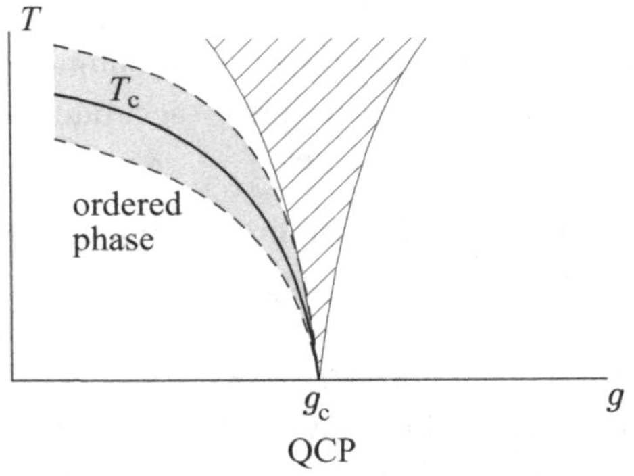


Fig. 2.14

may be disordered even at $T = 0$, not because of classical, but because of quantum fluctuations – we can speak of a quantum disordered phase.

There is here no contradiction with the Nernst theorem, which requires that the entropy of the system should tend to zero as $T \to 0$: the state of such a system at $T = 0$ and $P > P_c$ is a disordered one from the point of view of standard phase transitions (e.g. magnetic ordering may be suppressed for $P > P_c$), but it is a unique quantum state described by a, maybe very complicated, but unique wavefunction. Consequently the entropy of such state is zero. A simple example is given by an ordinary metal (Fermi liquid), as compared, for example, with the ferromagnetic metallic state: although, in contrast to the latter, a normal metal is paramagnetic (Pauli paramagnetism, see Chapter 6 below), it is a unique state – the filled Fermi surface, which has zero entropy. As discussed below, in Section 2.7.3, this determines, for example, the slope of the high-temperature insulator–metal transition in $V_2O_3$.

The behaviour of the system close to the point ($T = 0$, $g = g_c$) in Fig. 2.14 is determined by *quantum fluctuations*, and such a point is called a *quantum critical point* (QCP). One can show that quantum fluctuations dominate the behaviour of the system 'above' QCP – in the region of the phase diagram marked by hatching in Fig. 2.14 (of course the intensity of these fluctuations and consequently the amplitude of all the anomalies decreases with increasing temperature). The proximity to a quantum critical point may strongly influence not only thermodynamic, but also transport properties of the corresponding systems, leading, e.g. to a non-Fermi-liquid behaviour, see Chapter 10 below. There are even suggestions that the proximity to QCP may be important for the appearance of unconventional superconductivity, in particular for the high-$T_c$ superconductivity in cuprates. More detailed descriptions of these phenomena one can find, e.g. in Sachdev (1999) and in von Löhneisen *et al.* (2007).

## 2.7 General considerations

### 2.7.1 *Different types of order parameters*

There may exist different types of order parameters in different systems. For example, they may be scalars, e.g. periodic density in crystals as compared to a homogeneous liquid; vectors (or pseudovectors), e.g. magnetization in ferromagnets or spontaneous polarization in ferroelectrics. Order parameters may also be complex. An example is a complex scalar – the electron (condensate) wavefunction $\Psi(\boldsymbol{r})$ in superconductivity. There may also exist more complicated types of order parameters, for example tensor order parameters (liquid crystals, anisotropic superconductivity and superfluidity, e.g. in $^3$He and most probably in high-temperature and in heavy-fermion superconductors).

### 2.7.2 *General principle*

The general principle is that the free energy should contain *invariants* built in from the order parameter. Thus, for the vector order parameter $\boldsymbol{M}$ in ordinary magnets, $\Phi$ may contain $(\boldsymbol{M})^2$ or $\left(\text{div}\,\boldsymbol{M}\right)^2 = (\nabla \cdot \boldsymbol{M})^2$, but, in the absence of an external field, no terms linear in $\boldsymbol{M}$ or terms $\sim\boldsymbol{M}^3$ because they break the inversion symmetry (equivalence of the states $\boldsymbol{M}$ and $-\boldsymbol{M}$).[2] Similarly, for the complex scalar order parameter $\Psi$ in isotropic systems the free energy $\Phi$ should contain $|\Psi|^2$, but not $\Psi^2$, etc.

Symmetry considerations are very powerful, and they determine the form of the Landau free energy expansion. In the general treatment one should always keep all the terms allowed by symmetry, even if we do not know in detail the physical mechanisms responsible for their appearance. This was, for example, the reason why we have included the term $\sim\eta^2 u$ in (2.22), although we did not specify the type of the order parameter or its dependence on the lattice distortion. Of course, the coefficients of such terms *do depend* on the microscopic nature of the ordering, and to calculate these coefficients is a separate, often very difficult problem. However we can deduce many general conclusions even without such a microscopic treatment, using only very general properties such as symmetry.

[2] For magnetic states there is another, even more powerful restriction. All magnetic states break *time reversal invariance*; magnetic moments are odd with respect to time inversion. (This becomes clear when one remembers that the standard way to create a magnetic moment is by a current running in a coil; and time inversion means that the current would run in the opposite direction, changing accordingly $\boldsymbol{M}$ to $-\boldsymbol{M}$.) But the free energy should of course remain the same under time inversion, from which it follows that magnetic vectors should always enter in even combinations; e.g. $\boldsymbol{M}^2$, $\boldsymbol{M} \cdot \text{curl}\,\boldsymbol{M}$, or $\boldsymbol{M} \cdot \boldsymbol{L}$ where $\boldsymbol{L}$ is the antiferromagnetic vector, see Section 6.2.3.

### 2.7.3 Broken symmetry and driving force of phase transitions

At second-order phase transitions the symmetry changes (it decreases in the ordered phase). Thus, e.g. in a paramagnetic phase there exists spherical symmetry (free rotation of spins), whereas in a ferromagnet all spins point in the same direction, the spin orientation is fixed, i.e. we have broken spin rotation invariance.

The situation is similar in liquids compared with crystals: in liquids there is continuous translational symmetry (shift by an arbitrary vector), whereas in a crystal there remains only a shift by a vector equal to the lattice period. This means that we have broken the continuous translational symmetry. The low-temperature phase usually has lower symmetry than the high-temperature one.

One should keep in mind an important distinction: there exist cases of a broken *continuous* symmetry, or of a *discrete* one. The examples given above correspond to broken continuous symmetry. But there are also cases of a broken discrete symmetry. Such is, e.g. the case of strongly anisotropic magnets. Suppose that spins can take not an arbitrary orientation, but only two: $\uparrow$ or $\downarrow$. Again, in the high-temperature disordered phase there is equal probability of finding spins $\uparrow$ or $\downarrow$ at a given site. The low-temperature ordered phase would correspond, e.g. to spins being predominantly $\uparrow$ (at $T = 0$, only $\uparrow$). This is the so-called Ising ferromagnet. Here a *discrete symmetry* (spin inversion $\uparrow \Longleftrightarrow \downarrow$) is broken.

As we have already discussed, it is easy to understand why the system goes to a more disordered phase with increasing temperature. If we start from the low-temperature ordered phase, we see that the transition to a disordered phase is driven by the entropy. According to (1.10) the free energy (for fixed pressure) is $\Phi = E - TS$. At low temperatures, e.g. $T = 0$, to minimize the free energy we should make the (interaction) energy as low as possible, which is reached in the ordered phase. However at high enough temperatures to decrease $\Phi$ we should make the entropy nonzero, and this drives the transition to a disordered phase. We thus gain in entropy, losing in energy, which, according to (1.10), becomes favourable with increasing temperature.

We have to include *all* contributions to the entropy. Sometimes it is not easy to understand at first glance why the transition occurs in a certain way. A good example is the insulator–metal transition (Mott transition), e.g. in $V_2O_3$, see Chapter 12 below. Briefly, this is a transition between the state with electrons localized each at their own site (Mott–Hubbard insulator) and a metallic state with itinerant electrons. It seems that in this case we should expect the insulating state to be the stable state at low temperatures (electrons localized at their sites, an 'electronic crystal'), and the metallic phase to be the high-temperature phase (delocalized, moving electrons – an electron liquid), similar to *melting* of an ordinary crystal with increasing temperature. However the actual situation is not so simple.

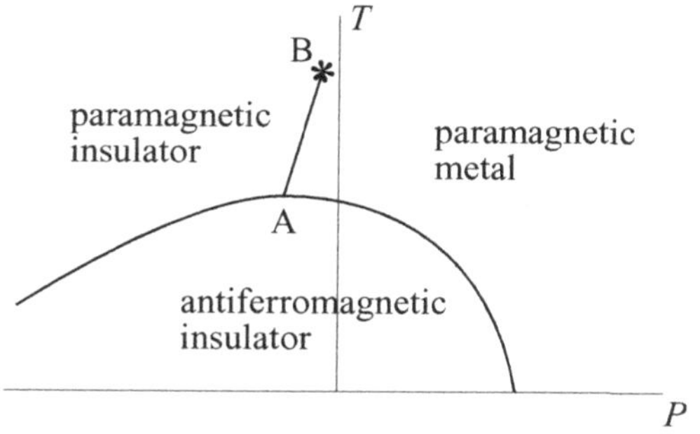


Fig. 2.15

The experimental phase diagram of $V_2O_3$ looks schematically as shown in Fig. 2.15. We see that the transition between an antiferromagnetic insulator and a metal does indeed occur with increasing temperature. However, the 'pure' Mott transition (line AB), not associated with the destruction of magnetic order, is such that with increasing temperature we go from a metal to an insulator! Why? The explanation is most probably that in the metallic phase we have a Fermi surface, and in a sense it is an 'ordered' state (a unique state, the entropy of which is small and at $T = 0$ would be zero). But the insulating state with localized electrons also has localized spins, which in the paramagnetic phase are *disordered*, i.e. the paramagnetic insulator has entropy (per site) $k_B \ln(2S + 1)$, higher than that of the metallic Fermi sea. Therefore, according to the general rule, with increasing temperature the system goes over from the state with lower entropy (metallic state) to the one with higher entropy (paramagnetic insulator), although at first glance this looks strange and counter-intuitive. This is not a unique situation, and in each such case we have to think *which entropy* drives the observed transition with increasing temperature.

### *2.7.4 The Goldstone theorem*

There exists one general result which is known as the *Goldstone theorem*. According to this, when there is a *broken continuous symmetry* at the phase transition, there should exist in the ordered state of the system (without long-range interaction) a collective mode, an excitation, with *gapless energy spectrum* (the energy starts continuously from 0). There exist many examples of such excitations. Here I give only a few: in an isotropic ferromagnet these are spin waves, with the spectrum shown qualitatively in Fig. 2.16. In crystals these are the usual phonons, Fig. 2.17. We will meet many other such examples later in this book.

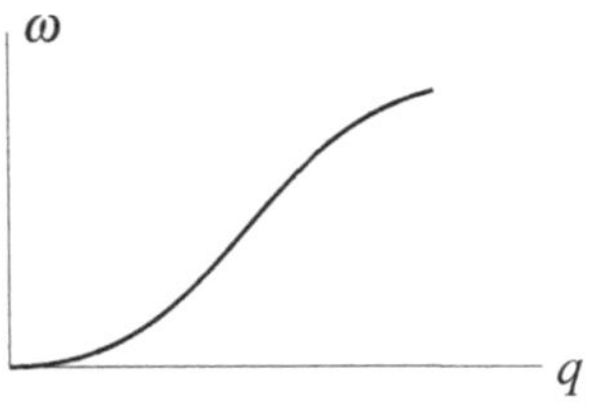


Fig. 2.16

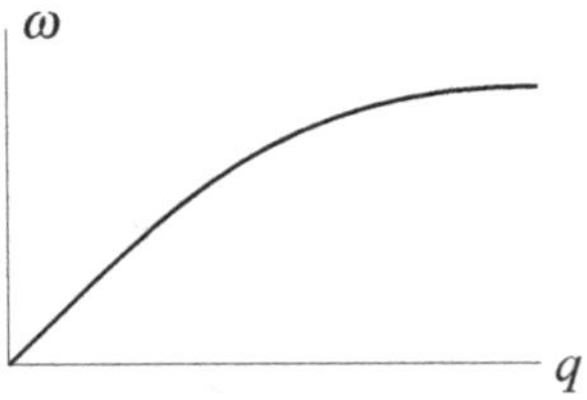


Fig. 2.17

*Qualitative explanation*

In the case of broken continuous symmetry in the ordered phase there exist infinitely many degenerate states (e.g. in a ferromagnet all directions of spontaneous magnetization are possible). All these states have the same energy and are equivalent. A collective mode with $q = 0$ describes a transition from one such state to another (e.g. rotation of the total magnetization of the sample as a whole) – and it should cost us no energy, hence the spectrum of such excitations $\omega(q)$ should start from $\omega = 0$, i.e. there exists gapless excitation. This is the content of the Goldstone theorem; the corresponding gapless modes are often called *Goldstone modes*.

### 2.7.5 Critical points

In principle there may exist phase transitions *without* a change of symmetry, but only the first-order ones. Such is, for example, the liquid–gas phase transition. The symmetry of both these phases, gas and liquid, is the same, but they differ in density (and of course in many other properties, not related to the phase transition itself). Such a transition can end at a *critical point* $(\tilde{P}, \tilde{T})$ (Fig. 2.18): here the thick dashed line is a first-order transition, and the thin solid lines are the limits of hysteresis. As in this case there is no change of symmetry across the first-order phase transition, one can in principle go from one phase to another continuously, e.g. from point A to point B in Fig. 2.18, moving around the critical point without crossing the transition line. For the second-order phase transitions this is impossible, because the symmetry of the two phases, the disordered one and the ordered one (or two different ordered phases) is different.

The same situation in $(P, V)$ coordinates has a familiar form, Fig. 2.19, where above the critical point (for $T > \tilde{T}$, $P > \tilde{P}$) the state of the system is unique, and below it the system with fixed total density, or fixed volume in the grey region, would phase-separate (decompose) into different phases (here gas and liquid), the relative volumes of which will be determined by the well-known Maxwell construction. The point $(\tilde{P}, \tilde{T})$ is the critical point. Above it the $P$–$V$ isotherms have only negative

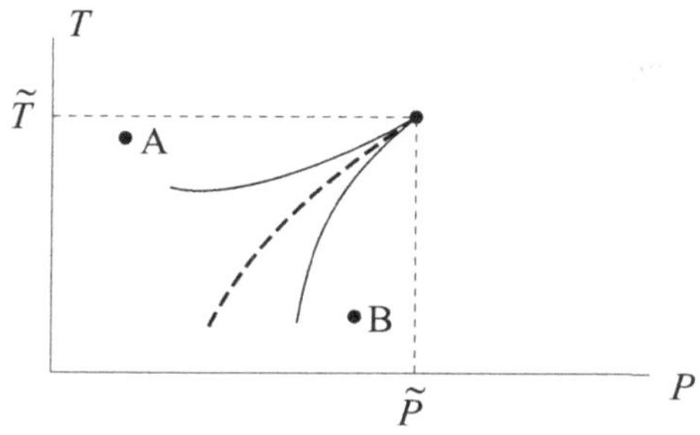


Fig. 2.18

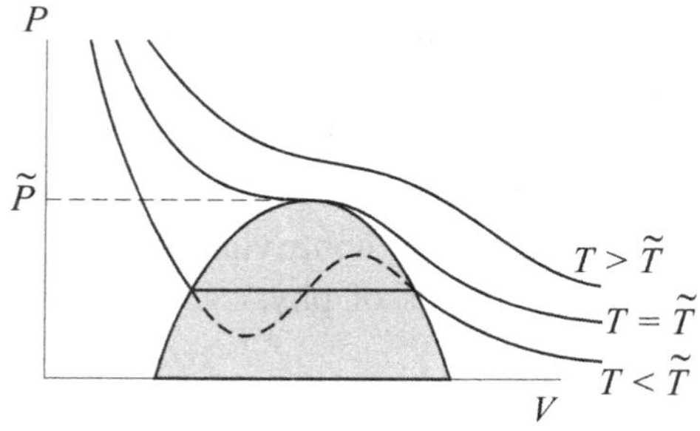


Fig. 2.19

slope, $dP/dV < 0$, the compressibility (1.21) is positive, and the transition (crossover), e.g. from a liquid to a gas state, is smooth. Inside the grey region in Fig. 2.19 there are parts of $P$–$V$ isotherms with $dV/dP > 0$, which would imply negative compressibility, i.e. an absolute instability of the corresponding homogeneous state. This is in fact the reason for the phase separation in this region.

Critical points can appear even in solids. This is the situation, e.g. for the isomorphous phase transitions like the $\gamma$–$\alpha$ transition in Ce or in a high-temperature metal–insulator transition in $V_2O_3$ shown in Fig. 2.15, where the point B is such a critical point.

---

Concluding this chapter, one general remark is in order. We have seen that the Landau approach to phase transitions, although conceptually and technically rather simple, is nevertheless very powerful in describing many quite different situations. It uses the most general arguments such as those of symmetry, etc. and if there are no special indications otherwise, makes the simplest assumptions possible, such as the use of a Taylor expansion in the small parameter $\eta$ or $\nabla\eta$, to give quite general and very successful descriptions of very complicated phenomena. Such an approach is often very fruitful also in many other fields of physics.

One good example is the treatment by Landau of the extremely complicated phenomenon of turbulence (see Landau and Lifshits, *Fluid Mechanics*, 1987). In this problem it is known that the flow of a liquid remains homogeneous, laminar, if the so-called Reynolds number $R$ is less than a certain critical value $R_c$, and the flow develops instability (transition to turbulence) for $R > R_c$. Landau again used here, in the absence of a complete theory (which still does not exist!) an expansion in terms of the amplitude $\Pi$ of a new mode of the motion, which appears for $R > R_c$ and breaks the laminar flow. The expansion again has the form $A\Pi^2 + B\Pi^4$, with $A = \gamma(R_c - R)$, so that the new mode $\Pi$ is absent for $R < R_c$ and the laminar flow is stable, and $\Pi \neq 0$ ($\sim\sqrt{R - R_c}$) for $R > R_c$. The analogy with the treatment of the second-order phase transition (cf. (2.1)–(2.6)) is of course apparent. Again, this approach is limited, and the real theory of turbulence should be much more complicated, e.g. it may resemble the theory of second-order phase transitions going beyond Landau's approach and briefly discussed in Section 2.5.1, with critical exponents, etc. or be even more complicated. But the Landau theory gives at least a first orientation in this nontrivial problem. Such an approach is widely used also in many other fields of physics, including elementary particle physics and cosmology.

# 3

# Bose and Fermi statistics

This chapter is a short reminder and a collection of basic formulae on Fermi and Bose statistics. For noninteracting particles with the spectrum $\varepsilon_k$, the energy of the quantum state, in which there are $n_k$ particles, is $E_{k,n_k} = \varepsilon_k n_k$ (we incorporated in the index $k$ also spin indices, and other indices if they exist). Then, according to (1.32), (1.33), the thermodynamic potential $\Omega$ for this quantum state is equal to

$$\Omega_k = -T \ln \sum_{n_k} \left( e^{(\mu-\varepsilon_k)/T} \right)^{n_k} . \tag{3.1}$$

*For bosons* (the occupation numbers $n_k$ can take any value, i.e. the summation in equation (3.1) goes from $n_k = 0$ to $\infty$), the expression (3.1) converges if $e^{(\mu-\varepsilon_k)/T} < 1$, i.e. we necessarily have $\mu < \varepsilon_k$ (or $\mu < 0$ if $\varepsilon_k \sim k^2/2m$), and after summation we obtain

$$\Omega_k^{\mathrm{B}} = T \ln \left( 1 - e^{(\mu-\varepsilon_k)/T} \right) . \tag{3.2}$$

Then from (1.31) the average occupation of the state $k$ is

$$\boxed{\bar{n}_k^{\mathrm{B}} = -\frac{\partial \Omega_k^{\mathrm{B}}}{\partial \mu} = \frac{1}{e^{(\varepsilon_k-\mu)/T} - 1}} . \tag{3.3}$$

*For fermions*, due to the Pauli principle, we can have no more than one particle in a given quantum state, i.e. $n_k = 0$ or 1. Then the thermodynamic potential (3.1) is

$$\Omega_k^{\mathrm{F}} = -T \ln \left( 1 + e^{(\mu-\varepsilon_k)/T} \right) , \tag{3.4}$$

and

$$\boxed{\bar{n}_k^{\mathrm{F}} = -\frac{\partial \Omega_k^{\mathrm{F}}}{\partial \mu} = \frac{1}{e^{(\varepsilon_k-\mu)/T} + 1}} . \tag{3.5}$$

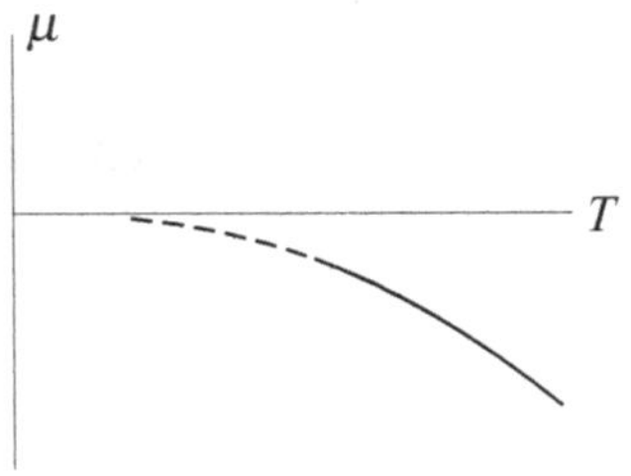


Fig. 3.1

In both cases the total number of particles $N = \sum_k n_k$; this condition determines the chemical potential $\mu(T)$.

**Problem:** Discuss the cross-over to Boltzmann statistics. Find the asymptotic behaviour of $\mu(T)$ at high temperatures.

**Solution:** Boltzmann statistics (the classical case) corresponds to the situation in which all $n_k \ll 1$ (no double occupancy, etc.), i.e.

$$\left.\begin{array}{c} e^{(\mu-\varepsilon_k)/T} \ll 1 \\ \left(\text{ or } e^{(\varepsilon_k-\mu)/T} \gg 1\right) \end{array}\right\} n_k \Longrightarrow e^{(\mu-\varepsilon_k)/T} . \tag{3.6}$$

The chemical potential $\mu$ is obtained from the condition

$$n = \frac{N}{V} = \int n_k \frac{d^3\boldsymbol{k}}{(2\pi\hbar)^3} . \tag{3.7}$$

Taking the energy spectrum $\varepsilon_k = k^2/2m$, and going over to the variable $\varepsilon$ (and then to the dimensionless variable $z = \varepsilon/T$), we obtain:

$$n = \frac{m^{2/3}}{\sqrt{2}\pi^2\hbar^3} \int_0^\infty e^{(\mu-\varepsilon)/T} \sqrt{\varepsilon}\, d\varepsilon \underset{(\varepsilon/T\equiv z)}{=} e^{\mu/T} \frac{m^{3/2}\,T^{3/2}}{\sqrt{2}\pi^2\hbar^3} \int_0^\infty \sqrt{z}\, e^{-z}\, dz . \tag{3.8}$$

The integral in (3.8) is the gamma function, $\Gamma\left(\frac{3}{2}\right) = \frac{\sqrt{\pi}}{2}$.

In effect $n = e^{\mu/T}(T/\tau)^{3/2}$, where $\tau$ is a constant (a combination of $m$, $\pi$, $\hbar, \ldots$). Inverting this equality, we find

$$e^{\mu/T} = n\left(\frac{T}{\tau}\right)^{-3/2} , \tag{3.9}$$

$$\mu = T \ln n - \frac{3}{2} T \ln\left(\frac{T}{\tau}\right) \tag{3.10}$$

which is valid at $T \gg \tau$. This behaviour is schematically shown in Fig. 3.1; we will need these results later on, in Chapter 5.

Special consideration is required for (quasi)particles, whose number is not conserved. Such is, for instance, the situation for phonons in crystals or for photons – quanta of the electromagnetic field. As mentioned in Chapter 1, in this case the number of particles $N$ itself is determined by the condition of minimization of the corresponding thermodynamic potential, e.g. the free energy, with respect to $N$. From the definition of the chemical potential (1.25) we then see that in this case the corresponding chemical potential has to be taken as zero for all temperatures.

# 4

# Phonons in crystals

In this chapter we will discuss the first, and probably the best-known example of bosonic systems – phonons in crystals. According to our general scheme, after briefly summarizing the basic facts about noninteracting phonons in a harmonic lattice, we will pay most attention to the next two factors: quantum effects in the lattice dynamics, and especially the interaction between phonons which leads, e.g. to such phenomena as thermal expansion, explains the features of melting, etc. But for completeness we give, at the beginning, a very short summary of the material well known from standard courses of solid state physics.

## 4.1 Harmonic oscillator

The classical equation for a harmonic oscillator is

$$M\ddot{x} = -Bx \; . \tag{4.1}$$

Its solution is:

$$x = x_0 e^{i\omega t} \; , \tag{4.2}$$

where the frequency is

$$\omega = \sqrt{B/M} \; . \tag{4.3}$$

The Hamiltonian of the harmonic oscillator has the form

$$\mathcal{H} = \frac{p^2}{2M} + \frac{Bx^2}{2} \qquad \left(\hat{p} = \frac{\hbar}{i}\frac{\partial}{\partial x}\right) . \tag{4.4}$$

In quantum mechanics the energy levels are quantized:

$$\varepsilon_n = \left(n + \tfrac{1}{2}\right)\hbar\omega \; . \tag{4.5}$$

## 4.2 Second quantization

The operators of the coordinate $\hat{x}$ and momentum $\hat{p}$ obey the commutation relation

$$[\hat{x}, \hat{p}]_- = \hat{x}\hat{p} - \hat{p}\hat{x} = i\hbar. \tag{4.6}$$

It is convenient to introduce the *annihilation and creation operators* of phonons (in the following we omit the sign ˆ for the operators):

$$\left.\begin{aligned} b &= \frac{1}{\sqrt{2\hbar M\omega}}(M\omega x + ip) \\ b^\dagger &= \frac{1}{\sqrt{2\hbar M\omega}}(M\omega x - ip). \end{aligned}\right\} \tag{4.7}$$

**Problem:** Check that $b$, $b^\dagger$ obey the commutation relation

$$[b, b^\dagger]_- = 1 \tag{4.8}$$

(in future we also omit the minus sign on the commutator $[\ \ ]_-$).

**Solution:**

$$\begin{aligned} [b, b^\dagger] &= \frac{1}{2\hbar M\omega}\Big\{(M\omega)^2[x, x] + iM\omega[p, x] - iM\omega[x, p] + [p, p]\Big\} \\ &= \frac{1}{2\hbar}\big(-2i[x, p]\big) = \frac{(-2i)i\hbar}{2\hbar} = 1\ . \end{aligned} \tag{4.9}$$

We can express $x$, $p$ through $b$, $b^\dagger$, using (4.7):

$$x = \sqrt{\frac{\hbar}{2M\omega}}\,(b^\dagger + b)\ , \qquad p = \frac{i\sqrt{2\hbar M\omega}}{2}\,(b^\dagger - b)\ . \tag{4.10}$$

From (4.4), (4.10) and using (4.8), we obtain

$$\mathcal{H} = \tfrac{1}{2}\hbar\omega(bb^\dagger + b^\dagger b) = \hbar\omega b^\dagger b + \tfrac{1}{2}\hbar\omega = \hbar\omega\Big(n + \tfrac{1}{2}\Big)\ , \qquad n = b^\dagger b\ . \tag{4.11}$$

For the eigenstates (the states with certain particular value $n$) equation (4.11) gives the energy levels (4.5); this is actually the simplest way to obtain the energy spectrum of the harmonic oscillator. Note the presence of $\frac{1}{2}$ in (4.11); this term corresponds to the so-called *zero-point oscillations* and it describes a real physical effect, which, as we will see later, often has very important physical implications.

The states with $n$ excitations $|n\rangle$ obey the relations

$$\begin{aligned} b\,|n\rangle &= \sqrt{n}\,|n-1\rangle \\ b^\dagger|n\rangle &= \sqrt{n+1}\,|n+1\rangle\ , \end{aligned} \tag{4.12}$$

that is, $b$ is indeed an annihilation operator and $b^\dagger$ is a creation operator; they respectively decrease and increase the number of phonons by 1. For the number

Fig. 4.1

operator $\hat{n}$ we have, as we should,

$$\hat{n}|n\rangle = b^\dagger b\,|n\rangle = b^\dagger\sqrt{n}\,|n-1\rangle = n\,|n\rangle\ . \tag{4.13}$$

For the ground state $|0\rangle$ (the state with zero phonons)

$$b\,|0\rangle = 0\ . \tag{4.14}$$

From (4.12) we then obtain

$$|n\rangle = \frac{1}{\sqrt{n!}}(b^\dagger)^n\,|0\rangle\ . \tag{4.15}$$

Let us now consider not an isolated oscillator, but a linear chain, consisting of atoms with the harmonic interaction: its Hamiltonian is

$$\mathcal{H} = \sum_n \left[\frac{M\dot{u}_n^2}{2} + \frac{B}{2}(u_n - u_{n+a})^2\right]\ . \tag{4.16}$$

(Here we have introduced the deviation of the $n$-th atom from its equilibrium position $u_n = x_n - x_n^0$.) The equations of motion have the form

$$M\ddot{u}_n = -B(2u_n - u_{n+1} - u_{n-1})\ . \tag{4.17}$$

We seek the solution in the form $u_n = u_q e^{iqn}$. Then from (4.17) we get

$$M\ddot{u}_q = -n(2 - e^{iqa} - e^{-iqa}) = -2B(1 - \cos qa)\,u_q\ . \tag{4.18}$$

This is also the equation for the harmonic oscillator, with frequency

$$\omega_q = 2\sqrt{\frac{B}{M}}\sin\frac{qa}{2} \underset{(q\to 0)}{\sim} \sqrt{\frac{B}{M}}\,qa \tag{4.19}$$

(compare with (4.1), (4.3)).

**Problem:** Consider a linear chain 'with the basis' (two atoms per unit cell – e.g. with different masses $M_1$ and $M_2$, see Fig. 4.1). Find the phonon spectrum.

**Solution:** For convenience we introduce two variables, $u_n$ and $v_n$, for the atoms of two kinds, see Fig. 4.2. The equations of motion for $u_n$ and $v_n$ are:

$$\begin{aligned} M_1\ddot{u}_n &= B(v_n + v_{n-1} - 2u_n) \\ M_2\ddot{v}_n &= B(u_{n+1} + u_n - 2v_n)\ . \end{aligned} \tag{4.20}$$

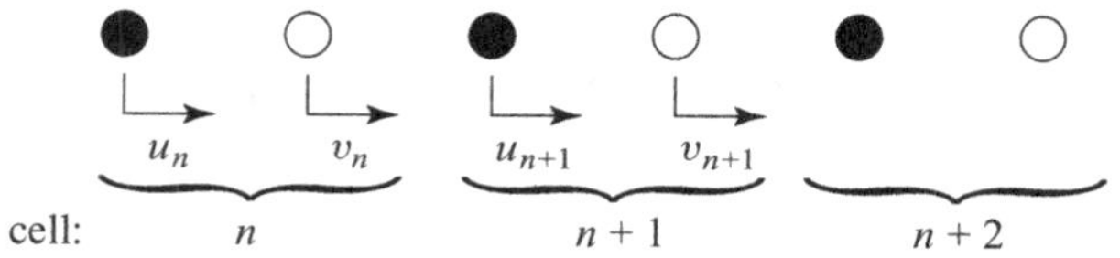


Fig. 4.2

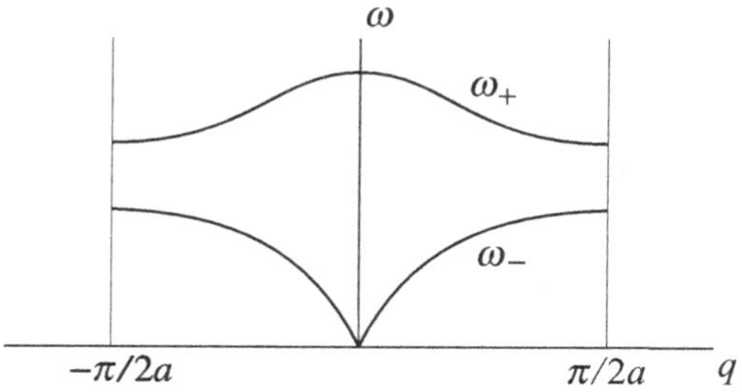


Fig. 4.3

Let us seek the solution in the form $u_n = ue^{iqna-i\omega t}$, $v_n = ve^{iqna-i\omega t}$. The solution of equation (4.20) is now reduced to the diagonalization of the matrix

$$\begin{vmatrix} 2B - M_1\omega^2 & -B(1+e^{-iqa}) \\ -B(1+e^{iqa}) & 2B - M_2\omega^2 \end{vmatrix} = 0 . \tag{4.21}$$

The eigenenergies are

$$\omega_\pm^2(q) = B\left(\frac{1}{M_1} + \frac{1}{M_2}\right) \pm B\sqrt{\left(\frac{1}{M_1} + \frac{1}{M_2}\right)^2 - \frac{4\sin^2 qa}{M_1 M_2}} . \tag{4.22}$$

Thus there exist in this case two branches of phonons, see Fig. 4.3: optical phonons with the spectrum $\omega_+$, which at $q = 0$ have finite frequency,

$$\omega_+^2(q=0) = 2B\left(\frac{1}{M_1} + \frac{1}{M_2}\right) , \tag{4.23}$$

and ordinary acoustic phonons, whose spectrum is obtained by taking the 'minus' sign in equation (4.22),

$$\omega_-(q=0) = \sqrt{2B/(M_1+M_2)}\,\sin qa , \tag{4.24}$$

which for $M_1 = M_2$ coincides with (4.19).

**Problem:** The the same for equal masses, but alternating spring constants, Fig. 4.4.

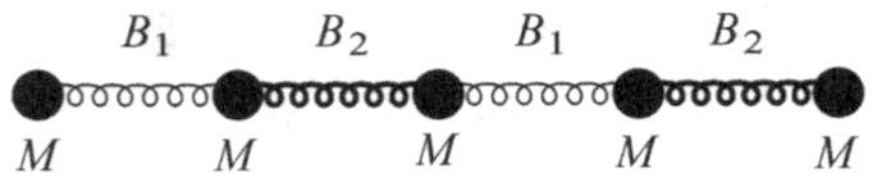


Fig. 4.4

**Solution:**

$$\omega_{\pm}^2(q) = \frac{B_1 + B_2}{M} \pm \frac{1}{M}\sqrt{B_1^2 + B_2^2 + 2B_1 B_2 \cos qa}\ . \tag{4.25}$$

In the general case the Hamiltonian of the harmonic lattice takes the form

$$\mathcal{H} = \sum_{\boldsymbol{q},\alpha} \hbar\omega_{\boldsymbol{q}\alpha}\left(b^\dagger_{\boldsymbol{q}\alpha} b_{\boldsymbol{q}\alpha} + \tfrac{1}{2}\right) + \text{const.} \tag{4.26}$$

where $\boldsymbol{q}$ is the (quasi)momentum (in the one-dimensional case $-\frac{\pi}{a} < q < \frac{\pi}{a}$; in general $\boldsymbol{q}$ lies in the first Brillouin zone), and $\alpha$ is the mode index (denoting acoustic or optical modes and the corresponding polarization – one longitudinal, two transverse modes).

## 4.3 Physical properties of crystals in the harmonic approximation

Phonons are *bosons*. The number of phonons is not fixed, which means that the phonon chemical potential should be taken as zero, see Chapter 3:

$$\mu_{\text{ph}} = 0\ . \tag{4.27}$$

Consequently, the phonon occupation number is (cf. (3.3))

$$\bar{n}_{\boldsymbol{q}} = \frac{1}{e^{\hbar\omega_{\boldsymbol{q}}/T} - 1}\ . \tag{4.28}$$

The energy of the mode $\boldsymbol{q}$ is $E_{\boldsymbol{q}} = (n_{\boldsymbol{q}} + \frac{1}{2})\hbar\omega_{\boldsymbol{q}}$. Thus the total energy (including the zero-point energy – the term with $\frac{1}{2}$ in (4.26)) is

$$E = \sum_{\boldsymbol{q}} \hbar\omega_{\boldsymbol{q}}\left(\frac{1}{e^{\hbar\omega_{\boldsymbol{q}}/T} - 1} + \frac{1}{2}\right) = \sum_{\boldsymbol{q}} \frac{\hbar\omega_{\boldsymbol{q}}}{2}\,\frac{e^{\hbar\omega_{\boldsymbol{q}}/T} + 1}{e^{\hbar\omega_{\boldsymbol{q}}/T} - 1}$$

$$= \sum_{\boldsymbol{q}} \frac{\hbar\omega_{\boldsymbol{q}}}{2} \coth\frac{\hbar\omega_{\boldsymbol{q}}}{2T}\ . \tag{4.29}$$

*General rule*

The transformation of the sum in $\boldsymbol{q}$ to an integral is done as follows:

$$\sum_{\boldsymbol{q}} \Longrightarrow \int \frac{d^d \boldsymbol{q}}{(2\pi)^d} \qquad \left(\begin{array}{l}\text{in 3d-case } d = 3;\\ \text{similarly in one-dimensional and in}\\ \text{two-dimensional cases } (d = 1, 2)\end{array}\right). \tag{4.30}$$

By (1.18) the specific heat (per unit volume) is

$$c_V = \left(\frac{\partial E}{\partial T}\right)_V = \frac{1}{T^2} \int \frac{d^3 \boldsymbol{q}}{(2\pi)^3} \frac{(\hbar\omega_{\boldsymbol{q}})^2 \, e^{\hbar\omega_{\boldsymbol{q}}/T}}{(e^{\hbar\omega_{\boldsymbol{q}}/T} - 1)^2}\,. \tag{4.31}$$

Later, in most cases we put $\hbar = 1$ (as well as the Boltzmann constant $k_B = 1$).

One can introduce the phonon density of states $D(\omega)$: $d^3\boldsymbol{q}/(2\pi)^3 = D(\omega)\,d\omega$, where $D(\omega)\,d\omega$ is the number of phonon states at energy $\omega$ in an interval $d\omega$. Normalization should be such that $\int d\omega\, D(\omega)$ over the Brillouin zone is equal to the total number of phonon modes per unit cell, i.e. equal to the total number of degrees of freedom. If there are $m$ atoms per unit cell and $N_c$ unit cells, there are $3mN_c = 3N$ modes. Then

$$c_V = 3N \int \frac{(\omega/T)^2 \, e^{\omega/T}}{(e^{\omega/T} - 1)^2} D(\omega)\, d\omega\,. \tag{4.32}$$

Different models for the phonon spectra $\omega(\boldsymbol{q})$ give different forms of the phonon density of states $D(\omega)$ and consequently in general different phonon specific heat:

(1) Einstein model, $\omega(\boldsymbol{q}) = \omega_0 = \text{const.}$ (this is not such a bad approximation for optical phonons). The phonon density of states is then a delta function, and the specific heat is

$$c_V^E = 3N \frac{(\omega_0/T)^2 \, e^{\omega_0/T}}{(e^{\omega_0/T} - 1)^2} \tag{4.33}$$

or, with $\hbar\omega_0 \equiv \Theta_E$, where $\Theta_E$ is the corresponding temperature (we then put $\Theta_E$ in (4.33) instead of $\omega_0$, or simply take $\omega_0$ in kelvin), it is

$$T \to 0 \;\; : \qquad c_V^E(T \ll \Theta_E) \simeq 3N \left(\frac{\Theta_E}{T}\right)^2 \exp\left(-\frac{\Theta_E}{T}\right) \tag{4.34}$$

$$T \gg \Theta_E : \qquad c_V^E(T \gg \Theta_E) \simeq 3N + O\left(\frac{\Theta_E}{T}\right). \tag{4.35}$$

The expression (4.35) is the Dulong–Petit law, well known from classical physics (we recall that above we put $k_B = 1$; the standard form of the Dulong–Petit law is $c = 3k_B N$).

(2) Debye model: we approximate the total phonon spectrum by the spectrum of acoustic phonons $\omega(\boldsymbol{q}) = sq$, with the appropriate upper cut-off, see the schematic picture in Fig. 4.5. The integration in $\int d^3\boldsymbol{q}$ is carried out not over

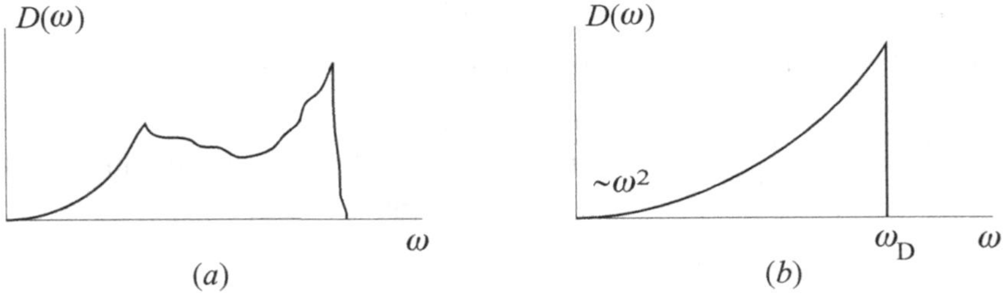


(a) Typical 'real' density of states for phonons

(b) Debye model (with the approximation $\omega = sq$)

Fig. 4.5

the real Brillouin zone, but over a *sphere*, with the volume equal to the volume of the Brillouin zone.

The number of $q$-points in the Brillouin zone is $N$; their density is $V/(2\pi)^3$; thus

$$N = \frac{V}{(2\pi)^3}\frac{4}{3}\pi q_0^3 \tag{4.36}$$

where $q_0$ – the Debye wavevector – is the maximum wavevector of the equivalent sphere in $q$-space.

The maximum frequency $\omega_D = sq_0$ is called the Debye frequency ($s$ is the sound velocity); $\Theta_D = \hbar\omega_D$ is the Debye temperature.

In effect

$$c_V = 3N\left(\frac{T}{\Theta_D}\right)^3 \cdot 3\int_0^{\Theta_D/T} \frac{x^4 e^x}{(e^x-1)^2}\,dx\ . \tag{4.37}$$

(Here we have used that $D(\omega)\,d\omega$ in (4.29) is

$$D(\omega)\,d\omega = \frac{4\pi q^2\,dq}{\frac{4}{3}\pi q_D^2} = \frac{3\omega^2}{\omega_D^2}\,d\omega\ , \tag{4.38}$$

and substituted $k_B\Theta_D$ for $\hbar\omega_D$.)

The limiting behaviour of the specific heat in the Debye model is

$$T \to 0 \quad (T \ll \Theta_D)\ : \qquad c_V \sim \frac{12\pi^4}{5}N\left(\frac{T}{\Theta_D}\right)^3, \tag{4.39}$$

$$T \to \infty \ (T \gg \Theta_D)\ : \qquad c_V \simeq 3N\ . \tag{4.40}$$

At high temperatures we have the Dulong–Petit law again. (Actually the crossover from the low-temperature behaviour $c_V \sim T^3$ to the high-temperature limit occurs not at $T \sim \Theta_D$, but at approximately $T \sim \frac{1}{4}\Theta_D$.) One has to remember that at low temperatures the phonon specific heat is $\sim T^3$; this relation is very important for many experiments.

**Problem:** Find the free energy of the harmonic crystal.

**Solution:** The energy of the harmonic crystal is

$$E = \sum_q \hbar\omega_q \left(n_q + \tfrac{1}{2}\right), \qquad n_q = 1, 2, 3, \dots . \tag{4.41}$$

The partition function is

$$Z = \prod_q \sum_{n_q} \exp\left\{-\frac{\hbar\omega_q}{T}\left(n_q + \tfrac{1}{2}\right)\right\} = \prod_q \frac{e^{-\hbar\omega_q/2T}}{1 - e^{-\hbar\omega_q/T}} . \tag{4.42}$$

The free energy (1.9) is then given by the expression

$$F = -T \ln Z = \tfrac{1}{2}\sum_q \hbar\omega_q + T\sum_q \ln(1 - e^{-\hbar\omega_q/T}) = T\sum_q \ln\left[2\sinh\frac{\hbar\omega_q}{2T}\right] . \tag{4.43}$$

Here we have used a simple transformation

$$\begin{aligned} \ln(1 - e^{-\hbar\omega/T}) &= \ln\left[2e^{-\hbar\omega/2T}\left(\frac{e^{\hbar\omega/2T} - e^{-\hbar\omega/2T}}{2}\right)\right] \\ &= -\frac{\hbar\omega}{2T} + \ln\left(2\sinh\frac{\hbar\omega}{2T}\right) . \end{aligned} \tag{4.44}$$

## 4.4 Anharmonic effects

In general the interatomic interaction, i.e. the potential energy of two atoms with coordinates $x_1$, $x_2$ is $\mathcal{V}(x_1 - x_2)$. Previously, in the harmonic approximation, we used an expansion in small deviations from the static positions

$$\mathcal{V}(x_1 - x_2) = \mathcal{V}_0 + \tfrac{1}{2}B\,(u_1 - u_2)^2 \tag{4.45}$$

$$\left(x_1 = x_1^0 + u_1, \quad x_2 = x_2^0 + u_2\right),$$

see, e.g. (4.16).

In general there exist *anharmonic terms*, proportional respectively to $(u_1 - u_2)^3$, $(u_1 - u_2)^4$, etc.:

$$\mathcal{V} = \mathcal{V}_0 + \frac{B}{2}(u_1 - u_2)^2 + \frac{\zeta}{3!}(u_1 - u_2)^3 + \frac{\nu}{4!}(u_1 - u_2)^4 + \cdots . \tag{4.46}$$

$$\left(B = \frac{\partial^2 \mathcal{V}}{\partial x^2}, \quad \zeta = \frac{\partial^3 \mathcal{V}}{\partial x^3}, \quad \nu = \frac{\partial^4 \mathcal{V}}{\partial x^4}, \quad \dots\right) . \tag{4.47}$$

The term with the coefficient $\zeta$ gives rise to cubic anharmonism, the one with $\nu$ to quartic, etc.

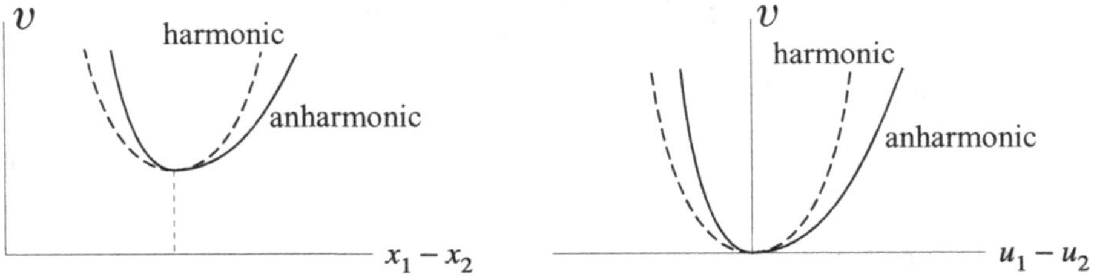


Fig. 4.6

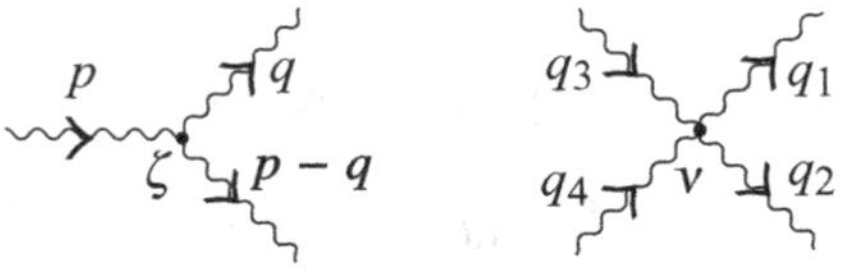


(*a*) (*b*)

Fig. 4.7

Usually the coefficient $\zeta$ is negative, $\zeta < 0$, and the potential looks like the one shown in Fig. 4.6 by the solid line, i.e. it is steeper for negative relative distortion (when two atoms approach one another) and less steep for positive $u_1 - u_2$, when the two atoms move further apart. (This is quite natural: the interaction between atoms becomes stronger when they approach – the overlap of atomic or ionic cores leads to a very strong, almost hard core repulsion; and this interaction is weaker when the distance between atoms becomes large.)

One can also write down the corresponding anharmonic terms in second quantization form, through the operators $b$, $b^\dagger$, and add them to the phonon Hamiltonian (4.26). Cubic anharmonism will give rise to terms with the structure $\zeta \sum_{p,q}(b^\dagger_{p-q} b^\dagger_q b_p + \text{h.c.})$, and quartic anharmonism, to terms of the type $\nu\big[\sum_{q_1 q_2 q_3 q_4}(b^\dagger_{q_1} b^\dagger_{q_2} b_{q_3} b_{q_4} + \text{h.c.}) + \sum_{k_1 k_2 k_3 k_4}(b^\dagger_{k_1} b_{k_2} b_{k_3} b_{k_4} + \text{h.c.})\big]$, where the momenta involved obey conservation laws $q_1 + q_2 = q_3 + q_4$, $k_1 = k_2 + k_3 + k_4$ (the total momentum of created phonons is equal to the total momentum of annihilated ones).

Anharmonic interactions have several consequences. One is that they lead to phonon–phonon interactions, schematically illustrated in Fig. 4.7. These processes are important, e.g. for thermal conductivity. These graphs, Fig. 4.7(*a*), 4.7(*b*), at this stage can be treated simply as pictorial representations of certain processes. Thus, if we depict the phonon by a wavy line, Fig. 4.7(*a*) corresponds to a process in which in the initial state there was one phonon with momentum $p$, which as a result of anharmonic interaction is transformed into two other phonons, with momenta $q$ and $p - q$; such processes are allowed by cubic anharmonism. Correspondingly,

Fig. 4.7(*b*) describes the process of scattering of two phonons with momenta $q_3$ and $q_4$ into two others, with momenta $q_1$, $q_2$; such processes are contained in the quartic term of the Hamiltonian. But one can give these 'pictures' much more meaning: they are actually *Feynman diagrams*, which allow one really to *calculate* the probabilities of corresponding processes. This method will be discussed in more detail in Chapter 8, and in Chapters 9–11 we will show how it works in many specific problems.

Another important consequence of anharmonic interactions is their role in thermal expansion and in the melting of crystals. This will be discussed in the following sections.

### 4.4.1 Thermal expansion

By (4.43) the free energy of a crystal in the harmonic approximation is (putting $\hbar = 1$)

$$F = T \sum_q \ln \left[ 2 \sinh \frac{\omega_q}{2T} \right] . \tag{4.48}$$

With the inclusion of anharmonicity, i.e. of the phonon–phonon interaction, this is no longer true. But one can still use this expression approximately in the so-called *quasiharmonic approximation*, accounting for anharmonicity in the following way. The phonon frequencies $\omega_q$ in the anharmonic crystal in general depend on the specific volume, $\omega_q(V)$ (see Fig. 4.6: the curvature of the potential $\upsilon(x)$, $d^2\upsilon/dx^2$, which according to (4.3), (4.4) determines phonon frequencies, in the anharmonic case depends on $x$). Usually this dependence is described by the phenomenological relation

$$\frac{V}{\omega} \frac{d\omega}{dV} = \frac{d \ln \omega}{d \ln V} = -\gamma , \tag{4.49}$$

which is called the Grüneisen approximation; $\gamma$ is the Grüneisen constant (usually, in ordinary crystals, $\gamma \sim 1$–$2$).

The total free energy as a function of volume can then be written as

$$F(V) = \frac{1}{2\kappa} \left( \frac{\delta V}{V} \right)^2 + T \sum_q \ln \left[ 2 \sinh \frac{\omega_q(V)}{2T} \right] . \tag{4.50}$$

The first term in (4.50) is the elastic energy after deformation $\delta V$, and $\kappa$ is the lattice compressibility (inverse bulk modulus). In equation (4.50) we considered the situation when we (artificially) fix the volume of the system $V$, which may differ from the equilibrium volume without phonons by the distortion $\delta V$. Then we indeed should include in the total energy the first term of equation (4.50) describing such deformation. The second term gives the phonon contribution to the free energy, which is calculated by treating the lattice as harmonic, but *at a given volume* $V$ (the expression (4.43) for the phonon contribution, strictly speaking, is valid only for the harmonic crystal). However, whereas in the purely harmonic case the phonon frequencies are constant, independent of the distance between atoms or of the volume, here we assume that they do depend on $V$, by the relation (4.49). Thus we effectively take into account anharmonic effects, albeit approximately, via the phenomenological Grüneisen relation. Such a scheme is often used in treating anharmonic lattices, and it is called the quasiharmonic approximation. The equilibrium volume $V$ in this scheme should be determined by minimizing the free energy (4.50); due to the presence of the second term the resulting volume would depend on temperature, and this dependence gives thermal expansion.

Differentiating (4.50) in $V$ and using (4.49), we obtain

$$\frac{1}{\kappa}\frac{\delta V}{V} = \sum_q \gamma \frac{\omega_q}{2} \coth\frac{\omega_q}{2T} = \gamma \bar{E}(T)\,, \tag{4.51}$$

where $\bar{E}(T)$ is the average energy of the lattice (cf. (4.29)).

Thus the volume thermal expansion $\beta$ is:

$$\beta = \frac{1}{V}\frac{\partial(\delta V)}{\partial T} = \kappa\gamma\frac{\partial \bar{E}(T)}{\partial T} = \gamma\kappa c_V = \gamma\frac{c_V}{B}\,, \tag{4.52}$$

where $B = 1/\kappa$ is the bulk modulus.

The linear thermal expansion coefficient is

$$\alpha = \frac{\partial l}{\partial T} = \frac{1}{3}\beta \left(\text{as } \beta = \frac{1}{V}\frac{\partial V}{\partial T} = \frac{1}{l^3}\frac{\partial(l^3)}{\partial T} = \frac{3}{l}\frac{\partial l}{\partial T} = 3\alpha\right).$$

Thus

$$\boxed{\alpha = \gamma\cdot\frac{c_V}{3B}}\,. \tag{4.53}$$

This is called the Grüneisen equation. We see that we get nonzero thermal expansion only when we include anharmonic effects (the dependence of the phonon frequencies on interatomic distance or on volume, which is phenomenologically described by the relation (4.49)). For constant phonon frequencies (Grüneisen constant $\gamma = 0$) thermal expansion would be absent.

The Grüneisen equation (4.53) is very useful, as it establishes the relation between different measurable quantities and allows us, e.g. to calculate the thermal expansion if the specific heat and bulk modulus are known. It is often used in practice, to check the consistency of different thermophysical data of real materials.

Why is anharmonism necessary for thermal expansion? The physics of the expansion is illustrated in Figs. 4.8 and 4.9. At low temperatures the system is

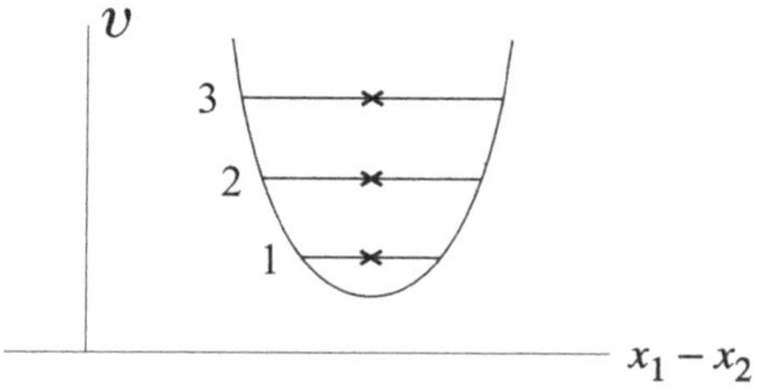


Fig. 4.8

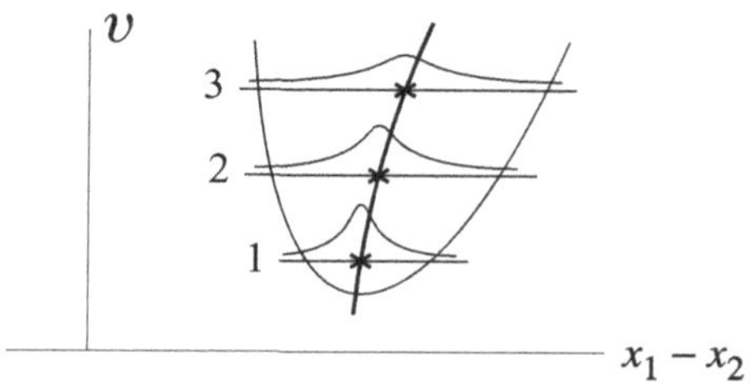


Fig. 4.9

at the lowest quantum level of the oscillator (state 1). At higher temperatures the average distance between atoms is given by averaging over all occupied higher-lying quantum levels. In the harmonic case we have the situation shown in Fig. 4.8, and the average displacement at each level, i.e. the average lattice parameter, is the same at each quantum level. Thus it does not change with temperature, and in the purely harmonic crystal we would not have any thermal expansion.

On the other hand, in an anharmonic crystal the interaction potential qualitatively looks as shown in Fig. 4.9 (cf. Fig. 4.6). Then with increasing temperature, when the higher levels become occupied, the average distance between atoms (lattice parameter) *increases* with temperature – and this is the conventional thermal expansion.

### 4.4.2 Melting

From (4.49) we see that the dependence of the phonon frequencies on the specific volume can be rewritten as

$$\omega = \omega_0 \left( \frac{V}{V_0} \right)^{-\gamma} . \tag{4.54}$$

As the change of volume (due to thermal expansion) is usually small, we can approximately write instead of (4.54) (and using the definition of thermal expansion $\beta = 3\alpha = \frac{1}{V}\frac{\partial V}{\partial T}$)

$$V \simeq V_0(1 + 3\alpha T) . \tag{4.55}$$

This means that

$$\omega(T) \simeq \omega_0(1 - 3\alpha\gamma T) . \tag{4.56}$$

Actually, as follows from a more complete treatment, see Section 8.6, the description of phonons always contains not the phonon frequency $\omega$, but rather $\omega^2$. Consequently, the equation describing the change of phonon frequencies with temperature

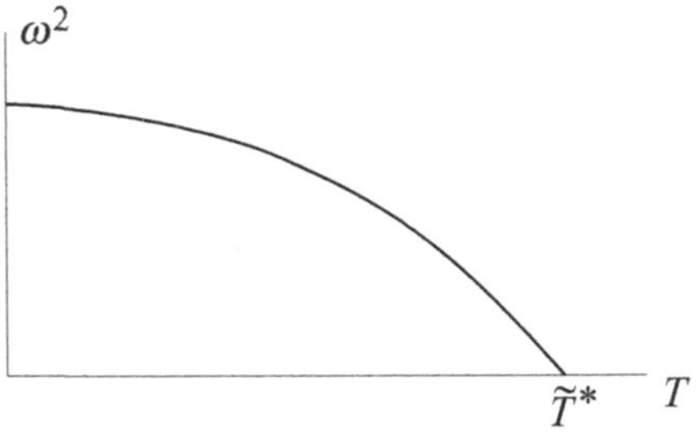


Fig. 4.10

has not the form (4.56), but is rather

$$\omega^2(T) = \omega_0^2(1 - 6\alpha\gamma T) . \tag{4.57}$$

The more general equation for $\omega(T)$ is

$$\omega^2(T) = \omega_0^2\big(1 - cn(T)\big) , \tag{4.58}$$

where $n(T) = 1/(e^{\omega/T} - 1)$ is the average number of excited phonons (this will be shown later); here $c$ is some constant ($\sim\zeta^2/B^3$). Then for $\omega \sim \omega_0 < T$

$$\omega^2 = \omega_0^2\left(1 - \frac{cT}{\omega_0}\right) , \tag{4.59}$$

i.e. for high temperatures the expression (4.58) gives the same temperature dependence as (4.56), (4.57).

These expressions give the dependence of $\omega$ on $T$ shown in Fig. 4.10. At a certain temperature $\tilde{T}^*$ the phonon frequency $\omega^2 = 0$, and for $T > \tilde{T}^*$ it becomes negative, which means an instability of our system and the transition to a new state, e.g. melting of the crystal.

I remind readers that in quantum mechanics the time dependence of the wavefunction is given by $\psi(t) = \psi(0) \exp i\omega t$. When $\omega^2$ becomes negative, i.e. $\omega t = \pm i|\omega|t$, this would give an exponential growth of the corresponding quantities. In our case, when the phonon frequencies cross zero, this would mean an exponential growth of the number of corresponding phonons, or of the respective distortion, which means absolute instability of the initial state, in this case a crystal. Whether such instability would indeed correspond to melting or to a structural transition to a different crystal structure, depends on which particular phonon mode becomes unstable. If this is the phonon with momentum $\boldsymbol{Q}$, then this implies the change of the crystal structure with the formation of a superstructure with this wavevector, $u(\boldsymbol{r}) \sim \exp(i\,\boldsymbol{Q}\cdot\boldsymbol{r})$. Melting corresponds in this language to the softening of transverse phonons at $\boldsymbol{q} = 0$ (in other words, the *shear modulus* becomes negative).

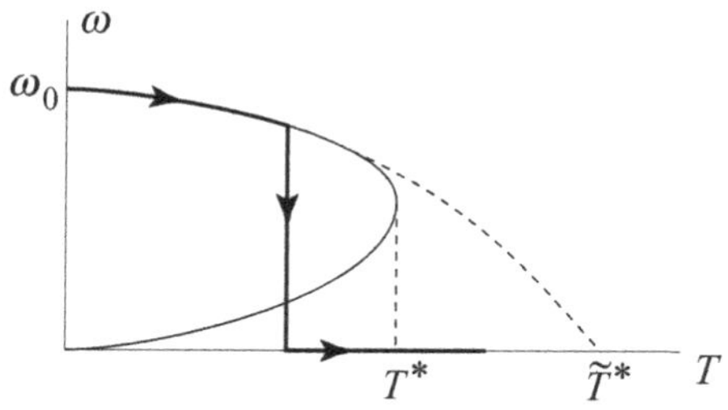


Fig. 4.11

In general, the description of structural phase transitions as an instability of the original crystal structure due to softening of particular phonon modes is known as the *soft mode* concept (Ginzburg, Cochran, Anderson). It is widely used in solid state physics and leads to definite predictions which can be checked experimentally.

Returning to melting, we see that in this approximation the transition would be continuous, i.e. second order (the phonon frequency goes continuously to zero). But on general grounds melting should be a first-order phase transition (one can show that in this case there exist cubic invariants in the Landau free energy expansion, see Section 2.2 above). How can we correct this drawback? It is more correct to take in (4.58) not $n(T) = 1/(e^{\omega_0/T} - 1)$ (which leads to (4.59)), but the number of phonons with an already renormalized, new frequency $\omega$, $1/(e^{\omega/T} - 1)$. Then instead of (4.59) we would obtain

$$\omega = \omega_0\big(1 - cn(\omega)\big) = \omega_0\left(1 - \frac{cT}{\omega}\right) \tag{4.60}$$

(here we have simplified the mathematics by writing the equation not for $\omega^2$, but for $\omega$; this is sufficient for our qualitative treatment).

The equation (4.60) is a self-consistent equation for $\omega$. Its solution is

$$\omega = \frac{\omega_0}{2} \pm \sqrt{\frac{\omega_0^2}{4} - cT\omega_0}\,, \tag{4.61}$$

see Fig. 4.11. We see that in contrast to the previous treatment, the self-consistent solution with real $\omega$ exists only up to a temperature $T^* = \omega_0/4c$ (smaller than $\tilde{T}^*$), after which it becomes complex, i.e. the time dependence of the vibration amplitude (4.2) $\sim e^{+i\omega t}$ is again diverging, which indeed signals an instability of the lattice, in our case melting. And in this theory melting would be a first-order phase transition, with a jump in $\omega$, as it should be. The results are qualitatively the same if we had proceeded from equation (4.58) for $\omega^2$.

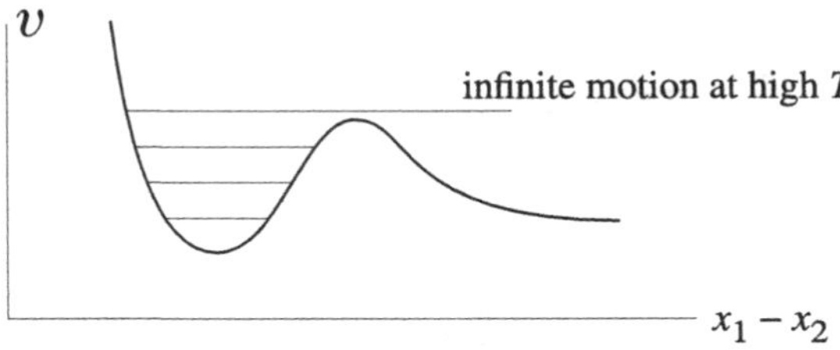


Fig. 4.12

The softening of phonons means at the same time a decrease of the bulk modulus of the lattice, $B$: as $\omega^2 = B/M$, we get from (4.57)

$$B(T) \simeq B_0(1 - 6\alpha\gamma T) . \tag{4.62}$$

But, from the Grüneisen relation (4.53), $\alpha \sim 1/B$, which again gives a self-consistent equation for $B$ (or for $\omega^2$), equivalent to (4.58):

$$B(T) = B_0 \left(1 - \frac{2\gamma^2 c_V T}{B(T)}\right) , \tag{4.63}$$

i.e. we again obtain a quadratic equation for $B(T)$ similar to (4.60), which will again give first-order melting.

Once again, strictly speaking an instability of a phonon does not mean necessary melting; it can signal, e.g. a transformation into another crystal structure. To check that we will indeed have melting, we have to show that the *shear modulus* (shear modes for $q \to 0$) becomes unstable. But qualitatively the picture described is correct, and it is also consistent with other approaches to melting, described below.

Why does melting occur at all? How can one explain it qualitatively? Let us extend Fig. 4.6 a bit. From (4.46), $v = v_0 + \frac{1}{2}Bu^2 - \frac{1}{3!}|\zeta|u^3$, i.e. this potential actually looks as in Fig. 4.12. We see that there exists an infinite motion at high enough temperature! The lattice is no longer stable, which means melting. Thus we again see that the melting is intrinsically connected with the anharmonicity of the lattice.

### *4.4.3 Another approach to melting. Quantum melting*

From equation (4.10) we see that the vibration amplitude can be expressed through the phonon operators as

$$u = \sqrt{\frac{\hbar}{2M\omega}}\,(b^\dagger + b) \tag{4.64}$$

(for one phonon mode). The average shift from the equilibrium position is of course zero, $\langle u\rangle = 0$, but the square of the average amplitude of vibrations is nonzero:

$$\begin{aligned}\langle u^2\rangle &= \frac{\hbar}{2M\omega}\big\langle(b^\dagger + b)^2\big\rangle = \frac{\hbar}{2M\omega}\langle b^\dagger b^\dagger + bb + b^\dagger b + bb^\dagger\rangle \\ &= \frac{\hbar}{2M\omega}\big(2\langle b^\dagger b\rangle + 1\big) = \frac{\hbar}{M\omega}\left(n + \tfrac{1}{2}\right)\end{aligned} \tag{4.65}$$

(we use the commutation relation $bb^\dagger - b^\dagger b = 1$).

Thus

$$\langle u^2\rangle = \frac{\hbar}{M\omega}\left(n + \tfrac{1}{2}\right) = \frac{\hbar^2}{M\Theta_{\rm D}}\left(n + \tfrac{1}{2}\right) \tag{4.66}$$

(there may enter some numerical factors such as 3 because of the presence of several phonon modes). The limiting values of this mean square vibration amplitude are:

$$T \to 0 \;:\; \langle u^2\rangle \sim \frac{\hbar^2}{2M\Theta_{\rm D}} \qquad \text{(important: note that } \langle u^2\rangle_{T=0} \neq 0\text{; these are the famous zero-point oscillations)} \tag{4.67}$$

$$T \gg \Theta_{\rm D} : \; \langle u^2\rangle \sim \frac{\hbar^2 T}{M\Theta_{\rm D}^2} \qquad \left(\text{actually it is equal to } \frac{9\hbar^2 T}{M\Theta_{\rm D}^2}\right). \tag{4.68}$$

When the vibration amplitude becomes comparable to the lattice spacing itself, $\langle u^2\rangle \sim a^2$ (actually when $\langle u^2\rangle/a^2 \sim 0.2$), melting occurs; this is the *Lindemann criterion* of melting.

The factor $\langle u^2\rangle$ also enters into the intensity of X-ray scattering in crystals, in the theory of the Mössbauer effect, etc. It enters through the factor $e^{-2W}$, where

$$W \sim \langle u^2\rangle, \qquad W = \begin{cases} \dfrac{3}{8}\dfrac{\hbar^2 K^2}{M\Theta_{\rm D}} & (T \to 0) \\[2ex] \dfrac{3}{2}\dfrac{\hbar^2 K^2 T}{M\Theta_{\rm D}^2} & (T \gg \Theta_{\rm D})\,. \end{cases} \tag{4.69}$$

$K$ is the Umklapp wavevector $\sim\hbar/a$; the quantity $W$ is called the *Debye–Waller* factor.

Thus, in melting an important parameter is

$$\Lambda = \frac{\langle u^2\rangle}{a^2}\,, \tag{4.70}$$

the so-called *de Boer* parameter (usually this term is used at $T = 0$, for zero-point vibrations; this is then the *quantum de Boer parameter*). At $T = 0$

$$\langle u^2 \rangle \sim \frac{\hbar}{M\omega} \qquad \text{with} \quad \omega = \sqrt{\frac{B}{M}} \, . \tag{4.71}$$

The stiffness of the lattice $B$ (the bulk modulus) can be estimated as follows: when an atom is shifted from its equilibrium position by a distance $\sim a$ ($a$ is the lattice parameter), the change in potential energy $\sim Ba^2$ is of the order of the typical interaction between atoms $v(a)$, i.e.

$$B \sim \frac{v}{a^2} \, . \tag{4.72}$$

As a result the quantum de Boer parameter $\Lambda$ (4.70) becomes

$$\Lambda \sim \frac{\hbar}{a} \frac{1}{\sqrt{Mv}} \, . \tag{4.73}$$

If $\Lambda \gtrsim 0.2$–$0.3$, the crystal is unstable with respect to *zero-point motion* even at $T = 0$, which means *quantum melting*. This will also be important for electrons (see the discussion of Wigner crystals and cold melting below, and in Section 11.8).

When will a substance melt by quantum fluctuations and remain liquid down to $T = 0$? According to (4.73), better chances for this exist if:

- the mass of the atom $M$ is small;
- the interaction $v$ is weak.

The best candidate for this is helium: it is light, and He atoms have filled 1s shells (inert atoms), so that the He–He interaction is weak. That is why helium remains liquid down to $T = 0$ (at normal pressure).

Another good candidate could have been hydrogen. It is even lighter than helium. But the effective interaction between hydrogen atoms and even between $H_2$ molecules is too strong. Nevertheless there have been active experimental attempts to stabilize hydrogen in a liquid phase down to the lowest temperatures, i.e. to prevent its crystallization, with the idea that it would then experience Bose condensation and possibly would become superfluid, as $^4$He. These attempts have not yet succeeded; instead Bose condensation was reached in completely different systems, in optically trapped and supercooled alkali atoms (Rb, Cs, etc.).

As we have said, the condition for quantum melting is determined by the value of the quantum de Boer parameter $\Lambda$. Typical values of $\Lambda$ are: $\Lambda_{^3\mathrm{He}} = 0.5$, $\Lambda_{^4\mathrm{He}} \approx 0.4$, $\Lambda_{\mathrm{H_2}} \approx 0.3$, $\Lambda_{\mathrm{Ne}} \approx 0.1$. (In the book by Ashcroft and Mermin (1976) other values for $\Lambda$ are given, which is due to the different normalization used there, but the ratio for different elements is the same.)

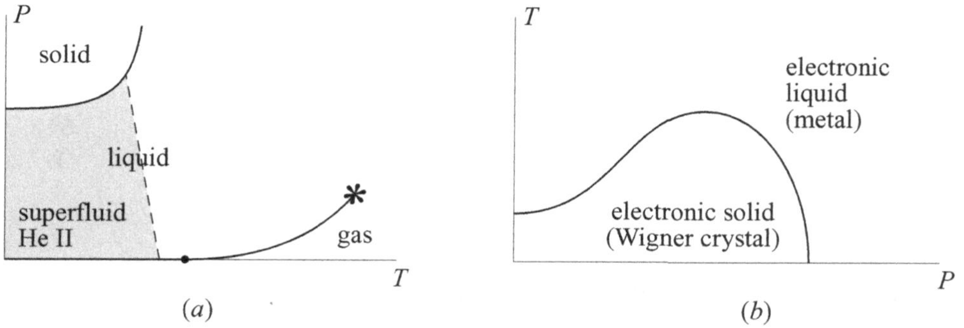


Fig. 4.13

Under pressure the average distance $a$ between atoms decreases, the atom–atom interaction $\upsilon(a)$ increases, and $\Lambda$ may decrease or increase depending on the behaviour of $\upsilon a^2$ (see (4.73)). For neutral particles (helium) the interaction $\upsilon$ increases faster, $\Lambda$ decreases, and He becomes solid under pressure (the phase diagram of $^4$He is shown schematically in Fig. 4.13($a$)). For electrons the opposite is true: the characteristic Coulomb interaction between electrons is $\upsilon = e^2/a$, and $\Lambda = \hbar/a\sqrt{me^2/a} \sim 1/\sqrt{a})$, where $a$ is the average distance between electrons. Under pressure this distance decreases, the de Boer parameter increases, and there occurs *cold melting* of the electronic (Wigner) crystal (Fig. 4.13($b$)).

When helium crystallizes under pressure, it still has a large value of $\Lambda$, i.e. large quantum fluctuations. This means that it is a *quantum crystal*, for which quantum effects are important. Thus, for instance, the vacancies in solid He are very mobile, they behave like quasiparticles. There is even discussion of a possible *superfluidity* of vacancies in solid He ('superfluidity in solid' – 'supersolid'). In 2004 there appeared the first experimental indications that it could indeed be true, although this question is still controversial.

Yet another manifestation of quantum effects in solid He is their importance for the exchange interaction, especially in solid $^3$He. The exchange interaction in solid $^3$He ($^3$He is a fermion!) is usually due to ring exchange in the presence of vacancies, see Fig. 4.14. As a result of this process, after three steps the atoms 1 and 2 interchange their positions. This can be shown to lead to a partial ferromagnetism of solid $^3$He.

### *4.4.4 Low-dimensional solids; why is our world three-dimensional?*

**Problem:** Using the approach described above (see, e.g. equation (4.65)), discuss what would be the situation with crystals in the one-dimensional (1d) and two-dimensional (2d) cases.

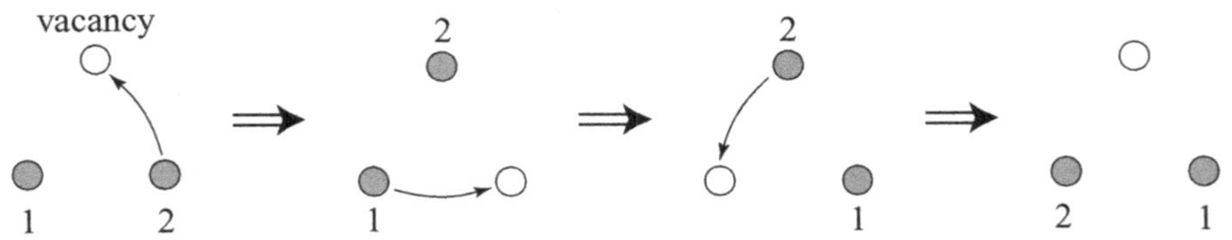


Fig. 4.14

**Solution:** Similar to the single oscillator case, see (4.65), in a crystal the average vibration amplitude $\langle u^2 \rangle$ is

$$\langle u^2 \rangle = \sum_{\boldsymbol{k}} \frac{\hbar}{M\omega_{\boldsymbol{k}}} \left( \langle b_{\boldsymbol{k}}^\dagger b_{\boldsymbol{k}} \rangle + \frac{1}{2} \right) = \sum_{\boldsymbol{k}} \frac{\hbar}{M\omega_{\boldsymbol{k}}} \left( n_{\boldsymbol{k}} + \frac{1}{2} \right)$$

$$= \int \frac{d^d \boldsymbol{k}}{(2\pi)^d} \frac{\hbar}{M\omega_{\boldsymbol{k}}} \left( \frac{1}{e^{\omega_{\boldsymbol{k}}/T} - 1} + \frac{1}{2} \right) . \tag{4.74}$$

1. First consider the case of zero temperature. At $T = 0$, $n_{\boldsymbol{k}} = 0$, and

$$\langle u^2 \rangle = \int \frac{d^d \boldsymbol{k}}{(2\pi)^d} \frac{\hbar}{M\omega_{\boldsymbol{k}}} \cdot \frac{1}{2} . \tag{4.75}$$

1($a$) At $k \to 0$, $\omega_{\boldsymbol{k}} = sk$ (acoustic phonons, $s$ is the sound velocity). Thus for $d = 1$ (one-dimensional system)

$$\langle u^2 \rangle \sim \int \frac{dk}{sk} , \tag{4.76}$$

i.e. $\langle u^2 \rangle$ is logarithmically divergent! This means that in a 1d system even at $T = 0$ $\langle u^2 \rangle \to \infty$, and as a result *there is no long-range crystalline order* in this case! Zero-point vibrations are so strong that they destroy the ordered state even at $T = 0$! Mathematically this divergence is due to the behaviour of the integral (4.75) at the lower limit of integration $k \to 0$, or $\omega \to 0$; this is what is called an infrared divergence. The upper limit of integration is determined by the upper edge of the spectrum, which in solids is finite, so that usually there are no divergences there. Therefore here and in the future we do not specify this upper limit of integration.

1($b$) $T = 0$, 2d system. Here $d^2\boldsymbol{k} \sim k\, dk$, and everything is OK,

$$\langle u^2 \rangle \sim \int \frac{k\, dk}{sk} , \tag{4.77}$$

the integral is convergent, the mean square vibration amplitude $\langle u^2 \rangle$ is finite, and in general there may exist long-range crystalline order at $T = 0$ in 2d case.

1($c$) And of course everything is fine in 3d systems, where $d^3\boldsymbol{k} \sim k^2\, dk$.

2. $T \neq 0$. Again the behaviour as $k \to 0$ is critical; thus we consider the region $\omega_k = sk < T$. This part of the spectrum exists if the spectrum is gapless. In this region

$$n_k = \frac{1}{e^{\omega_k/T} - 1} \simeq \frac{1}{\omega_k/T} = \frac{T}{sk} ,$$

and from (4.74) we get

$$\langle u^2 \rangle \sim \int \frac{d^d\mathbf{k}}{\omega_k}\left(\frac{T}{\omega_k} + \frac{1}{2}\right) . \tag{4.78}$$

In this expression again the most dangerous part is that close to $\mathbf{k} = 0$ or $\omega = 0$, and we can ignore the term with $\frac{1}{2}$ in the integrand.

2(*a*) 1d case: $\langle u^2 \rangle$ was divergent already at $T = 0$. At $T \neq 0$ it is even more divergent – the corresponding expression for $\langle u^2 \rangle$ would be proportional to $T \int \frac{dk}{s^2k^2}$ and would diverge not logarithmically, but linearly.

2(*b*) 2d case: The most dangerous term has the form

$$\langle u^2 \rangle \sim \int \frac{k\,dk \cdot T}{\omega_k^2} \sim T \int \frac{k\,dk}{s^2k^2} , \tag{4.79}$$

i.e. in the 2d case, $\langle u^2 \rangle$ is logarithmically divergent *at any finite T*. Thus *at finite temperature there is no long-range order either in 1d or in 2d systems.*

2(*c*) 3d case. Here all is 'quiet' even at nonzero temperatures:

$$\langle u^2 \rangle \sim T \int \frac{k^2\,dk}{s^2k^2} , \tag{4.80}$$

which is convergent, so that the fluctuations are finite. Luckily for us! Otherwise everything surrounding us, and maybe we ourselves, would not be stable. Our bones, and all other tissues would 'melt'. (If you like, this may be the physical explanation of why we exist in a three-dimensional world.)

There exists a general theorem – the *Mermin–Wagner theorem* – which states that whenever an ordering corresponds to a breaking of *continuous* symmetry, there is *no long-range order* in one-dimensional and two-dimensional cases *at any nonzero temperature*. Actually this theorem is intrinsically connected with the Goldstone theorem about the presence of gapless Goldstone excitations for a broken continuous symmetry, mentioned above in Section 2.7.4: we saw above that for the divergence of the mean square vibration amplitude $\langle u^2 \rangle$ and consequently for the instability of the crystal, it is crucial that the energy spectrum $\omega_k$ which stands in the denominator in equations (4.74)–(4.80) should be gapless, $\omega_k \to 0$ for $k \to 0$. All these features are especially important in different magnetic systems, see below, Chapter 6, but also in low-dimensional superconductors, etc.

# 5

# General Bose systems; Bose condensation

## 5.1 Bose condensation

There exist in nature different kinds of bosons. These may be phonons or photons. Their number is not conserved, and consequently their chemical potential is $\mu = 0$, see the previous chapter.

We meet a different situation in the case of systems of bosons with conserved particle number. These are, for example, atoms or molecules with an *even spin*, such as $^4He$. For these cases in general the chemical potential is $\mu \neq 0$; it is determined by the condition (1.25), that is by the requirement that the total number of particles, or particle density, is fixed. In these cases we meet the phenomenon of *Bose condensation.*

Consider ideal noninteracting bosons with the spectrum $\varepsilon_{\boldsymbol{p}} = \boldsymbol{p}^2/2m$. From equation (3.3) we obtain the number of bosons in a unit of the phase space $d^3\boldsymbol{p}/(2\pi\hbar)^3$:

$$dn_{\boldsymbol{p}} = \frac{d^3\boldsymbol{p}}{(2\pi\hbar)^3} \frac{1}{e^{(\varepsilon_{\boldsymbol{p}}-\mu)/T} - 1} , \tag{5.1}$$

or

$$n = \frac{m^{3/2}}{\sqrt{2}\pi^2\hbar^3} \int_0^\infty \frac{\sqrt{\varepsilon}\, d\varepsilon}{e^{(\varepsilon-\mu)/T} - 1} . \tag{5.2}$$

To show that the phenomenon of Bose condensation considered below is intrinsically a quantum phenomenon, we keep here the Plank constant $\hbar$ in an apparent way, and do not put it equal to 1, as elsewhere in this book. For a *given density* (given number of particles) equation (5.2) is an equation for the chemical potential $\mu(T)$. At high temperatures, $\mu(T) < 0$ (cf. (3.10) and Fig. 3.1). As discussed in the problem in Chapter 3, the chemical potential $\mu$ increases with decreasing temperature, and it tends to zero at a certain temperature $T_0$, given by the condition

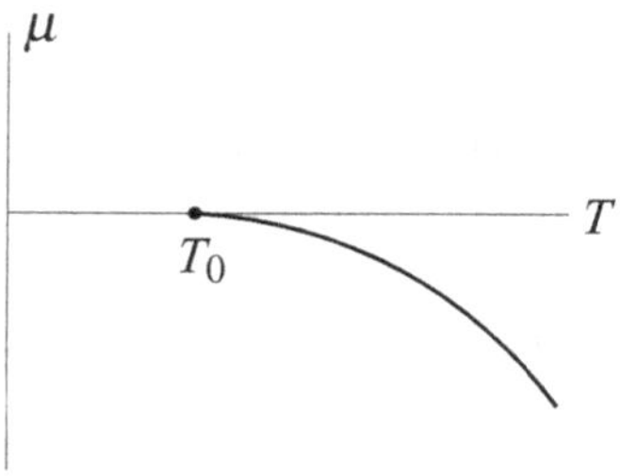


Fig. 5.1

(we use below the dimensionless variable $z = \varepsilon/T$)

$$n = \frac{m^{3/2}\,T^{3/2}}{\sqrt{2}\pi^2\hbar^3}\int_0^\infty \frac{\sqrt{z}\,dz}{e^z - 1}\,, \tag{5.3}$$

see Fig. 5.1. The integral in (5.3) is finite; it can be expressed through the Riemann $\zeta$ function,

$$\int_0^\infty \frac{z^{x-1}\,dz}{e^z - 1} = \Gamma(x)\,\zeta(x)\,, \tag{5.4}$$

where $\Gamma(x)$ is the gamma function.

Similarly, more general integrals of this type are

$$\int_0^\infty \frac{z^{x-1}\,dz}{e^z + 1} = (1 - 2^{1-x})\,\Gamma(x)\,\zeta(x)\,. \tag{5.5}$$

These formulae will also be useful for fermions later on.

In our case ($x = \frac{3}{2}$) $\Gamma\left(\frac{3}{2}\right) = \frac{\sqrt{\pi}}{2}$, $\zeta\left(\frac{3}{2}\right) \simeq 2.61$. This finally gives

$$T_0 = \frac{3.31\,\hbar^2}{m}\,n^{2/3}\,. \tag{5.6}$$

As follows from the general principles of statistical mechanics, the chemical potential $\mu(T)$ for Bose particles should always be negative (or, in general, should lie below the bottom of the corresponding band); otherwise the sum over $N$ (from 0 to $\infty$) in (1.33), (1.32) would not converge, see Chapter 3 (for free bosons we have to put in (1.33) $E_{nN} = \varepsilon_{\boldsymbol{p}} N$, where $\varepsilon_{\boldsymbol{p}} = \boldsymbol{p}^2/2m$).

For $T < T_0$ there is no solution of equation (5.2) with $\mu < 0$, but that is a *necessary requirement* for a Bose system! Indeed the left-hand side of equation (5.3) is constant, whereas the right-hand side goes to zero as $T \to 0$.

The solution of this apparent paradox is the following: there is a *macroscopic occupation* of the state with $\boldsymbol{p} = 0$ or with $\varepsilon = 0$. The transition from summation over discrete values of $\boldsymbol{p}$, $\sum_{\boldsymbol{p}}$, to the integral $\int d^3\boldsymbol{p}/(2\pi)^3$, which we usually do when going from a finite system to a system with infinite volume and which we 'automatically' did in writing equation (5.2), is not valid in this case. Due to the macroscopic occupation of one particular state, here the state with $\boldsymbol{p} = 0$ (which is allowed for bosons!), the number of particles in this state is infinite, and in the corresponding summation over $\boldsymbol{p}$ the first term ($\boldsymbol{p} = 0$, $\varepsilon = 0$) tends to infinity. The remaining sum $\sum_{\boldsymbol{p}\neq 0}$ can then be transformed into an integral in (5.3), which then can be finite. There is no contradiction any more: the number, or density of particles for all $\varepsilon > 0$, given by the expression (5.3), can indeed go to zero as $T \to 0$, but the *total* number of particles can still be conserved: the 'missing' particles are now *in the condensate*, in the state with $\boldsymbol{p} = 0$ and $\varepsilon = 0$.

Thus at $T < T_0$, where the chemical potential is identically zero, we have for the states with $\varepsilon > 0$,

$$dN_\varepsilon\Big|_{\varepsilon>0} = \frac{V\, m^{3/2}}{\sqrt{2}\pi^2\hbar^3}\frac{\sqrt{\varepsilon}\, d\varepsilon}{e^{\varepsilon/T} - 1}\,, \tag{5.7}$$

and the total number of particles with $\varepsilon > 0$ is

$$N_{\varepsilon>0} = \int dN_\varepsilon = \frac{V(mT)^{3/2}}{\sqrt{2}\pi^2\hbar^3}\int_0^\infty \frac{\sqrt{z}\, dz}{e^z - 1} = N\left(\frac{T}{T_0}\right)^{3/2}. \tag{5.8}$$

The remaining

$$N_{\varepsilon=0} = N - N_{\varepsilon>0} = N\left(1 - \left(\frac{T}{T_0}\right)^{3/2}\right) \tag{5.9}$$

particles are in the state $\boldsymbol{p} = 0$ – in a condensate. This is the *Bose condensation* (or Bose–Einstein condensation).

**Problem:** Check what would be the situation with Bose condensation of an ideal Bose gas in one-dimensional and two-dimensional systems.

**Solution:** Why does Bose condensation occur in 3d systems? In equations (5.2), (5.3) (with the minimal possible value of the chemical potential $\mu = 0$) the integral *converges*. Then as $T \to 0$ we cannot fulfil the condition (5.3), and we have to put an infinite number of particles in one particular state – in a condensate.

It turns out that in 1d and 2d cases the corresponding integrals diverge, and there exists a solution of similar equations with $\mu \neq 0$ *for all* $T > 0$. As a result there occurs *no Bose condensation* at any nonzero temperature in 1d and 2d cases!

Indeed, the equation similar to (5.2) has, in general, schematically the form

$$n = \text{const.}\int_0^\infty \frac{\rho_d(\varepsilon)\, d\varepsilon}{e^{(\varepsilon-\mu)/T} - 1} \tag{5.10}$$

where at small $\varepsilon$ the density of states $\rho_d(\varepsilon) \sim \sqrt{\varepsilon}$ for the 3d case ($d = 3$), $\rho_2(\varepsilon) \sim \rho_2 = \text{const.}$, and $\rho_1(\varepsilon) \sim 1/\sqrt{\varepsilon}$.[1]

Let us make a change of variables, $\varepsilon/T = z$, $\mu/T = \tilde{\mu}$. We then obtain

$$n = Tc \int \frac{dz}{e^{z-\tilde{\mu}} - 1} \qquad \text{(2d case)} \qquad (5.11)$$

and

$$n = c\, T^{1/2} \int \frac{dz}{\sqrt{z}\left(e^{z-\tilde{\mu}} - 1\right)} \qquad \text{(1d case)} \qquad (5.12)$$

($c$ is a certain constant). The integral in (5.3) is finite even for $\tilde{\mu} = 0$, and when $T \to 0$ we have a contradiction: the left-hand side of equation (5.3) is finite, and the right-hand side goes to zero. The resolution of this paradox leads to Bose condensation. However, for the 2d case the integral in (5.11) is logarithmically divergent for $\tilde{\mu} = 0$. Thus when $T \to 0$ we can compensate the small factor $T$ in (5.11) by the corresponding increase of the integral, choosing the appropriate dependence $\tilde{\mu}(T)$ ($\neq 0$) so that the product $Tc \int dz/(e^{z-\tilde{\mu}} - 1)$ remains finite (equal to $n$).

The same is true also for the 1d case: the integral in (5.12) diverges even more strongly than in the 2d case, which means that there should be no Bose condensation in one-dimensional systems either.

In an ideal Bose gas one can calculate all thermodynamic functions at the Bose condensation transition. It turns out that thermodynamic functions $E$, $F$, $\Phi$, $S$, $c_V$ are continuous, i.e. this transition is not even a second-order phase transition (at the second-order phase transition there is a jump in $c$). Here, not $c_V$ but $dc_V/dT$ has a jump, i.e. it is a 'third-order' phase transition.

But:

(1) The situation would be different if we were to work not *at fixed volume* (or fixed density of particles), as we have until now, but at *fixed pressure*. In this case Bose condensation becomes a real second-order phase transition even for an ideal Bose gas.

(2) The interaction between bosons is especially important – it will also change the order of the transition. This we will consider in the next section.

[1] Why do we have this form of the density of states $\rho_d(\varepsilon)$ at the edge of the spectrum $\varepsilon \to 0$? Say, for the 3d case we had initially $d^3\boldsymbol{p}$, with $d^3\boldsymbol{p} \sim p^2 dp$ which we transformed to $\rho(\varepsilon)d\varepsilon$. In the 3d case, with the spectrum $\varepsilon(p) = p^2/2m$, this gives

$$\rho(\varepsilon)\Big|_{3d} \sim \frac{p^2}{d\varepsilon/dp} \sim \frac{p^2}{p} \sim p \sim \sqrt{\varepsilon}\,.$$

Similar considerations show that:
In the 2d case: $d^2\boldsymbol{p} \sim p\, dp = \rho(\varepsilon)\, d\varepsilon$, $\rho(\varepsilon) \sim p/(d\varepsilon/dp) \sim \text{const.}$
In the 1d case: $d^1 p \sim dp = \rho(\varepsilon)\, d\varepsilon$, $\rho(\varepsilon) \sim 1/(d\varepsilon/dp) \sim 1/p \sim 1/\sqrt{\varepsilon}$.
We will often use these asymptotics of the density of states later on.

## 5.2 Weakly interacting Bose gas

Let us include an interaction between bosons

$$v(\boldsymbol{r}-\boldsymbol{r}')\,n(\boldsymbol{r})\,n(\boldsymbol{r}') = v(\boldsymbol{r}-\boldsymbol{r}')\,\Psi^*(\boldsymbol{r})\,\Psi(\boldsymbol{r})\,\Psi^*(\boldsymbol{r}')\,\Psi(\boldsymbol{r}')\,, \tag{5.13}$$

which we will treat as weak.

In the second quantization form it is convenient to work in the momentum representation. The Hamiltonian in this representation is

$$\mathcal{H} = \mathcal{H}_0 + \mathcal{H}_{int} = \sum_{p} \frac{p^2}{2m}\hat{a}_p^\dagger \hat{a}_p + \sum_{pp'q} v(\boldsymbol{q})\,\hat{a}_{p+q}^\dagger \hat{a}_{p'-q}^\dagger \hat{a}_{p'}\hat{a}_p \tag{5.14}$$

(or $\sum\limits_{p_1+p_2=p_3+p_4} v\hat{a}_{p_1}^\dagger \hat{a}_{p_2}^\dagger \hat{a}_{p_3}\hat{a}_{p_4}$). Here, in the beginning, we keep the notation ˆ for operators.

For simplicity we take the interaction as constant, $v = U/2V$ (the volume $V$ in the denominator is needed for normalization). Physically this corresponds to the assumption that the interaction is point-like, and that it does not depend on the angle, so-called $s$-wave scattering.

In the Bose condensed state there is a macroscopic occupation of the state $p = 0$ by $N_0$ particles. This means that for the momentum $p = 0$,

$$\hat{a}_0^\dagger \hat{a}_0 = N_0 \sim N\,, \tag{5.15}$$

i.e. $a_0 \sim \sqrt{N} \gg 1$. In this case the commutator

$$[\hat{a}_0, \hat{a}_0^\dagger] = \hat{a}_0\hat{a}_0^\dagger - \hat{a}_0^\dagger\hat{a}_0 = 1 \tag{5.16}$$

is small compared with $a_0$ itself, and therefore commutation relations for the zero momentum operators $a_0$ are not important. This means that for this particular state we may treat $a_0$, $a_0^\dagger$ as ordinary c-numbers and not as operators.

*Note*: $a_0$, $a_0^\dagger$ are *complex numbers*, that is

$$\begin{aligned} a_0 &= \sqrt{N_0}\,e^{i\varphi}\,, \\ a_0^\dagger &= \sqrt{N_0}\,e^{-i\varphi}\,. \end{aligned} \tag{5.17}$$

The phase $\varphi$ is in general very important, as we will discuss later. In our present discussion, however, we always have bilinear combinations of the type $a^\dagger a$, and at this stage we do not have to worry about the phase.

Thus we can treat $a_0$, $a_0^\dagger$ as (large, $\sim \sqrt{N}$) c-numbers, and $\hat{a}_p$, $\hat{a}_p^\dagger$ as (small) operators. Let us make this substitution and keep the leading terms in the Hamiltonian, of order $N^2$ and $N$:

$$\mathcal{H}_{\text{int}} = \frac{U}{2V}\Big[a_0^\dagger a_0^\dagger a_0 a_0 + \sum_{p\neq 0}(2\hat{a}_p^\dagger a_0^\dagger \hat{a}_p a_0 + 2\hat{a}_{-p}^\dagger a_0^\dagger \hat{a}_{-p} a_0 + \hat{a}_p^\dagger \hat{a}_{-p}^\dagger a_0 a_0 + a_0^\dagger a_0^\dagger \hat{a}_p \hat{a}_{-p})\Big]$$

$$= \frac{U}{2V}\Big[a_0^4 + a_0^2 \sum_{p\neq 0}(4\hat{a}_p^\dagger \hat{a}_p + \hat{a}_p \hat{a}_{-p} + \hat{a}_p^\dagger \hat{a}_{-p}^\dagger)\Big]. \tag{5.18}$$

For noninteracting particles at $T = 0$ all particles are in the condensate, $N_0 = N$. For weakly interacting bosons *almost* all particles will be there, $N_0 \sim N$, $N - N_0 \ll (N, N_0)$. In the second term in (5.18) we can put $a_0^2 = N$ (the second term is already of first order in the small parameter, due to the presence of $a_p^\dagger a_p$). The term $a_0^4$ should be treated more accurately, keeping all terms of the same order. As

$$a_0^2 + \sum_{p\neq 0} \hat{a}_p^\dagger \hat{a}_p = N\,, \tag{5.19}$$

we should write

$$a_0^4 = \Big(N - \sum_{p\neq 0} \hat{a}_p^\dagger \hat{a}_p\Big)^2 = N^2 - 2N\sum_p \hat{a}_p^\dagger \hat{a}_p + \sum_{p,p'\neq 0} \hat{a}_p^\dagger \hat{a}_p \hat{a}_{p'}^\dagger \hat{a}_{p'}\,. \tag{5.20}$$

The last term here is only of order 1, thus it can be omitted. In effect equation (5.18) becomes

$$\mathcal{H}_{\text{int}} = \frac{U}{2V}\Big[N^2 + N\sum_{p\neq 0}(2\hat{a}_p^\dagger \hat{a}_p + \hat{a}_p^\dagger \hat{a}_{-p}^\dagger + \hat{a}_p \hat{a}_{-p})\Big], \tag{5.21}$$

and the full Hamiltonian (5.14) takes the form

$$\mathcal{H} = \frac{UN^2}{2V} + \frac{1}{2}\sum_{p\neq 0}\Big[\Big(\frac{p^2}{2m} + \frac{UN}{V}\Big)(\hat{a}_p^\dagger \hat{a}_p + \hat{a}_{-p}^\dagger \hat{a}_{-p}) + \frac{UN}{V}(\hat{a}_p^\dagger \hat{a}_{-p}^\dagger + \hat{a}_p \hat{a}_{-p})\Big]. \tag{5.22}$$

The expression (5.22) is nondiagonal in the operators $\hat{a}_p^\dagger$, $\hat{a}_p$. But it is a quadratic form and can be easily diagonalized using the so-called *Bogolyubov canonical transformation*, or *u–v* transformation:

$$\begin{aligned} \hat{a}_p &= u_p b_p + v_p b_{-p}^\dagger \\ \hat{a}_p^\dagger &= u_p b_p^\dagger + v_p b_{-p} \end{aligned} \tag{5.23}$$

(here we assume the coefficients $u_p$, $v_p$ to be real).

**Problem:** Write down $b^\dagger$, $b$ in terms of $a^\dagger$, $a$.

**Solution:** $b_p = u_p a_p - v_p a^\dagger_{-p}$ ($u_p$, $v_p$ are even functions of $p$), and the corresponding equation for $b_p^\dagger$ is obtained from this one by Hermitian conjugation.

We require that (we now omit the notation ^ for operators) $b^\dagger$, $b$ are bosons:

$$b_p b^\dagger_{p'} - b^\dagger_{p'} b_p = \delta_{pp'} , \qquad b_p b_{p'} - b_{p'} b_p = 0 . \tag{5.24}$$

In order for this to hold, the coefficients $u_p$, $v_p$ should satisfy the condition

$$u_p^2 - v_p^2 = 1 . \tag{5.25}$$

Thus we can write

$$\begin{aligned} a_p &= \frac{1}{\sqrt{1 - A_p^2}} (b_p + A_p b^\dagger_{-p}) \\ a_p^\dagger &= \frac{1}{\sqrt{1 - A_p^2}} (b_p^\dagger + A_p b_{-p}) , \end{aligned} \tag{5.26}$$

where $u_p = 1/\sqrt{1 - A_p^2}$, $v_p = A_p/\sqrt{1 - A_p^2}$ (or, equivalently, we can write down $u_p = \cosh \alpha_p$, $v_p = \sinh \alpha_p$, and express everything through $\alpha_p$). We can determine the coefficients $u_p$, $v_p$, or $A_p$ from the condition that the nondiagonal terms in (5.22), after making the canonical transformation, drop out. This gives the equation

$$\left( \frac{p^2}{2m} + nU \right) 2A_p + nU(1 + A_p^2) = 0 , \tag{5.27}$$

or

$$A_p = \frac{1}{nU} \left( -\frac{p^2}{2m} - nU + \sqrt{\left( \frac{p^2}{2m} + nU \right)^2 - (nU)^2} \right) , \tag{5.28}$$

where we have introduced the density $n = N/V$. The Hamiltonian then takes the form

$$\begin{aligned} \mathcal{H} = \frac{UN^2}{2V} - \frac{1}{2} \sum_{p \neq 0} \left[ \left( \frac{p^2}{2m} + nU \right) - \sqrt{\left( \frac{p^2}{2m} + nU \right)^2 - (nU)^2} \right] \\ + \sum_{p \neq 0} \sqrt{\left( \frac{p^2}{2m} + nU \right)^2 - (nU)^2} \; b_p^\dagger b_p . \end{aligned} \tag{5.29}$$

The first two terms give the energy of the ground state at $T = 0$, and the third term describes *elementary excitations* (cf. phonons, (4.26)),

$$\sum_{p \neq 0} \varepsilon_p b_p^\dagger b_p , \tag{5.30}$$

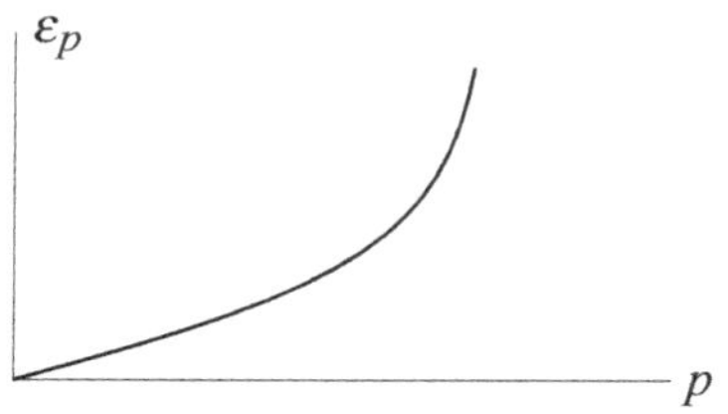


Fig. 5.2

with the spectrum

$$\varepsilon_p = \sqrt{\left(\frac{p^2}{2m} + nU\right)^2 - (nU)^2} = \begin{cases} \sqrt{\dfrac{nU}{m}} \cdot p & (p \to 0) \\ \dfrac{p^2}{2m} & (\text{large } p). \end{cases} \tag{5.31}$$

Thus the elementary excitations in a weakly interacting Bose condensed system have the character of *sound* at small momenta $p$ (Bogolyubov sound), and they continuously go over to *free particles* with the spectrum $p^2/2m$ at large $p$, see Fig. 5.2.

We should make here several remarks:

- The sound velocity (5.31) coincides with the standard sound velocity of a gas with density $n = N/V$ and interaction $U$.
- One can find the coefficients of the canonical transformation (5.23) or (5.26) not from the condition that nondiagonal terms in the Hamiltonian cancel, but from the condition of the *minimum* of the ground state energy.

**Problem:** Check this: put (5.26) in (5.22), collect the terms without operators, and minimize the resulting expression in $A_p$.

- One can calculate the total energy of this weakly interacting Bose gas. Usually one expresses all the quantities through the scattering length $a$ given by

$$U = \frac{4\pi}{m}a \tag{5.32}$$

($a$ is the scattering amplitude for energy tending to zero). The resulting energy is given by the expression

$$\frac{E}{V} = \frac{2\pi a}{m}n^2\left[1 + \frac{128}{15\sqrt{\pi}}\, a^{3/2}n^{1/2}\right] \tag{5.33}$$

(T. D. Lee, C. N. Yang, 1957).

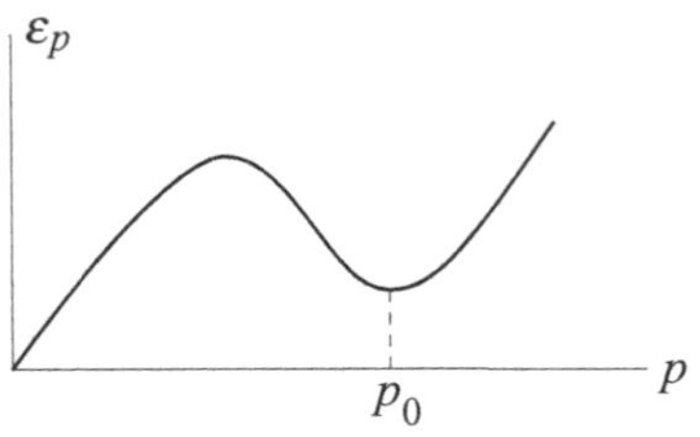


Fig. 5.3

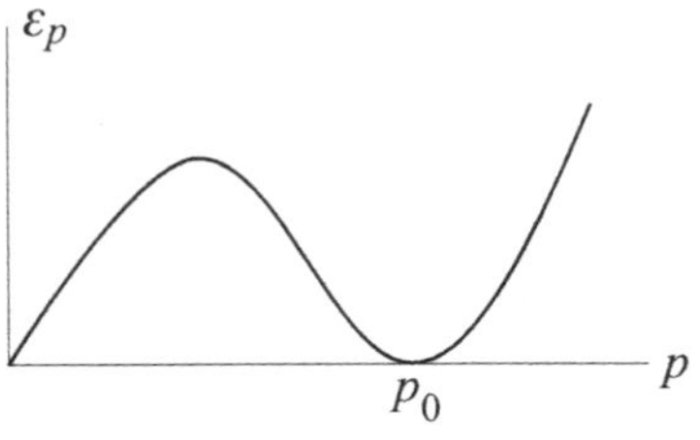


Fig. 5.4

From (5.33) one can find the sound velocity in the usual way, as we are doing in gases:

$$s = \sqrt{\frac{V^2}{mN}\frac{\partial^2 E}{\partial V^2}} = \frac{\sqrt{4\pi a n}}{m}\,, \tag{5.34}$$

which coincides with (5.31) (taking into account (5.32)).

## 5.3 Bose condensation and superfluidity

The main system for which, for a long time, one applied the concept of Bose condensation (before the recently observed Bose condensation of optically trapped supercooled atoms) is superfluid $^4$He. $^4$He atoms are *bosons*, and helium remains liquid down to $T = 0$ (see Section 4.4.3). It goes over to a superfluid state at $T_c = 2.4$ K (P. L. Kapitza). Superfluidity in $^4$He is attributed to a Bose condensation of He atoms into the state with $p = 0$.

There exists in real helium one important difference relative to the previous treatment: He atoms interact strongly, as a result of which there is a number of modifications and because of which there exists actually no microscopic theory of superfluid $^4$He (there is no small parameter – weak interaction – which we have used in the theoretical treatment above). Nevertheless, the basic concepts described above apply, with some modifications, also to this case. The main modifications are:

1. The actual number of atoms in the condensate $N_0$ is finite, but even at $T = 0$ it is rather small, $N_0/N \sim$ 8–10% at most.
2. The spectrum of elementary excitations (originally postulated in a slightly different form by Landau) has the form shown in Fig. 5.3. As $p \to 0$ the excitations remain phonons (spectrum linear in $p$). At larger $p$ there exists an extra minimum – the so-called *rotons*. Physically the roton minimum is a consequence of *strong interaction* in the liquid; it reflects the tendency to crystallization (under pressure the roton minimum becomes deeper, and when $\varepsilon_{p_0}$ approaches zero, Fig. 5.4, the corresponding mode becomes *unstable*, and there appears a positive increment, i.e. a large standing wave with wavevector $p_0$, or with period $a = \hbar/p_0$, will develop; this signals the formation of a crystal). Cf. the discussion of the reverse process of melting in Section 4.4.3, Fig. 4.13($a$).

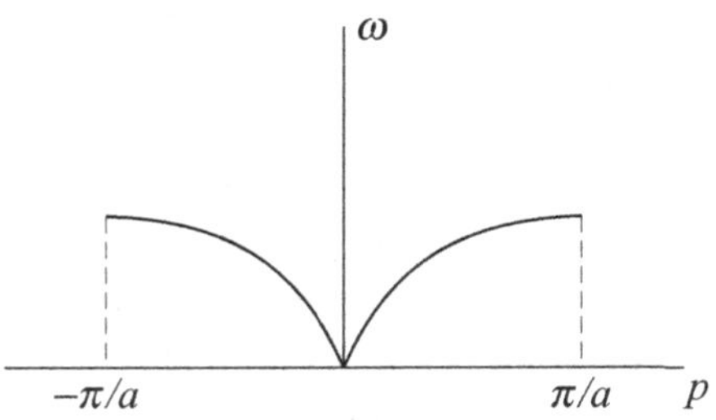


Fig. 5.5

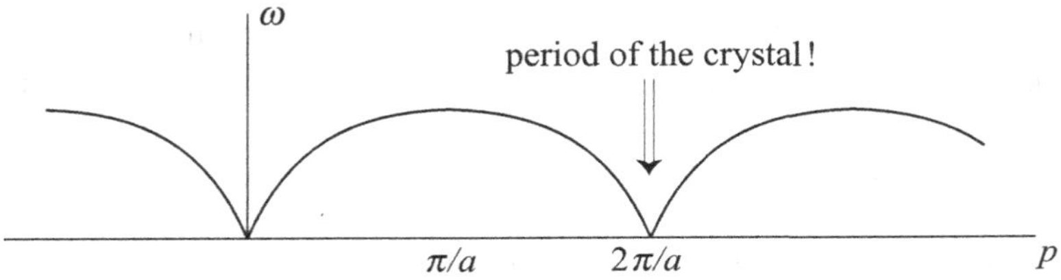


Fig. 5.6

**Problem:** Compare this situation with ordinary phonons in crystals.

**Solution:** For an ordinary crystal with period $a$ the phonon spectrum in the first Brillouin zone has the form schematically shown in Fig. 5.5. In the extended zone scheme it may be represented as shown in Fig. 5.6, with the plot extended to larger $p$. If we do not have real long-range crystalline order, but only short-range order typical for liquids, the spectrum would not go to zero at $p = 2\pi/a$, but would retain a minimum at this wavevector (and of course would acquire an imaginary part, i.e. would be strongly damped). Thus there is a close analogy between this situation and that of rotons in liquid helium shown in Figs. 5.3 and 5.4; that is, the spectrum of Fig. 5.4 with the roton minimum is not something specific to liquid $^4$He, but in principle such a minimum could exist in any liquid close to the melting point. An important difference is that in normal liquids these excitations with short wavelengths are usually extremely strongly damped, so that there is not much sense in speaking about them there, whereas they are well-defined excitations in the superfluid phase of $^4$He.

These arguments are of course only qualitative. But one can make them more accurate. R. Feynman has shown (1954) that the spectrum of elementary excitations in $^4$He can be written as

$$\varepsilon_p = \frac{p^2}{2mS(p)} , \tag{5.35}$$

where $S(p)$ is the static structure factor describing spatial correlation of atoms in a liquid. At small $p$ $S(p) = p/2ms$, where $s$ is the velocity of Bogolyubov sound, $s = \sqrt{nU/m}$, see (5.31), so that the spectrum (5.35) goes over to (5.31). For large $p$, however, $S(p)$ has a maximum at the value $p_0 \sim \hbar/d_0$ where $d_0$ is the typical distance between He atoms (of the order of the lattice parameter of solid $^4$He at high pressures); this tells us that the probability of finding an atom at distance $d$ from a given atom is a maximum at $d \sim d_0$. Correspondingly, the spectrum (5.35) would develop a minimum at $p \sim p_0$ which is nothing else but the roton minimum of Fig. 5.3.

3. An important point: in Bose condensation, and in superfluidity, the order parameter $\eta$ (cf. (2.1)) is $\eta \sim \langle \hat{a}_0 \rangle$ (it is zero above $T_c$, and nonzero in the Bose condensed phase). This is a *complex* scalar (see the remark (5.17) in Section 5.2):

$$\langle \hat{a}_0 \rangle = a_0 = \sqrt{N_0}\, e^{i\varphi} \ . \tag{5.36}$$

At the phase transition, at $T < T_c$, its *phase* becomes fixed, i.e. the superfluid state is a *coherent* state (phase coherence). But the *number* of particles in the condensate $N_0$ fluctuates. The operators $\hat{N}$ and $\hat{\varphi}$ are conjugate variables in quantum mechanics, like $\hat{x}$ and $\hat{p}$; they obey the uncertainty relation (Heisenberg relation)

$$\Delta N_0 \, \Delta\varphi \simeq \hbar \ . \tag{5.37}$$

Bose condensation is thus a phase transition with breaking of a continuous symmetry – gauge symmetry (fixing of the phase $\varphi$ which is a continuous variable). Correspondingly, Bogolyubov sound is the *Goldstone mode* for this broken symmetry. This also gives another explanation of the absence of Bose condensation at any $T \neq 0$ in the 1d and 2d systems discussed in the Problem in Section 5.1: phase fluctuations caused by the excitation of this gapless mode are so strong that in the 1d and 2d cases they destroy the long-range order at any finite temperature. This is yet another application of the Mermin–Wagner theorem mentioned at the end of Chapter 4.

4. The fact that the order parameter in Bose condensed and superfluid systems is a complex scalar (5.36) permits one also to establish the correspondence with certain other systems with the same symmetry, notably the anisotropic spin system – the so-called $xy$ model, in which spins are confined to the $xy$-plane. In this case the order parameter takes the form $s = |s|e^{i\varphi}$, where the angle $\varphi$ determines the orientation of spin in the $xy$-plane (see below, Section 6.4.3). One can use this analogy by 'borrowing' the concepts and results from one field and applying them to the other. Thus one often speaks now about Bose condensation of magnons, although one must be careful in applying this

concept. Also, the concept of vortices, first introduced in superfluid $^4$He, is now 'translated' to magnetic vortices, see Section 6.4.3(*b*), p. 118.

5. Generalizing the description of a Bose condensed state to the case with spatial inhomogeneities (cf. Section 2.4 above), we should treat the order parameter (5.36) as a function of position. The most important is the change in space of the phase $\varphi(\boldsymbol{r})$. It turns out that the gradient of the phase determines the local superfluid velocity in the system:

$$\boldsymbol{v}_s = \frac{\hbar}{m}\nabla\varphi \,. \tag{5.38}$$

Actually the collective mode we have described above – Bogolyubov sound – is predominantly the oscillations of the *phase* $\varphi$ and velocity $\boldsymbol{v}_s$ (or local currents) in the liquid.

### 5.3.1 Landau criterion of superfluidity

Why is $^4$He superfluid? The answer to this question is given by the *Landau criterion of superfluidity*: it turns out that for the excitation spectrum of Fig. 5.3 there is no dissipation at velocities not exceeding a certain critical value. The corresponding arguments are straightforward, but require a bit of concentration.

Suppose that a liquid flows with velocity $\boldsymbol{v}$ through a capillary (thin tube). The elementary processes leading to *friction* are the creation of excitations in the liquid, one after another. These excitations should reduce the total momentum of the flowing liquid. If such processes are allowed, friction would appear, and there would be no superfluidity.

Let us first consider this situation in the system of coordinates where the liquid is at rest but the walls of the tube move with velocity $-\boldsymbol{v}$. Suppose that one (the first) such an excitation appears, with momentum $\boldsymbol{p}$ and energy $\varepsilon(\boldsymbol{p})$. Now let us go back to the laboratory coordinate frame, in which the liquid moves with velocity $\boldsymbol{v}$. In this coordinate system the total energy of the liquid with one excitation is[2]

$$E = \frac{Mv^2}{2} + \varepsilon(\boldsymbol{p}) + \boldsymbol{p}\cdot\boldsymbol{v} = \frac{Mv^2}{2} + \delta E\,, \quad \text{where} \quad \delta E = \varepsilon(\boldsymbol{p}) + \boldsymbol{p}\cdot\boldsymbol{v}\,. \tag{5.39}$$

Here $Mv^2/2$ is the initial kinetic energy of the moving liquid, and $\delta E$ is the change in the total energy due to the creation of the excitation. For such a process to occur we need $\delta E < 0$, otherwise the excitation would cost us energy, and such

[2] This follows from the well-known formulae of mechanics: energy and momentum are transformed from one reference frame to another, moving with velocity $\boldsymbol{v}$, as $E = E_0 + \boldsymbol{p}_0\cdot\boldsymbol{v} + Mv^2/2$, $\boldsymbol{p} = \boldsymbol{p}_0 + M\boldsymbol{v}$, where $E_0$ and $p_0$ are the energy and momentum in the frame with the liquid at rest. In our case $E_0 = \varepsilon(\boldsymbol{p})$, $\boldsymbol{p}_0 = \boldsymbol{p}$, which gives (5.39).

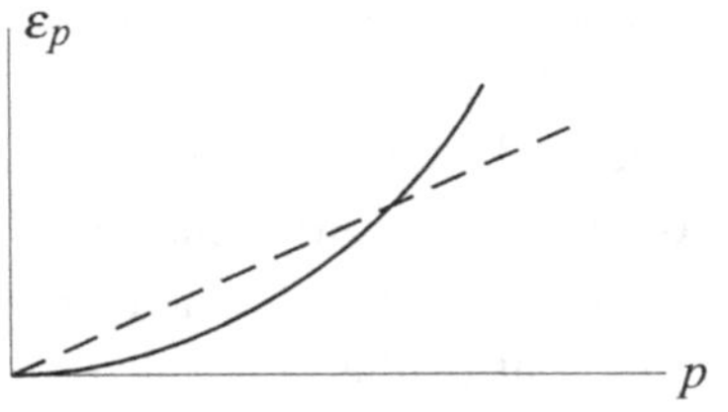


Fig. 5.7

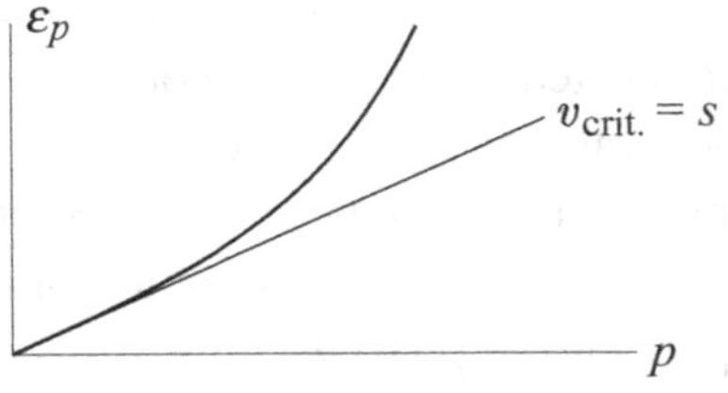


Fig. 5.8

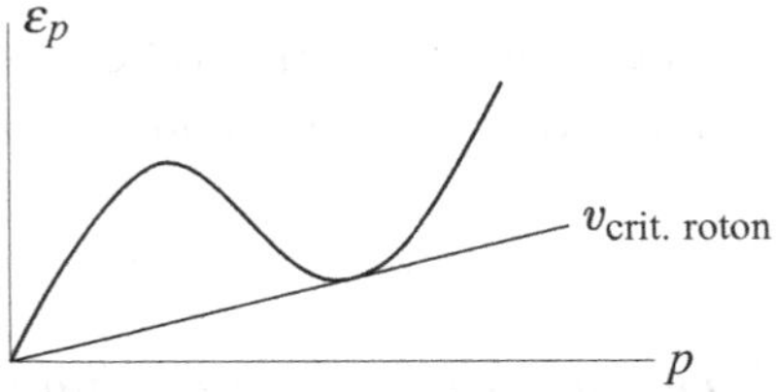


Fig. 5.9

an excitation would not be created spontaneously. But for that $\boldsymbol{p}$ should, first of all, be antiparallel to $\boldsymbol{v}$, and, most important, the change of the energy $\delta E$ (5.39) should be negative, i.e. $\varepsilon - pv < 0$. Thus for such a process (friction) to begin, the velocity of the liquid should satisfy the condition

$$v > \frac{\varepsilon(\boldsymbol{p})}{p} \tag{5.40}$$

(at least for some $p$). That is, the excitations can be created only if the velocity of the flow exceeds the *critical velocity*,

$$v \geq v_{\text{crit}} = \left(\frac{\varepsilon}{p}\right)_{\text{min}} , \tag{5.41}$$

starting from which the flow can slow down, and dissipation, or friction, appears.

If we have a usual liquid or gas, consisting of noninteracting particles, the excitation spectrum is $\varepsilon_p = p^2/2m$, Fig. 5.7, and there is *no* superfluidity at any velocity (the critical velocity $v_{\text{crit}} = 0$). But if the spectrum has the form shown in Fig. 5.8, see equation (5.31), then there exists a finite critical velocity $v_{\text{crit}} = s$, which in this case is equal to the sound velocity; for smaller $v$ such excitations cannot be formed, and the motion is dissipationless, i.e. superfluid.

For real $^4$He the sound velocity $s = 2.4 \times 10^4$ cm/sec, and this condition would give the critical velocity which is too large. Experimentally dissipation starts much earlier. One of the possible explanations is that the real spectrum looks as shown in Fig. 5.9 and the critical velocity could be determined by *rotons*. Actually even this velocity is too large, and the excitations determining the critical velocity in the bulk helium in most of the actual experiments are special topological excitations – vortices (similar to smoke rings from a pipe) – see below, Section 5.3.2. Only in cases when the formation of vortices is suppressed due to a restricted geometry, for example in thin films of $^4$He or in very thin capillaries, can we reach the critical velocity determined by rotons, which is indeed much higher than in bulk $^4$He.

We have seen that if the velocity is below a certain critical value, new excitations cannot be spontaneously created in the moving liquid. This conclusion is valid not only at $T = 0$ and not only for the ground state. However, at finite temperature there are always present thermally excited elementary excitations in the liquid. And when the liquid flows through a capillary, these excitations can collide with the walls and can change their momentum. Therefore these excitations, which initially 'flow' with the liquid, will gradually slow down, exactly like ordinary gas flowing through a tube. In effect it looks as though a part of the liquid experiences friction, whereas the remaining part moves without any resistance. In other words, it seems as though there exist two components in the liquid; a normal component and a superfluid one, with the total density $\rho = \rho_{\text{norm}} + \rho_s$. Such a two-fluid picture (L. Tisza) gives a very useful phenomenological description of many properties of superfluid helium. It is also widely used for the description of many properties of superconductors. But one has to realize that it is only a way to interpret the properties of these systems; in no way should we take this picture too literally and think that indeed some of the atoms are moving without dissipation, whereas the others experience friction. In fact, it is the same atoms which display both types of behaviour, and the real meaning of these two 'fluids' is the one explained above: the normal 'fluid' consists of elementary excitations – collective modes of the liquid as a whole.

### 5.3.2 *Vortices in a superfluid*

As mentioned in the previous section, there is yet another very important type of excitation in a superfluid liquid: topological excitations, or vortices. If one starts to rotate a vessel containing a superfluid, initially the liquid remains at rest. However, starting from a certain critical angular velocity, a vortex will be formed in the liquid: a circular motion of the superfluid around a certain line which is called the vortex core. As a result some circulation is transferred to the liquid, i.e. the liquid starts to participate in the rotation. In a cylindrical vessel vortices start at the bottom and go all the way up to the upper surface of the liquid, Fig. 5.10; they cannot be 'interrupted' and cannot simply end inside the liquid.

Using equation (5.38), we can show that the circulation of velocity around a vortex should be quantized. Let us integrate (5.38) over a contour $c$ surrounding

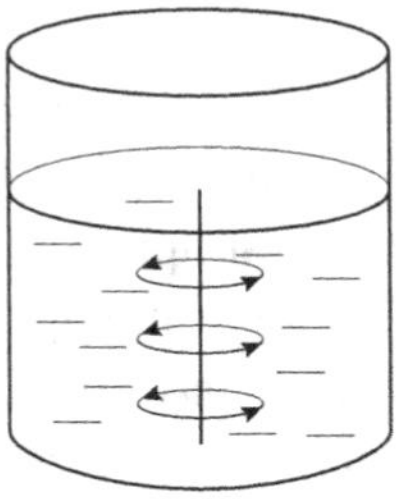

Fig. 5.10

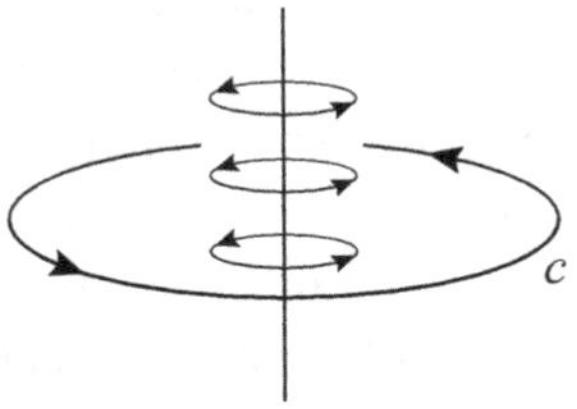


Fig. 5.11

the vortex, Fig. 5.11. The circulation is

$$\oint \boldsymbol{v}_s \cdot d\boldsymbol{l} = \frac{\hbar}{m} \Delta\varphi \,, \tag{5.42}$$

where $\Delta\varphi$ is the total change of phase along the contour $c$. However the total wavefunction of the superfluid has to be a single-valued function, which means that $\Delta\varphi = 2\pi n$, where $n = 0, \pm 1, \pm 2, \ldots$. Thus the circulation is

$$\oint \boldsymbol{v}_s \cdot d\boldsymbol{l} = \frac{\hbar}{m} 2\pi n \,, \tag{5.43}$$

i.e. it is indeed quantized. Such quantization is a manifestation of the quantum nature of superfluidity and confirms its interpretation as a 'macroscopic quantum phenomenon' (the whole system is described by one quantum wavefunction $\Psi(r) = \sqrt{N_0}\, e^{i\varphi(r)}$). The above-mentioned fact that the vortex cannot end inside the liquid is actually connected with this property: if it were to, we could continuously deform the contour $c$ in equation (5.43) in such a way that it would be 'above' the end of the vortex, after which we could contract it to zero, with zero circulation – in contrast to the finite (quantized) value (5.43) we started with. This is why the vortices are actually *topological excitations*.

The situation may be much more intricate in superfluid systems with a more complicated order parameter, e.g. such as that of liquid $^3$He. In these cases the space of the order parameter is different from the case of $^4$He (which lives 'on the circle' $\eta = |\eta| e^{i\varphi}$ with $0 \leq \varphi \leq 2\pi$). Consequently the types and properties of

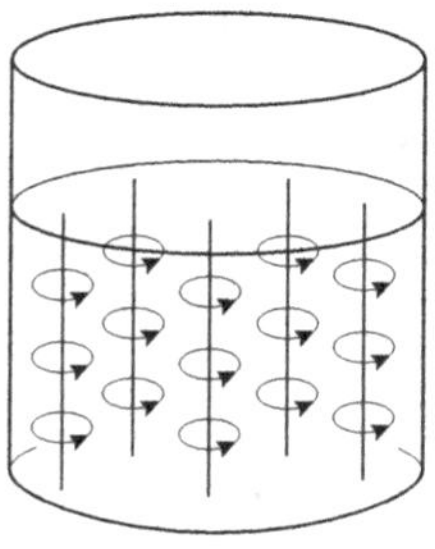

Fig. 5.12

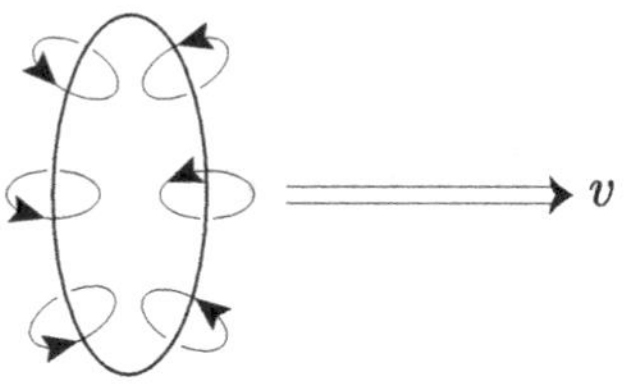


Fig. 5.13

topological excitations in such systems, including vortices, may be very different from those of $^4$He.

As mentioned above, in most of the real experiments the critical velocity in liquid helium is determined not by the sound velocity and not even by rotons, but by excitation of vortices in moving liquid He. Vortices are created before the critical velocity for excitations of sound quanta or rotons is reached, and their creation and motion leads to dissipation. In a rotating cylinder vortices are parallel to the rotation axis, Fig. 5.12. In a liquid flowing through a capillary the vortices form vortex rings, Fig. 5.13, very much like smoke rings from a pipe (as we mentioned before, vortices cannot be simply interrupted inside a liquid: they either end at the surface of the liquid, or form closed loops, as in these rings).

Many features and phenomena described above are also met (and are even much better known) in superconductors. Thus, the general description of superconductivity as a macroscopic quantum phenomenon is rather similar to that given above; the Landau criterion of the existence of supercurrents without dissipation works there as well, and the notion of quantized vortices leading to dissipation plays a very important role in the physics and application of superconductivity (in the so-called type-II superconductors, to which all practically important superconducting materials, including high-temperature superconductors, belong). Theoretical work by Abrikosov and Ginzburg laying the foundations of the description of these systems earned them a Nobel Prize in 2003.

# 6

# Magnetism

## 6.1 Basic notions; different types of magnetic response

I will again start by briefly recalling the basic notions from general physics and quantum mechanics about different sources and types of magnetic response, which are actually covered in most of the corresponding textbooks. For simplicity I will mostly ignore in this chapter the details of atomic structure leading to different contributions to atomic magnetic moments (e.g. orbital contributions, with the often important role of spin–orbit coupling), see e.g. Goodenough (1963) and Kugel and Khomskii (1982), and will consider only the spin contribution, mostly illustrating the results on the example of spin $\frac{1}{2}$. More detailed treatments can be found, e.g. in the books by White (2006) and Yosida (1996), and in many others.

The Hamiltonian of electrons in a magnetic field has the form

$$\mathcal{H} = \frac{1}{2m}\left(\hat{\boldsymbol{p}} - \frac{e}{c}\boldsymbol{A}\right)^2 - \frac{\mu}{|\boldsymbol{S}|}\boldsymbol{S}\cdot\boldsymbol{H} \; . \tag{6.1}$$

The first term in this Hamiltonian describes the response due to the orbital motion of electrons, and the second one is due to the spin of the electron. Here $\mu$ is the magnetic moment of spin $\boldsymbol{S}$. [Often the last term in the Hamiltonian (6.1) is written as $g\mu_{\mathrm{B}}\boldsymbol{S}\cdot\boldsymbol{H}$, where $\mu_{\mathrm{B}} = e\hbar/2mc$ is the Bohr magneton, and $g$ is the so-called $g$-factor, which for free electron spins is $g_{\mathrm{spin}} = 2$ (and for orbital moments $g_{\mathrm{orb}} = 1$).] In general the vector potential $\boldsymbol{A}$ is a function of the coordinate, $\boldsymbol{A} = \boldsymbol{A}(\hat{\boldsymbol{x}})$, and it does not commute with the momentum $\hat{\boldsymbol{p}}$, $\hat{\boldsymbol{p}}\cdot\boldsymbol{A} - \boldsymbol{A}\cdot\hat{\boldsymbol{p}} = -i\hbar\,\mathrm{div}\,\boldsymbol{A}$. However, we can make a gauge transformation of $\boldsymbol{A}$, and if we choose a gauge such that $\mathrm{div}\,\boldsymbol{A} = 0$, the commutator of $\hat{\boldsymbol{p}}$ and $\boldsymbol{A}$ will be zero. For example, we can take

$$\boldsymbol{A} = \tfrac{1}{2}\,[\boldsymbol{H}\times\boldsymbol{r}] \; . \tag{6.2}$$

Then

$$\mathcal{H} = \frac{1}{2m}\hat{\boldsymbol{p}}^2 - \frac{e}{mc}\hat{\boldsymbol{p}}\cdot\boldsymbol{A} + \frac{e^2}{2mc^2}\boldsymbol{A}^2 - \frac{\mu}{|\boldsymbol{S}|}\boldsymbol{S}\cdot\boldsymbol{H} \; . \tag{6.3}$$

As to the magnetic response of various systems, there exist different situations:

(1) Closed shells with orbital moment $L = 0$ and spin $S = 0$. In the ground state $\langle 0|\boldsymbol{p} \cdot \boldsymbol{A}|0\rangle = 0$, and here first of all the third term in (6.3) works (see also (2) below); the energy is

$$\Delta E = \langle \mathcal{H} \rangle = \frac{e^2}{2mc^2}\langle \boldsymbol{A} \rangle^2 = \frac{e^2}{8mc^2}\sum_a [\boldsymbol{H} \times \boldsymbol{r}_a]^2 = \frac{e^2}{12mc^2}H^2 \sum_a \overline{r_a^2} \qquad (6.4)$$

after averaging over angles, i.e. over all directions of $\boldsymbol{r}$.

The energy of a system in a magnetic field is

$$\Delta E = -\boldsymbol{M} \cdot \boldsymbol{H}, \quad \text{i.e.} \quad \boldsymbol{M} = -\frac{\partial \Delta E}{\partial \boldsymbol{H}} . \qquad (6.5)$$

As the magnetic moment is $\boldsymbol{M} = \chi \boldsymbol{H}$, the susceptibility for $N$ such atoms with charge $Z$ is, using (6.4),

$$\chi = -\frac{e^2}{6mc^2}\sum_a \overline{r_a^2} \quad \left( = -\frac{Ze^2 N \langle r^2 \rangle}{6mc^2} \right) . \qquad (6.6)$$

This is the standard *diamagnetism* – classical diamagnetic screening of the external field. Note again that it is due to the term $\boldsymbol{A}^2$ in the Hamiltonian (6.3); this is also the case in other situations, e.g. in the famous Meissner effect, the ideal diamagnetism of superconductors.

(2) The term in (6.3) linear in $\boldsymbol{A}$ can admix the excited states with $L \neq 0$ to the ground state: in second order in perturbation theory (using the second term in (6.3) as a perturbation) we then have the change of the energy in the magnetic field

$$\Delta E = -\sum_n \frac{\left|\langle 0|\boldsymbol{H} \cdot \boldsymbol{M}_L|n\rangle\right|^2}{\varepsilon_n - \varepsilon_0} \qquad (6.7)$$

(here we have used that $\boldsymbol{M} = \frac{e}{2mc}\hat{\boldsymbol{L}}$, $\hat{\boldsymbol{L}} = \boldsymbol{r} \times \boldsymbol{p}$, and we have used the gauge (6.2)).

This would give *positive* susceptibility, which in a first approximation does not depend on temperature. This is the temperature-independent *Van Vleck paramagnetism.*[1]

The two terms described above, the usual diamagnetism and Van Vleck paramagnetism, are always present in all materials, even in those with much stronger magnetic response due to localized spins; these contributions are responsible

[1] Note that in certain cases, where there exist low-lying magnetic excited states close to the nonmagnetic ground state, the Van Vleck paramagnetism may become temperature dependent. This, for example, is the case for many compounds containing $Eu^{3+}$.

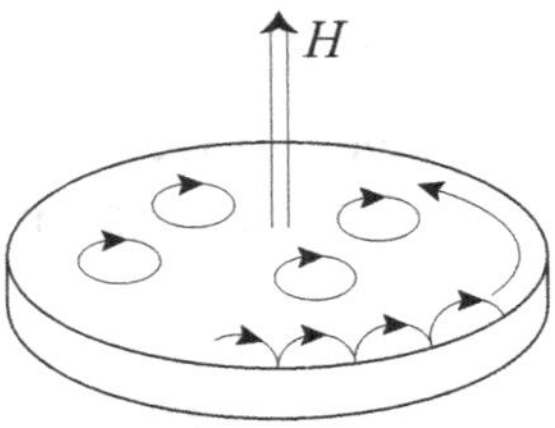


Fig. 6.1

for the temperature-independent 'background' in magnetic susceptibility. In metals there are two extra contributions to the magnetic response.

(3) Free electrons in metals. There exists a diamagnetic contribution to the susceptibility due to the orbital motion of electrons, which is called *Landau diamagnetism*:

$$\chi_{\text{Landau}} = -\frac{(N/V)e^2}{4mc^2 p_{\text{F}}^2} = -\frac{e^2 p_{\text{F}}}{12\pi^2 mc^2} \,. \tag{6.8}$$

The calculations giving this expression are not simple; the physics is connected with the boundary effects. In *classical physics* one can show that there is no magnetism in thermodynamic equilibrium (Bohr–van Leeuwen theorem): if we apply an external field $H$ to a metal with classical electrons, the diamagnetic currents created inside the sample will be compensated by the surface current along the boundary flowing in the opposite direction, see Fig. 6.1. In *quantum mechanics* there is no such compensation, which results in Landau diamagnetism.

The physical reasons underlying the Bohr–van Leeuwen theorem may be understood if we recall that classically the force acting on an electron in a magnetic field is the Lorentz force $\boldsymbol{F} \sim \boldsymbol{H} \times \boldsymbol{v}$, i.e. it is perpendicular to the velocity $\boldsymbol{v}$ and consequently it does not change the energy of the electron. In effect the magnetic field does not enter the thermodynamic potentials and does not induce a magnetic response in thermodynamic equilibrium. In quantum mechanics this is no longer true.

(4) Spins of electrons in metals give rise to *Pauli paramagnetism*. Its origin is the splitting and shift of the spin-up and spin-down subbands in a magnetic field, $\varepsilon(H) = \varepsilon_0 \pm \mu H$. This leads to the redistribution of electrons between these subbands, see Fig. 6.2, because the chemical potential of both these components should be the same. As a result there appears a net polarization of conduction electrons, proportional to the magnetic field, $\boldsymbol{M} = \chi_{\text{Pauli}} \boldsymbol{H}$. Straightforward calculations give for $\chi_{\text{Pauli}}$ the expression

$$\chi_{\text{Pauli}} = \mu_{\text{B}}^2 \, \rho(\varepsilon_F) = \frac{e^2 p_{\text{F}}}{4\pi^2 mc^2} \,. \tag{6.9}$$

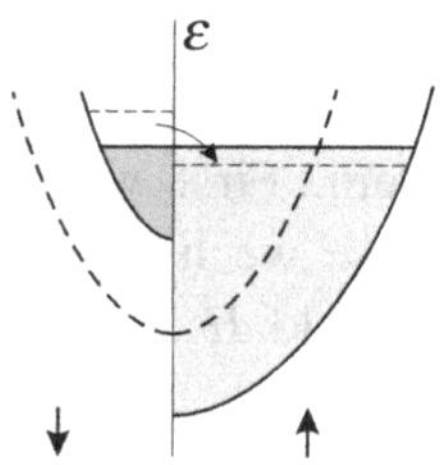


Fig. 6.2

It turns out that the Landau diamagnetism discussed above is determined by the same parameters as in equation (6.9), i.e. by the density of states at the Fermi level $\rho(\varepsilon_F)$, and is equal to

$$\chi_{\text{Landau}} = -\tfrac{1}{3}\chi_{\text{Pauli}} \,. \tag{6.10}$$

(5) Localized electrons (localized spins, localized magnetic moments). Why localized electrons exist in certain systems will be discussed later, especially in Chapters 12 and 13; briefly speaking, this is connected with the existence of partially filled inner shells and with strong electron–electron interactions. Typical systems of this type are those containing the following:

- *Transition metals with partially filled d shells.* In the 3d series (Mn, Fe, Co, Ni, ...) the d electrons are relatively strongly localized, especially in compounds such as oxides, e.g. NiO, MnO, $Fe_2O_3$, etc. In the 4d and 5d elements (Ru, Ir, Pd, Pt, ...) the d electrons are usually less localized than those of the 3d series.
- *Rare earth elements*, containing 4f electrons (Gd, Eu, Dy, ...). The 4f electrons are almost always very strongly localized (see, however, Section 13.3).
- *Actinides* with 5f electrons (U, Np, Am, ...). From the point of view of electron localization these are rather analogous to the 3d transition metals. However, there is one important difference here: due to the much larger atomic mass the relativistic effects in actinides, notably the spin–orbit interaction, are much stronger than in transition metals (this also applies to rare earths).

The main interaction of localized electrons with the magnetic field is the Zeeman term,

$$-\boldsymbol{M} \cdot \boldsymbol{H} = -g\mu_B \boldsymbol{S} \cdot \boldsymbol{H} \,. \tag{6.11}$$

This interaction gives rise to *paramagnetism of localized spins*; see the next section.

### 6.1.1 Susceptibility of noninteracting spins

Here we consider the susceptibility of localized magnetic moments without an exchange interaction between them. First we present the classical treatment, using Boltzmann statistics. Suppose we have a collection of atoms with magnetic moment $\boldsymbol{M}$. In the magnetic field $\boldsymbol{H}$ the probability of finding the moment $\boldsymbol{M}$ is

$$n(\boldsymbol{M}) \sim \exp\left(\frac{\boldsymbol{M}\cdot\boldsymbol{H}}{T}\right) \qquad (k_{\mathrm{B}}=1) \tag{6.12}$$

(see (1.1), with the energy (6.5)). From this we find the average moment

$$\langle \boldsymbol{M}\rangle = \frac{\int \boldsymbol{M} e^{\boldsymbol{M}\cdot\boldsymbol{H}/T}\, d\Omega}{\int e^{\boldsymbol{M}\cdot\boldsymbol{H}/T}\, d\Omega}\,, \tag{6.13}$$

where $\Omega$ is the solid angle. From (6.13), the magnetic susceptibility is

$$\chi_0 = N\left\langle\frac{\partial \boldsymbol{M}}{\partial \boldsymbol{H}}\right\rangle = \frac{N}{T}\langle \boldsymbol{M}^2\rangle = \frac{1}{3}\frac{Ng^2\mu_{\mathrm{B}}^2\, S(S+1)}{T}\,, \tag{6.14}$$

which is the well-known Curie law; the index '0' denotes the fact that we are dealing with noninteracting spins. Here we have already used the connection $\boldsymbol{M} = g\mu_{\mathrm{B}}\boldsymbol{S}$, see equation (6.11), and also the fact that in quantum mechanics $\langle \boldsymbol{S}^2\rangle = S(S+1)$. The factor $\frac{1}{3}$ comes from averaging over all directions, see e.g. Kittel (1987).

The susceptibility per unit volume is thus

$$\chi_0 = \frac{N}{V}\frac{g^2\,\mu_{\mathrm{B}}^2\, S(S+1)}{3T} \equiv \frac{\mathcal{C}}{T}\,, \tag{6.15}$$

where $\mathcal{C}$ is called the Curie constant.

Consider now the quantum case. Instead of the integral over all orientations of $\boldsymbol{M}$ in (6.11) we have to take a *sum* over possible quantum states $S^z$ ($M^z = g\mu_{\mathrm{B}}S^z$, $-S \le S^z \le S$).

Thus for $S = \frac{1}{2}$ we would have (for the field in the $z$-direction and for the moment parallel to the field; we omit the index '$z$' below):

$$M = g\mu_{\mathrm{B}}\langle S\rangle\,, \qquad \langle S\rangle = \frac{1}{2}\frac{e^{g\mu_{\mathrm{B}}\frac{1}{2}H/T} - e^{-g\mu_{\mathrm{B}}\frac{1}{2}H/T}}{e^{g\mu_{\mathrm{B}}\frac{1}{2}H/T} + e^{-g\mu_{\mathrm{B}}\frac{1}{2}H/T}} = \frac{1}{2}\tanh\left(\frac{g\mu_{\mathrm{B}}H}{2T}\right). \tag{6.16}$$

Taking into account that for electrons $g = 2$, we can rewrite equation (6.16) as

$$M = \mu_{\mathrm{B}}\tanh\left(\frac{\mu_{\mathrm{B}}H}{T}\right). \tag{6.17}$$

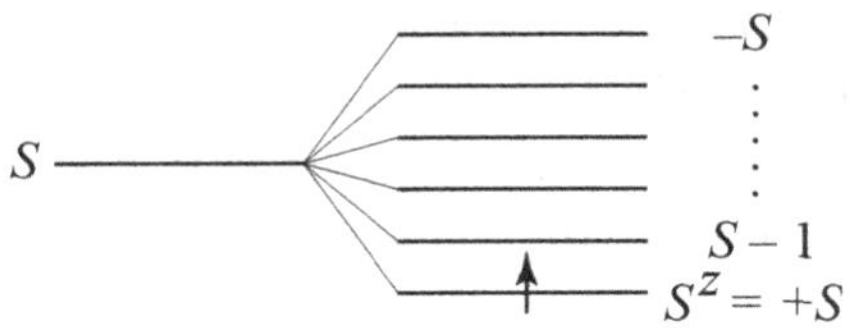


Fig. 6.3

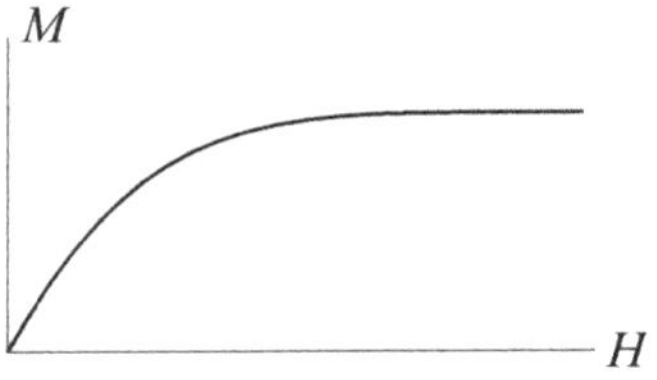


Fig. 6.4

For small $H$ ($g\mu_B H \ll T$) the moment $M$ is linear in the field,

$$M = \frac{N}{V}\frac{g^2\mu_B^2 H}{4T} ,$$

and

$$\chi_0 = \frac{\partial M}{\partial H} = \frac{N}{V}\frac{g^2\mu_B^2}{4T} , \tag{6.18}$$

which coincides with (6.15) for $S = \frac{1}{2}$, $S(S+1) = \frac{3}{4}$.

For the general case (arbitrary $S$, or total angular momentum $\mathcal{J}$):

$$M(H) = g\mu_B\, \mathcal{J}\, B_{\mathcal{J}}(g\mu_B \mathcal{J} H/T) , \tag{6.19}$$

where

$$B_{\mathcal{J}}(x) = \left(1 + \frac{1}{2\mathcal{J}}\right)\coth\left[\left(1 + \frac{1}{2\mathcal{J}}\right)x\right] - \frac{1}{2\mathcal{J}}\coth\frac{x}{2\mathcal{J}} \tag{6.20}$$

is the so-called Brillouin function (used here instead of tanh in equation (6.16), which was valid for $\mathcal{J} = S = \frac{1}{2}$).

The physics leading to (6.14)–(6.20) is illustrated in Fig. 6.3. In a magnetic field there occurs the Zeeman splitting of the levels, and with decreasing temperature more and more spins in our system will accumulate at the lowest level, with the spin parallel to the field, which leads to the increase of total magnetization (6.13), (6.16) and to the susceptibility (6.14).

Both equation (6.16) and equations (6.19)–(6.20) describe the *saturation* of magnetization in strong fields $g\mu_B \mathcal{J} H \gg T$, see Fig. 6.4. For very large spins the quantum expressions go over to the classical ones, with the substitution $S(S+1) \to S^2$. This is a general property of any spin system: it becomes classical for $S \to \infty$, because for $S \to \infty$ the noncommutativity of spin operators is irrelevant, and quantum effects disappear (cf. the treatment of Bose condensation in Chapter 5).

## 6.2 Interacting localized moments; magnetic ordering

The main interaction between localized spins is the exchange interaction:

$$\mathcal{H} = \sum_{ij} J_{ij} \boldsymbol{S}_i \cdot \boldsymbol{S}_j - g\mu_B \boldsymbol{H} \cdot \sum_i \boldsymbol{S}_i \,, \tag{6.21}$$

where we have introduced also the interaction with the external field $\boldsymbol{H}$. This is the Heisenberg exchange interaction, the simplest form of spin–spin interaction. In general exchange may be anisotropic, e.g. $J_\parallel S^z S^z + J_\perp(S^x S^x + S^y S^y)$; short-range ($J_{ij} = J\delta_{j,i\pm1}$) or long-range, etc. For higher spins $S > \frac{1}{2}$ the exchange interaction may also contain higher-order terms, for example biquadratic exchange $S_i^2 S_j^2$; there may also exist other, more special terms. We will treat below predominantly the simplest interaction (6.21) or its anisotropic generalizations; the general case is considered in specialized monographs or reviews on magnetism, e.g. in White (2006) and Yosida (1996).

*Note*: often the exchange interaction (our equation (6.21)) is defined differently: sometimes with the opposite sign, $-\sum J_{ij} \boldsymbol{S}_i \cdot \boldsymbol{S}_j$, and sometimes as $\pm 2\sum J_{ij} \boldsymbol{S}_i \cdot \boldsymbol{S}_j$; this corresponds to different definitions of the exchange integral $J_{ij}$. For example, the definitions of exchange integrals in two of the most popular textbooks, those by Kittel (1987) and by Ashcroft and Mermin (1976), differ by a factor of 2. It is also important to know whether in the summation in the Hamiltonian (6.21) each pair $ij$ is counted only once or the summation is carried out for all $i$ and $j$ independently, i.e. each pair enters twice (we use the latter convention below); this is in fact the reason for the difference in the factor of 2 mentioned above. Thus you have to be careful when someone cites the value of the exchange constant for a particular system; you should always check which definition of the exchange integral is being used.

Depending on the sign and the detailed distance dependence of $J_{ij} = J(\boldsymbol{R}_i - \boldsymbol{R}_j)$ the exchange interaction can give different types of magnetic ordering:

- ferromagnetic: all spins parallel, $\uparrow\uparrow\uparrow\uparrow\uparrow\uparrow$;
- antiferromagnetic: in the simplest case, alternating spins, $\uparrow\downarrow\uparrow\downarrow\uparrow\downarrow$ (so-called *Néel ordering*);
- spiral, e.g. helicoidal or cycloidal state: $\uparrow\nearrow\nearrow\rightarrow\searrow\searrow\downarrow\swarrow\swarrow\leftarrow\nwarrow\nwarrow\uparrow$;

etc.

### 6.2.1 Mean field approximation

The simplest treatment of the influence of the exchange interaction (6.21) on magnetic properties of the system, in particular on magnetic ordering, is the *mean field*, or self-consistent field approximation. In this method we consider each spin as being in an average field created by other spins of the system, which has to be determined self-consistently. This is equivalent to the *decoupling*

$$J_{ij}\boldsymbol{S}_i \cdot \boldsymbol{S}_j \Longrightarrow J_{ij}\Big(\boldsymbol{S}_i \cdot \langle \boldsymbol{S}_j \rangle + \langle \boldsymbol{S}_i \rangle \cdot \boldsymbol{S}_j - \langle \boldsymbol{S}_i \rangle \cdot \langle \boldsymbol{S}_j \rangle\Big) , \tag{6.22}$$

where

$$\langle \boldsymbol{S}_i \rangle = \langle \boldsymbol{S} \rangle \tag{6.23}$$

is the average spin, in the ferromagnetic case independent of the site. The last term in equation (6.22) is necessary to avoid double counting; this is important in the calculation of the total energy. If we take $z$ as the quantization axis, only $z$-components of the spin remain. Consequently, later on we will omit vector notation and denote the average spin simply as $\langle S \rangle$. This mean field approximation (the decoupling (6.22)) means that we consider each spin, say spin $S_i$, as being in a *molecular field* (also called the internal, or effective field) $H_{\text{intern}}$ created by its neighbours and given by the expression

$$g\mu_{\text{B}} H_{\text{intern}} = -2 \sum_j J_{ij} \langle S \rangle . \tag{6.24}$$

(We have used here the standard form of the coupling (6.11); the factor of 2 comes from the fact that, according to our definition of the exchange Hamiltonian (6.21), we sum over all indices $ij$ independently, i.e. each pair of spins is counted twice.) For the nearest-neighbour coupling $\sum_j J_{ij} = Jz$, where $z$ is the number of nearest neighbours. Having in mind ferromagnetic interactions and introducing for convenience the notation $\tilde{J} = -J$ (so that $\tilde{J} > 0$), we get for the molecular field the expression

$$g\mu_{\text{B}} H_{\text{intern}} = 2\tilde{J}z\langle S \rangle . \tag{6.25}$$

Putting this expression into equation (6.16), we now obtain the *self-consistency* equation (the *mean field* equation) for the average magnetization $M = g\mu_{\text{B}}\langle S \rangle$, which for the case of $S = \frac{1}{2}$ ($g = 2$) with ferromagnetic interaction takes the form

$$\langle S \rangle = \frac{1}{2} \tanh\left(\frac{\tilde{J}z\langle S \rangle}{T}\right) \tag{6.26}$$

or

$$M/\mu_{\text{B}} = \tanh \frac{\tilde{J}z(M/\mu_{\text{B}})}{2T} . \tag{6.27}$$

In the presence of an external field $H$ that field would enter in the numerator in tanh in (6.27) together with the molecular field (6.24).

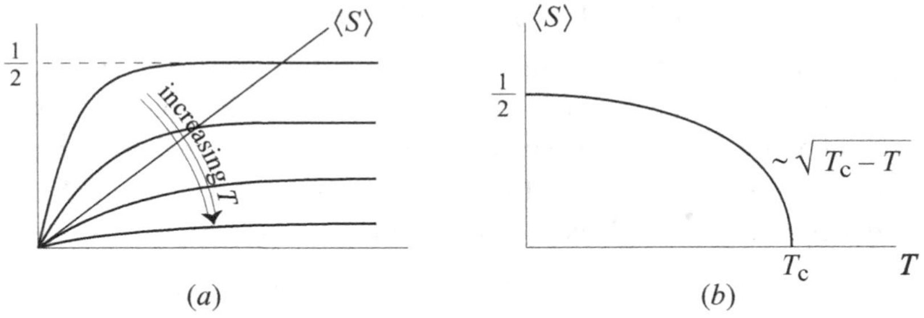


Fig. 6.5

The easiest way to analyse the mean field equations (6.26), (6.27) is the graphical solution, see Fig. 6.5($a$): we plot the curves for the left- and right-hand sides of these equations and look at their crossings. As we see, for zero external field $H = 0$ there is always a zero (nonmagnetic) solution $M = 0$, or $\langle S\rangle = 0$. But for low enough temperatures there also exists a nonzero solution with finite value $M$ (and also $-M$). One can show that these nontrivial solutions, when they exist, correspond to the minimum of the free energy, whereas the zero solution becomes the maximum. The temperature dependence of the spontaneous magnetization $M$, or the average spin $\langle S\rangle$, given by equation (6.27), is shown in Fig. 6.5($b$). The value of the critical temperature in this case ($S = \frac{1}{2}$) is

$$T_c = \tfrac{1}{2}\tilde{J}z \,. \tag{6.28}$$

For an arbitrary spin $S$ with nearest-neighbour interaction the corresponding expression is

$$T_c = \tfrac{2}{3}\tilde{J}zS(S+1)\,, \tag{6.29}$$

which for $S = \frac{1}{2}$ reduces to (6.28).

Using equations (6.24)–(6.27) we can also easily obtain an expression for the magnetic susceptibility of interacting electrons. We can write the magnetic moment as

$$M = \chi_0(H + H_{\text{intern}})\,, \tag{6.30}$$

where $\chi_0$ is given by the expression (6.15). Using the expression (6.24) for $H_{\text{intern}}$, we finally obtain for a short-range interaction

$$M = \frac{\mathcal{C}H}{T - T_c}\,, \tag{6.31}$$

where we took into account the expressions (6.15) for $\mathcal{C}$ and (6.29) for $T_c$.

The full susceptibility $\chi$, defined by the relation $M = \chi H$, is given in this case by the expression

$$\chi = \frac{\mathcal{C}}{T - T_c}\,. \tag{6.32}$$

This is the famous *Curie–Weiss* law, which is usually written as

$$\chi = \frac{\mathcal{C}}{T - \Theta} . \tag{6.33}$$

In the mean field approximation in the case of only a nearest-neighbour interaction $J$, the *Weiss temperature* is

$$\Theta = T_c = \tfrac{2}{3} S(S+1)\tilde{J} z = -\tfrac{2}{3} S(S+1) J z , \tag{6.34}$$

where $J = -\tilde{J}$ is the exchange integral as introduced in equation (6.21) for the nearest-neighbour interaction. The sign of $\Theta$ tells us whether the corresponding exchange interaction is ferromagnetic ($\Theta > 0$) or antiferromagnetic ($\Theta < 0$).

In the general case the Weiss temperature is given by the expression

$$\Theta = -\tfrac{2}{3} S(S+1) \sum_j J_{ij} , \tag{6.35}$$

where the summation goes over all neighbours with which a given spin interacts. This general expression (6.35) has broader applicability than the nearest-neighbour version (6.34). It can happen, for example, that for anisotropic and especially for frustrated magnets, or for systems with long-range interactions, the value of $\Theta$ is very different from $T_c$ or even has the opposite sign. It is important to realize that the sign of $\Theta$ is determined by the *sum of all exchange interactions*, with all neighbours of a given site. If these interactions are predominantly antiferromagnetic, the sign of $\Theta$ would be negative, and the magnetic ordering would be antiferromagnetic. There may be situations, however, when the strongest interactions are ferromagnetic, e.g. a strong ferromagnetic interaction in a magnetic layer, but if the interlayer coupling is antiferromagnetic (and, e.g. weaker), the resulting state would be also antiferromagnetic (ferromagnetic planes stacked antiferromagnetically). In this case at high temperatures we would have a 'ferromagnetic' susceptibility ($\Theta > 0$), although the ground state is actually antiferromagnetic. On the other hand, for frustrated systems with antiferromagnetic interactions the value of the critical temperature (the Néel temperature $T_N$, see below) may be much smaller than the typical absolute values of the exchange interaction. In this case $|\Theta|$ may be much larger than $T_N$; the small value of $T_N/|\Theta|$ (or large $|\Theta|/T_N$) is now often taken as an empirical criterion, a fingerprint of strong frustrations.

A word of caution again: for a different definition of the exchange constant, when, in contrast to our convention, each pair $ij$ in the summation in the exchange Hamiltonian (6.21) is counted only once, one would use the exchange integral $J'$ which is twice our $J$, $J' = 2J$; this is, e.g. the definition used by Ashcroft and Mermin (1976) and Ziman (1979). Then the results (6.29), (6.34) for the critical temperature $T_c$ or for the Weiss temperature $\Theta$, written through this $J'$, would contain, instead of the factor $\frac{2}{3}$, the factor $\frac{1}{3}$, and the value of the exchange constant

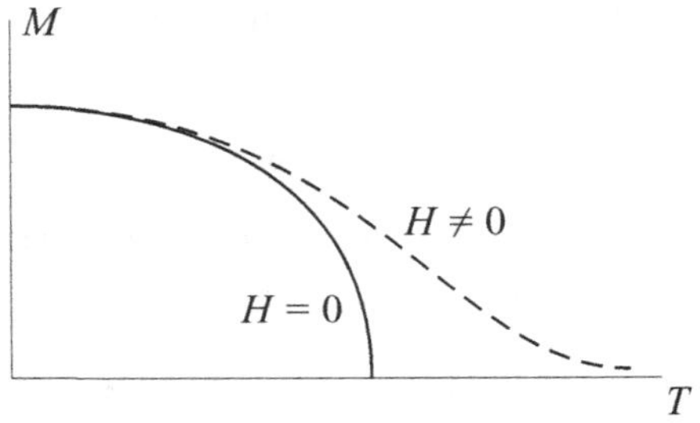


Fig. 6.6

determined, e.g. from the high-temperature susceptibility, would differ from the more conventionally determined one by a factor of 2.

### 6.2.2 Landau theory for ferromagnets

The mean field approximation gives results equivalent to the general Landau theory of second-order phase transitions. The Landau expansion of the free energy has here the form

$$\Phi = A\boldsymbol{M}^2 + B\boldsymbol{M}^4 - \boldsymbol{H}\cdot\boldsymbol{M}\,. \tag{6.36}$$

We have introduced in (6.36) the interaction with the external magnetic field $\boldsymbol{H}$ to point out two consequences:

1. We can obtain from this approach the behaviour of the magnetic susceptibility at the phase transition, which turns out to be the same as that obtained above in the mean field approximation. The equation for $\boldsymbol{M}$ is

$$\frac{\partial\Phi}{\partial\boldsymbol{M}} = 2A\boldsymbol{M} + 4B\boldsymbol{M}(\boldsymbol{M})^2 - \boldsymbol{H} = \boldsymbol{0}\,. \tag{6.37}$$

   As $\boldsymbol{M} \parallel \boldsymbol{H}$, we may omit vector notation here:

$$2AM + 4BM^3 - H = 0\,. \tag{6.38}$$

   It is clear that, for finite $H$, $M$ is always nonzero, even above $T_c$, Fig. 6.6. Close to $T_c$ ($T > T_c$), where $M \ll 1$, we have (with $A = a(T - T_c)$, see (2.3))

$$M = \frac{H}{2A} = \frac{H}{2a(T - T_c)}\,, \tag{6.39}$$

   i.e. with $M = \chi H$, we get

$$\chi = \frac{1}{2a(T - T_c)}\,, \tag{6.40}$$

   the standard Curie–Weiss law, cf. (6.32).

The expression (6.39) is valid for $T > T_c$. Below $T_c$ there exists spontaneous magnetization in the absence of the magnetic field $H$, so that one cannot define the susceptibility as $\chi = \lim_{H\to 0} M/H$, as is usually done. However one can always define the differential susceptibility $dM/dH$. From Fig. 6.6, where the temperature dependence of the magnetization of a ferromagnet is shown in the presence and in the absence of an external magnetic field, it is clear that in the finite external field the magnetization increases both above and below $T_c$. From equation (6.38) we find for $T < T_c$ (where $M_{H=0} \neq 0$):

$$\frac{d}{dH}\left(\frac{\partial\Phi}{\partial M}\right) = \frac{\partial}{\partial H}\left(\frac{\partial\Phi}{\partial M}\right) + \frac{\partial}{\partial M}\left(\frac{\partial\Phi}{\partial M}\right)\frac{dM}{dH} = 0\,, \tag{6.41}$$

i.e.

$$-1 + (2A + 12BM^2)\frac{dM}{dH} = 0\,, \tag{6.42}$$

or

$$\chi_{\text{diff}} = \frac{dM}{dH} = \frac{1}{2A + 12BM^2}\,. \tag{6.43}$$

For $T < T_c$ the coefficient $A = a(T - T_c) < 0$, $M^2 = -A/2B = a(T_c - T)/2B$, and

$$\chi = \frac{1}{2A - 6A} = -\frac{1}{4A} = +\frac{1}{4a(T_c - T)}\,. \tag{6.44}$$

Thus the differential susceptibility also diverges as $1/(T_c - T)$ from below, for $T \to T_c - 0$, but with the coefficient two times smaller than above $T_c$, cf. (6.44) and (6.40) – see also (2.37); as mentioned there, this is a real effect, measurable experimentally.

In obtaining equations (6.39)–(6.44) we have taken into account only terms linear in $M$ and $H$. However, the full equation (6.38) contains also nonlinear terms. They become important for higher fields. One can use equation (6.38) as a very convenient way to determine experimentally the critical temperature of a ferromagnetic phase transition.

The point is that by just measuring the magnetization in weak fields it is very difficult to get an accurate value of $T_c$. As is well known, there always appear in such systems ferromagnetic domains with different spin orientations, which change with field and temperature and which dominate the magnetic response at small fields. One can get rid of this problem if one works at high enough fields, sufficient to orient all spins in one direction. But in this case one has to take into account nonlinear effects in the $M(H)$ dependence, described by the cubic term

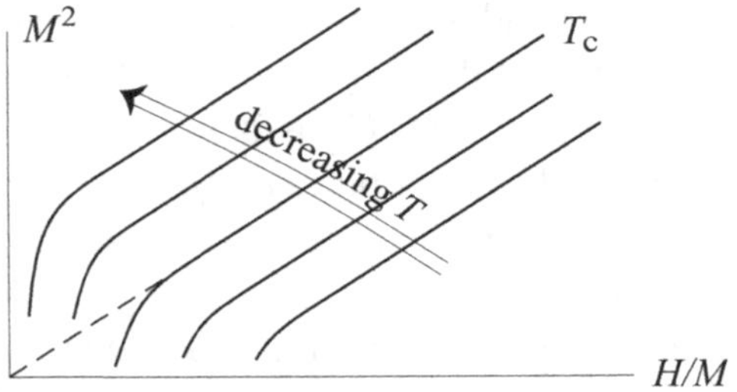


Fig. 6.7

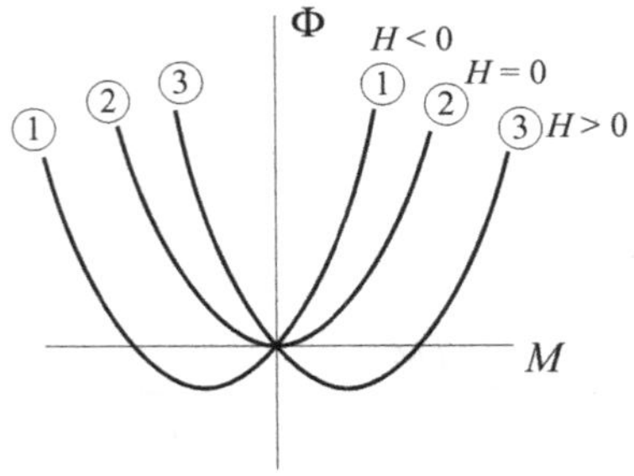


Fig. 6.8

in equation (6.38). Keeping all terms there, one can rewrite equation (6.38) in the form

$$2BM^2 = H/2M - A\,. \tag{6.45}$$

Thus if we plot $M^2$ vs. $H/M$ at different temperatures, we should have a series of straight lines, see Fig. 6.7. The slope of these lines would give us the value of the coefficient $B$ in the Landau expansion of the free energy, and the offset, the value at $H = 0$, would give the coefficient $A$. Remembering that in the Landau theory $A = a(T - T_c)$, we see that the value of the temperature at which such a straight line passes through zero would give us the value of $T_c$.

In real situations, due to domain effects, at small fields this dependence could strongly deviate from that of equation (6.45), but it is usually satisfied at large enough fields. Thus to determine $T_c$ one has to use the extrapolation of the curves of Fig. 6.7 measured at high fields to $H = 0$. This method is widely used in practice. The plot (6.45), Fig. 6.7, is called the *Arrott plot* (or sometimes the Belov–Arrott plot).

2. Let us now consider the free energy (6.36) as a function of $M$ for different temperatures *and fields*. For $T > T_c$ it has the form shown in Fig. 6.8, i.e. as a

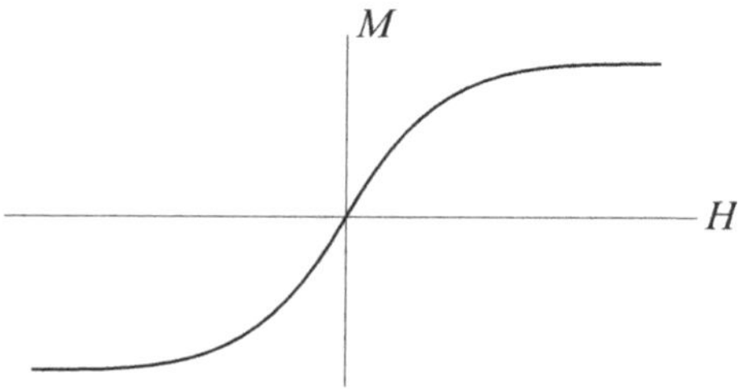


Fig. 6.9

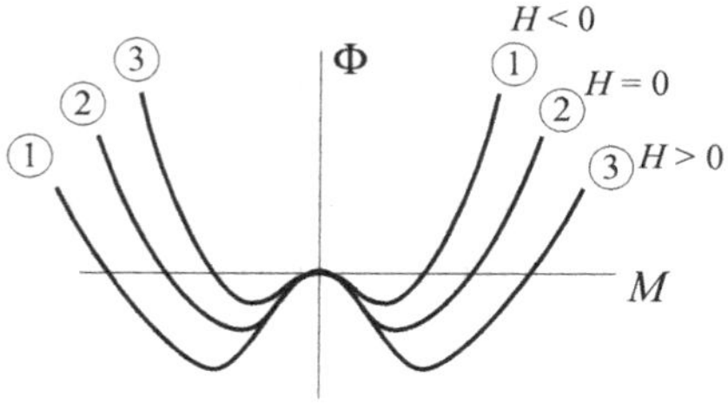


Fig. 6.10

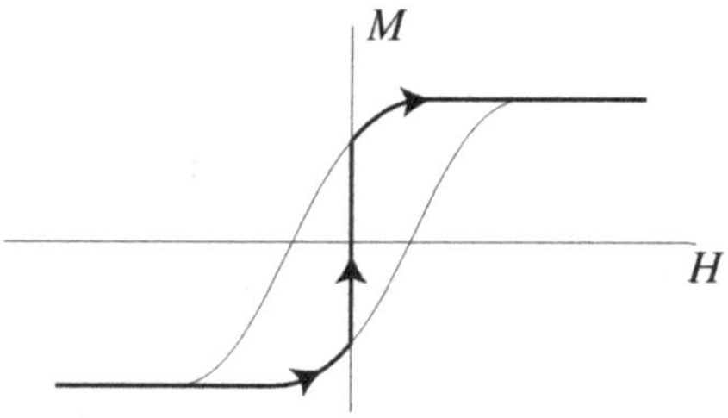


Fig. 6.11

function of $H$ the magnetization changes continuously, Fig. 6.9. Now, consider the situation for $T < T_c$. The set of curves for $\Phi(M)$ is now given in Fig. 6.10. We see that as the field changes sign, the minima with positive and negative $M$ change absolute values; for $H > 0$ the minimum with $M > 0$ lies deeper, and for $H < 0$ – that with $M < 0$. Thus there should be a jump of $M$, when $H$ passes zero, very much like in the first-order transitions discussed in Chapter 2! (cf. Figs. 2.5, 2.8). This is nothing else but the well-known *hysteresis* of ferromagnets, Fig. 6.11. And it really *is* a first-order transition. However, not as a function of temperature, but as a function of field. Thus, the phase diagram of a ferromagnet in the $(T, H)$ plane looks as shown in Fig. 6.12. There is indeed a first-order transition line, and the *end point* of this line (critical point,

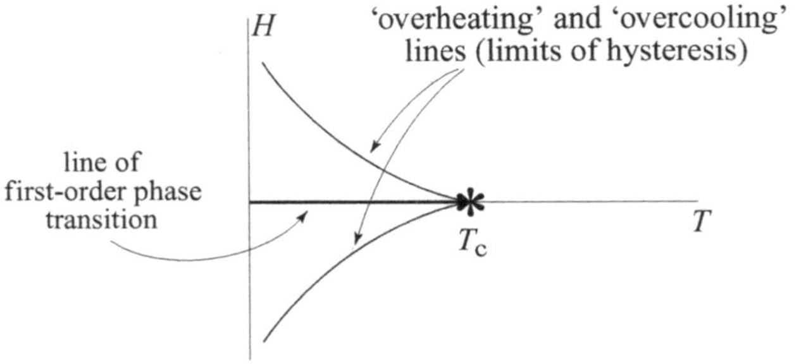


Fig. 6.12

in the terminology of first-order phase transitions) is our old familiar $T_c$ – the critical point of the second-order phase transition. All the anomalies of different quantities at the second-order phase transition point are actually the same as in the critical point of first-order phase transitions, cf. e.g. Fig. 2.9.

### *6.2.3 Antiferromagnetic interactions*

The mean field description of antiferromagnetic ordering coincides with that of ferromagnetic ordering, with the only difference that we change the spin direction in one of the sublattices, together with the opposite sign of the exchange interaction. The corresponding self-consistency equations, analogous to equations (6.26), (6.27), will now be written for a sublattice magnetization. Below we discuss the response of an antiferromagnet to an external field, which is different from that of a ferromagnet.

Let us write down the Landau free energy for an antiferromagnet. In an antiferromagnet in the absence of external field there are two sublattices with antiparallel spins, $\boldsymbol{M}_1 = -\boldsymbol{M}_2$. The order parameter here is

$$\boldsymbol{L} = \boldsymbol{M}_1 - \boldsymbol{M}_2 \,; \tag{6.46}$$

it is zero above the critical temperature (usually denoted as $T_N$, the *Néel temperature*), and nonzero for $T < T_N$. The free energy of an antiferromagnet in the presence of an external field has the form

$$\Phi = \Phi_0 + A\boldsymbol{L}^2 + B\boldsymbol{L}^4 + D(\boldsymbol{H}\cdot\boldsymbol{L})^2 - \tfrac{1}{2}\chi_p \boldsymbol{H}^2 \,. \tag{6.47}$$

It is important that in contrast to ferromagnets the external field $\boldsymbol{H}$ does not couple to the order parameter $\boldsymbol{L}$ linearly (linear couplings with different sublattices cancel, because $\boldsymbol{M}_1 = -\boldsymbol{M}_2$), so that the lowest nonzero coupling allowed by symmetry is quadratic. At the same time there would definitely appear a certain magnetization in

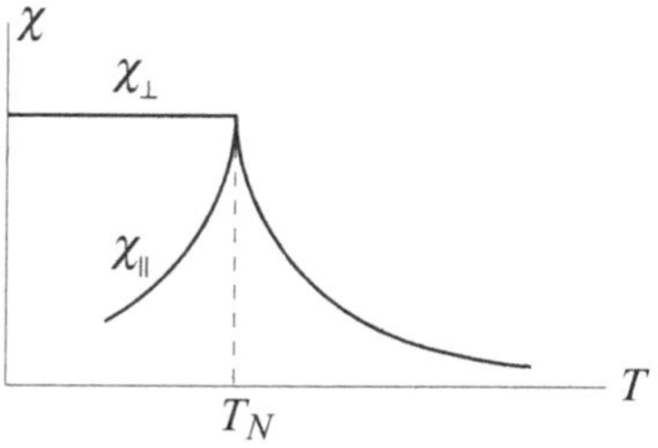


Fig. 6.13

the external field, even in the absence of antiferromagnetic ordering, e.g. above $T_{\rm N}$; this effect is described by the last term in equation (6.47). (We recall that the moment $M = -\partial\Phi/\partial H$, i.e. here it is equal to $\chi_{\rm p}H$, as it should be.) Here $\chi_{\rm p}$ is the magnetic susceptibility of the sample in the paramagnetic phase, which is given by the Curie–Weiss law (6.33) with $\Theta = -\frac{2}{3}S(S+1)Jz$ (6.34).

On the other hand, the presence of antiferromagnetic order below $T_{\rm N}$ should modify the susceptibility. Let us consider the susceptibility below the Néel temperature. We have to discriminate two situations: the field parallel to $\boldsymbol{L}$ (or to the sublattice magnetization) or perpendicular to it. In general the moment is

$$\boldsymbol{M} = -\frac{\partial\Phi}{\partial\boldsymbol{H}} = \chi_{\rm p}\boldsymbol{H} - 2D\boldsymbol{L}(\boldsymbol{L}\cdot\boldsymbol{H})\,. \qquad (6.48)$$

(*a*) For the perpendicular field, $\boldsymbol{H} \perp \boldsymbol{L}$, we obtain from (6.48) that

$$\chi_\perp = \chi_{\rm p}\,, \qquad (6.49)$$

i.e. below $T_{\rm N}$ the perpendicular susceptibility is constant, independent of temperature.

(*b*) For the field parallel to the sublattice magnetization, $\boldsymbol{H} \parallel \boldsymbol{L}$, the second term in equation (6.48) also contributes. We then obtain

$$\chi_\parallel = \chi_{\rm p} - 2DL^2 = \chi_\perp - \frac{Da}{B}(T_{\rm N} - T) \qquad (6.50)$$

(by (6.49) $\chi_{\rm p} = \chi_\perp$; we also took the standard expression for the order parameter of an antiferromagnet $L^2 = \frac{a}{2B}(T_{\rm N} - T)$, see equation (2.6)).

Thus we see that below $T_{\rm N}$, $\chi_\parallel$ and $\chi_\perp$ are different, see Fig. 6.13. The moment in the ordered phase is smaller for the field parallel to the sublattice magnetization than for the perpendicular field. In the parallel field we have to *invert* some spins, Fig. 6.14(*a*), and this costs a large amount of energy, and the resulting susceptibility is small, whereas for the field *perpendicular* to the sublattice magnetization moments the sublattices can *cant*, Fig. 6.14(*b*), which is much easier. And we

Fig. 6.14

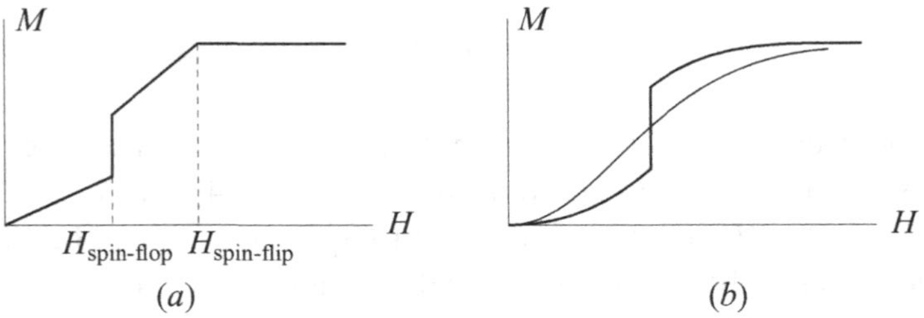


Fig. 6.15

gain more energy during this process. Therefore in an *isotropic* antiferromagnet the sublattice magnetization always orients *perpendicular* to the external field: in this case the direction of the sublattice magnetization does not matter and the antiferromagnetic exchange energy is the same, whereas we gain more energy due to interaction with the external field for the case of sublattices oriented perpendicular to the field ($\chi_\perp > \chi_\parallel$, see (6.50)).

What if there is a certain *anisotropy* in a material? For instance there can exist uniaxial anisotropy which favours all spins being oriented, say, along the $z$-axis. (This could be caused for example by single-site anisotropy, described by terms in the Hamiltonian of the type $-K(S^z)^2$ with $K > 0$, or by exchange anisotropy $J_\parallel \sum_{ij} S_i^z S_j^z + J_\perp \sum_{ij}(S_i^x S_j^x + S_i^y S_j^y)$, with $J_\parallel > J_\perp$.) Then at zero field and at small field along the $z$-axis, when anisotropy dominates, spins will still be parallel to $z$, $\uparrow\downarrow\uparrow\downarrow\uparrow\downarrow \Uparrow^{\boldsymbol{H}}$. However when the external field $\boldsymbol{H} \parallel z$ increases, at a certain moment, when it exceeds the anisotropy field, it will become favourable to turn all spins into the $xy$-plane $\perp \boldsymbol{H}$ (and cant them in the direction of the field). We gain in this process more energy on the interaction with the field than we lose on anisotropy. Such a transition is called a *spin-flop* transition: instead of having the structure of the type $\uparrow\downarrow\uparrow\downarrow\uparrow\downarrow \Uparrow^{\boldsymbol{H}}$, the structure now looks as $\nwarrow\nearrow\nwarrow\nearrow\nwarrow\nearrow \Uparrow^{\boldsymbol{H}}$, i.e. it is still antiferromagnetic, but with the sublattices perpendicular to the field and canted in the direction of the field. At still larger fields the canting angle increases till all the spins become parallel to the field. This transition to a 'forced' ferromagnetic arrangement is called a *spin-flip* transition. Thus altogether the behaviour of the total magnetization in this case has the form shown in Fig. 6.15($a$) (at low temperatures

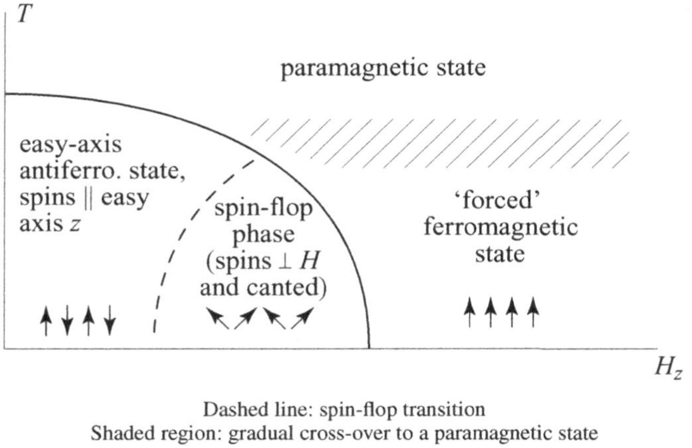


Fig. 6.16

$T \ll T_{\mathrm{N}}$). At higher temperatures this behaviour can be smeared out, Fig. 6.15(*b*). The phenomenon illustrated here (a sharp superlinear increase of magnetization at certain fields) is called *metamagnetism*; the situation described above gives one of the possible mechanisms of metamagnetic behaviour. A typical phase diagram of an easy axis antiferromagnet is shown in Fig. 6.16.

### 6.2.4 General case

Let us consider now the general case, with the exchange interaction $J_{ij}$ extending over larger distances and possibly of different sign. To study magnetic ordering in this case it is convenient to write down the Hamiltonian in the momentum representation,

$$\mathcal{H} = \sum_{ij} J_{ij} \boldsymbol{S}_i \cdot \boldsymbol{S}_j \implies \sum_{\boldsymbol{q}} J(\boldsymbol{q})\, \boldsymbol{S}(\boldsymbol{q}) \cdot \boldsymbol{S}(-\boldsymbol{q}) \,, \tag{6.51}$$

where

$$\boldsymbol{S}(\boldsymbol{q}) = \sum_n e^{-i\boldsymbol{q}\cdot\boldsymbol{R}_n} \boldsymbol{S}_n \tag{6.52}$$

is the Fourier transform of $\boldsymbol{S}_n$, and similarly for $J(\boldsymbol{q})$.

The mean field approximation can also be done in the momentum representation: $S(\boldsymbol{q})S(-\boldsymbol{q}) \Longrightarrow S(\boldsymbol{q})\langle S(-\boldsymbol{q})\rangle$. The energy $E = \sum_{\boldsymbol{q}} J(\boldsymbol{q})|S(\boldsymbol{q})|^2$ reaches the minimum at a certain value of $\boldsymbol{q} = \boldsymbol{Q}$, the value for which $J(\boldsymbol{q})$ is minimal. Then we can leave only these values of $\boldsymbol{q} = \boldsymbol{Q}$ in $S(\boldsymbol{q})$:

$$\langle S(\boldsymbol{q})\rangle = \langle S\rangle\, \delta(\boldsymbol{q} - \boldsymbol{Q}) \,, \tag{6.53}$$

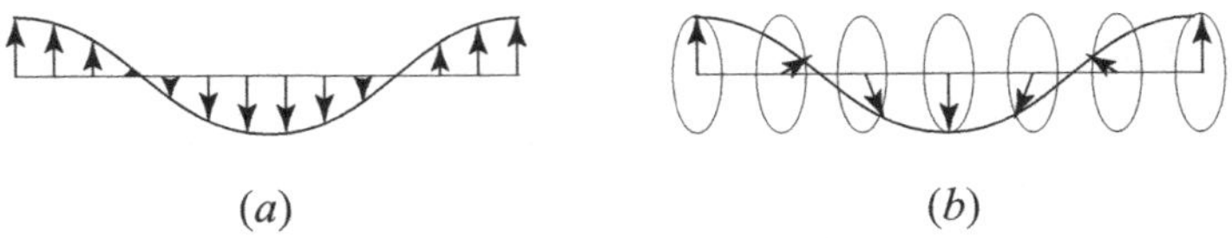

Fig. 6.17

$$\langle S_n \rangle \sim \int d^3\boldsymbol{q}\, e^{i\boldsymbol{q}\cdot\boldsymbol{R}_n} \langle S(\boldsymbol{q})\rangle \Longrightarrow \langle S \rangle e^{i\,\boldsymbol{Q}\cdot\boldsymbol{R}_n} . \tag{6.54}$$

Thus we see that for a general case the type of the resulting magnetic ordering (the one giving maximum energy gain) will be the modulated spin structure (e.g. spiral) with momentum $\boldsymbol{Q}$ corresponding to the minimum of $J(\boldsymbol{q})$.

Let us consider several examples, taking the simplest one-dimensional case. For the nearest-neighbour interaction

$$J_{ij} = J\,\delta_{j,i\pm 1} \Longrightarrow J(q) = 2J\cos qa . \tag{6.55}$$

(1) Let us first take $J$ negative, $J < 0$. $J(q)$ has a minimum at $q = Q = 0$. By (6.54)

$$\langle S_n \rangle = \langle S \rangle , \tag{6.56}$$

i.e. we automatically obtain a ferromagnetic solution, $\uparrow\uparrow\uparrow\uparrow\uparrow\uparrow$.

(2) For $J > 0$ the minimum of $J(q)$ is at $q = Q = \pi/a$. This gives

$$\langle S_n \rangle = \langle S \rangle e^{iQR_n} = \langle S \rangle e^{i\pi n} = \langle S \rangle (-1)^n , \tag{6.57}$$

which corresponds to antiferromagnetic ordering, $\uparrow\downarrow\uparrow\downarrow\uparrow\downarrow$, as it should. Thus in these two cases this method reproduces the known results.

(3) In general (for longer-range exchange $J_{ij}$) $J(q)$ can have a minimum at a certain arbitrary $Q$, and $\langle S(q)\rangle$ will be given by (6.53). In the coordinate representation this corresponds to the magnetization changing as

$$\langle S_n \rangle = \langle S \rangle e^{iQR_n} , \tag{6.58}$$

so that it describes a periodic (e.g. spiral) spin structure with period $d = 2\pi/Q$.

One should be more careful here in treating different projections of the spin; there may exist different types of structures with period $d$. Thus, this may be the so-called *sinusoidal* structure, with the magnetization at each site parallel say to the $z$-axis, with $\langle S_n^z \rangle = \langle S \rangle \cos \boldsymbol{Q}\cdot\boldsymbol{R}_n$, see Fig. 6.17($a$). Here *the absolute value* of the average magnetization changes in space. This will be, e.g. the situation for strong uniaxial (easy-axis) anisotropy. (We stress that we are speaking here about the thermodynamic average of the spin at site $n$, thus the fact that $\langle S_n^z \rangle$ is less than the nominal spin $S$, and can even be zero, does not imply any contradiction: one can visualize such a state as that with a finite average $z$-projection of $S$, $S^x$ and $S^y$ components remaining fluctuating.)

Another possibility is the *helicoidal* structure $\langle S_n^x \rangle = \langle S \rangle \cos \boldsymbol{Q} \cdot \boldsymbol{R}_n$, $\langle S_n^y \rangle = \langle S \rangle \sin \boldsymbol{Q} \cdot \boldsymbol{R}_n$. In this structure the value of the average momentum at each site remains constant, but the *spin rotates* for example in the $xy$ (or in another) plane, see Fig. 6.17($b$). If spins are rotating in the plane perpendicular to the wavevector of the spiral $\boldsymbol{Q}$, e.g. the spins are rotating in the $xy$-plane and $\boldsymbol{Q} = Q_z$, one can speak about a *proper screw* structure. If, however, the spins rotate in a plane containing $\boldsymbol{Q}$, e.g. with the same $\langle S_n^x \rangle$ and $\langle S_n^y \rangle$ as written above but with the vector $\boldsymbol{Q}$, e.g. in the $x$-direction, we will have a *cycloidal* spin structure. (This one, in particular, can give rise to ferroelectricity, which is very important in *multiferroics* – materials combining magnetic and ferroelectric properties.)

All such structures are often called *spin-density waves* (SDW), see also Section 11.6 below.

Note that in general the period of corresponding structures may have nothing to do with the periodicity of the underlying crystal lattice. Thus, for example, it can be *incommensurate* with respect to the lattice period.

Typical cases in which such magnetic structures can occur are those with competing (long-range) interactions. They are commonly met in metals with special shapes of the Fermi surface, e.g. in many rare earth metals and compounds, or in the metal Cr. Another typical example in which such spiral magnetic structures can appear are the *frustrated* magnets, with *geometrically frustrated* lattices (triangular, kagome, etc.) or with competing interactions.

**Problem:** Find the most favourable type of magnetic order in a one-dimensional system with competing nearest-neighbour and next-nearest-neighbour antiferromagnetic interactions

$$\mathcal{H} = J \sum_i \boldsymbol{S}_i \cdot \boldsymbol{S}_{i+1} + J' \sum_i \boldsymbol{S}_i \cdot \boldsymbol{S}_{i+2} \,. \tag{6.59}$$

**Crude solution:** Let us find the value of $q$ which minimizes the total $\tilde{J}(q)$:

$$\tilde{J}(q) = 2J \cos q + 2J' \cos 2q \tag{6.60}$$

$$\frac{d\tilde{J}}{dq} = -2J \sin q - 4J' \sin 2q = -2J \sin q - 8J' \sin q \cos q$$
$$= -2J \sin q \left(1 + 4\frac{J'}{J} \cos q\right) = 0 \,. \tag{6.61}$$

The solutions are:

(1) $\sin q = 0$, i.e. $q = 0$ or $q = \pi$.

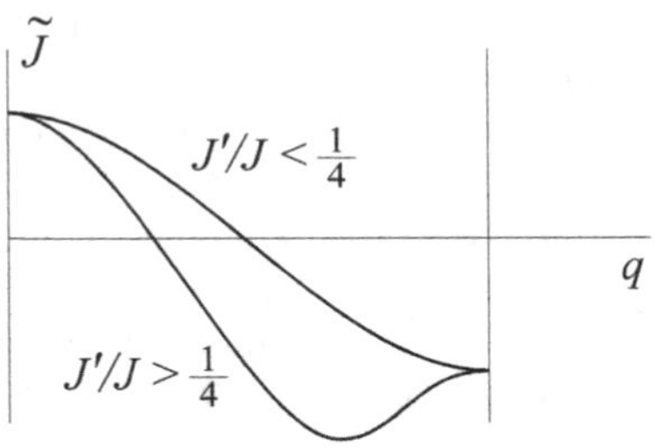


Fig. 6.18

From (6.60) we see that $q = 0$ corresponds to the *maximum* of $\tilde{J}(q)$, and $q = \pi$ may correspond to a minimum (if $\alpha = J'/J < \frac{1}{4}$). The solution with $q = \pi$ gives an antiferromagnetic state (the doubled period $l = \frac{2\pi}{\pi} = 2, \uparrow\downarrow\uparrow\downarrow\uparrow\downarrow$).

(2) For $\alpha = J'/J > \frac{1}{4}$ there exists yet another solution, which gives the absolute minimum of $\tilde{J}(q)$ in this case: this is given by the solution of the equation

$$1 + 4\frac{J'}{J}\cos q = 0\,, \qquad \cos q_0 = -\frac{1}{4J'/J}\,. \tag{6.62}$$

The dependence $\tilde{J}(q)$ has the form shown in Fig. 6.18. Indeed, as $|\cos q| < 1$, this solution exists only when $J'/J > \frac{1}{4}$.

Thus one may expect that the simple two-sublattice antiferromagnetic structure (the Néel configuration) will become *unstable* at $J'/J \geq \frac{1}{4}$ and could be transformed into a spiral with wavevector $q_0$. (Note that a similar spiral solution exists also for a ferromagnetic nearest-neighbour interaction $J < 0$ if $J'/|J| > \frac{1}{4}$.)

**Real solution:** The actual situation for $S = \frac{1}{2}$ and for $J, J' > 0$ is more complicated and much more interesting than in our mean field treatment. Indeed the two-sublattice Néel-like configuration becomes unstable for $(J'/J)_{\text{crit}} \simeq \frac{1}{4}$ (numerical calculations give the accurate critical value $(J/J')_{\text{crit}} = 0.241$, not so far from our mean field result $\frac{1}{4}$!). But due to the quantum nature of spins the actual state for $J'/J > (J'/J)_{\text{crit}}$ does not have real long-range order, it consists predominantly of singlets and is of spin-liquid type with a gap in the singlet–triplet excitations (this state is sometimes called a Majumdar–Ghosh state). But our simple treatment at least correctly predicts an instability of the simple Néel-like state and gives the value of the critical coupling close to the actual one. And this, in general, incommensurate, spiral state can indeed be realized in real quasi-1d materials when one takes into account the weaker interchain interaction. A similar treatment may be done for other types of lattices with competing interactions, not necessarily one-dimensional.

## 6.3 Quantum effects: magnons, or spin waves

Again first I give a short reminder:

The commutation relations for spin operators are

$$[S_i^x, S_j^y] = i S_i^z \delta_{ij} , \qquad [S_i^y, S_j^z] = i S_i^x \delta_{ij} , \qquad [S_i^z, S_j^x] = i S_i^y \delta_{ij} . \quad (6.63)$$

It is convenient to introduce the *spin raising* and *spin lowering* operators

$$\begin{aligned} S^+ &= S^x + i S^y , \\ S^- &= S^x - i S^y , \end{aligned} \qquad \text{so that} \qquad \begin{aligned} S^x &= \frac{S^+ + S^-}{2} , \\ S^y &= \frac{S^+ - S^-}{2i} . \end{aligned} \quad (6.64)$$

The commutation relations for these operators are

$$[S_i^+, S_j^-] = 2S_i^z \delta_{ij} , \qquad [S_i^z, S_j^+] = S_i^+ \delta_{ij} , \qquad [S_i^z, S_j^-] = -S_i^- \delta_{ij} . \quad (6.65)$$

Thus at *different sites* the operators $S^+$, $S^-$ *commute*, i.e. they behave as *bosons*. *But* we cannot put two spin-excitations on one site for $S = \frac{1}{2}$!

$$S^+|-\tfrac{1}{2}\rangle = |+\tfrac{1}{2}\rangle ; \qquad (S^+)^2|-\tfrac{1}{2}\rangle = S^+(S^+|-\tfrac{1}{2}\rangle) = S^+|\tfrac{1}{2}\rangle = 0 , \quad (6.66)$$

i.e. $(S^+)^2 = 0$, as for Fermi operators!

**Problem:** Check that at one site *for* $S = \frac{1}{2}$ the operators $S^+$, $S^-$ indeed obey anticommutation relation for Fermi operators.

**Solution:**

$$\begin{aligned} \{S^+, S^-\}_+ &= S^+S^- + S^-S^+ = 2\left((S^x)^2 + (S^y)^2\right) = 2\left(\mathbf{S}^2 - (S^z)^2\right) \\ &= 2\left(\frac{1}{2}\left(\frac{1}{2}+1\right) - \left(\frac{1}{2}\right)^2\right) = 2\left(\frac{3}{4} - \frac{1}{4}\right) = 1 , \end{aligned} \quad (6.67)$$

as it should be for Fermi operators. (Note that this is not the case for $S > \frac{1}{2}$.)

Thus the spin operators, strictly speaking, are neither Bose nor Fermi operators. For spin $\frac{1}{2}$ they behave as Bose operators at different sites and as Fermi operators at the same site. Sometimes they are called *paulions* (Pauli statistics, vs. Bose or Fermi statistics).

But it is often sufficient to treat spin excitations as *approximately bosons*, when their density is small and, on average, they are far from each other.

### 6.3.1 Magnons in ferromagnets

Let us now consider elementary excitations in a ferromagnet – spin waves, or magnons. It is convenient to rewrite the Heisenberg exchange Hamiltonian (6.21), using (6.64), as

$$\mathcal{H} = \sum J_{ij} \left[ S_i^z S_j^z + \tfrac{1}{2}(S_i^+ S_j^- + S_i^- S_j^+) \right] . \tag{6.68}$$

We start with the state $\left| \uparrow\uparrow\uparrow\uparrow\uparrow\uparrow \right\rangle = |0\rangle$, and act on this state by the operator $S_i^-$: $S_i^- \, |0\rangle = \left| \uparrow\uparrow\uparrow \cdots \underset{i}{\downarrow} \cdots \uparrow\uparrow\uparrow \right\rangle$, i.e. in this state one spin is reversed. The term $S_i^z S_j^z$ in (6.68) leaves the reversed spin at the same place, and only gives an energy cost to reverse it. The terms $S_i^+ S_j^- + S_i^- S_j^+$, on the other hand, *move* the reversed spin to another site, i.e. they create a *spin wave* with a certain dispersion.

The traditional way to treat this problem is the Holstein–Primakoff transformation. We introduce new operators $a^\dagger$, $a$ instead of $S^+$, $S$:

$$\begin{aligned} S_i^+ &= \sqrt{2S}\,(1 - a_i^\dagger a_i/2S)^{\frac{1}{2}}\, a_i \\ S_i^- &= \sqrt{2S}\, a_i^\dagger\,(1 - a_i^\dagger a_i/2S)^{\frac{1}{2}} \end{aligned} \tag{6.69}$$

so that $[a_i, a_j^\dagger] = \delta_{ij}$, the standard commutation relation for bosons.

One can then show that

$$S_i^z = S - a_i^\dagger a_i \, . \tag{6.70}$$

**Problem:** Check this, starting from the expression $(S^z)^2 = (\boldsymbol{S})^2 - (S^x)^2 - (S^y)^2$, using equation (6.69), the commutation relations for $a$, $a^\dagger$ and expanding $(1 - a^\dagger a/2S)^{1/2}$; use the fact that $\hat{n} = a^\dagger a$ commute, $[a^\dagger a, a^\dagger a] = 0$, and $[\hat{n}, a] = [a^\dagger a, a] = -a$. See Kittel (1987).

Usually it is sufficient to keep only the lowest terms in the expansion of (6.69), i.e.

$$S_i^+ \Longrightarrow \sqrt{2S}\, a_i, \qquad S_i^- \Longrightarrow \sqrt{2S}\, a_i^\dagger \, ; \tag{6.71}$$

other terms are small at least for large $S$. For $S = \frac{1}{2}$ we simply replace $S_i^+$ and $S_i^-$ by $a_i$ and $a_i^\dagger$. For convenience changing the sign of the exchange constant as in Section 6.2, $\tilde{J}_{ij} = -J_{ij}$, we obtain

$$\begin{aligned} \mathcal{H} &= -\sum \tilde{J}_{ij} \left[ (S - a_i^\dagger a_i)(S - a_j^\dagger a_j) + \tfrac{1}{2} \cdot 2S(a_i^\dagger a_j + a_i a_j^\dagger) \right] \\ &= -\sum \tilde{J}_{ij} \left[ S^2 + S(a_i^\dagger a_j + a_i a_j^\dagger - a_i^\dagger a_i - a_j^\dagger a_j) \right] + O(a^4) \, . \end{aligned} \tag{6.72}$$

Here we have omitted quartic terms in the Hamiltonian, originating from the term $S_i^z S_j^z$ and describing the interaction between magnons. The resulting expression

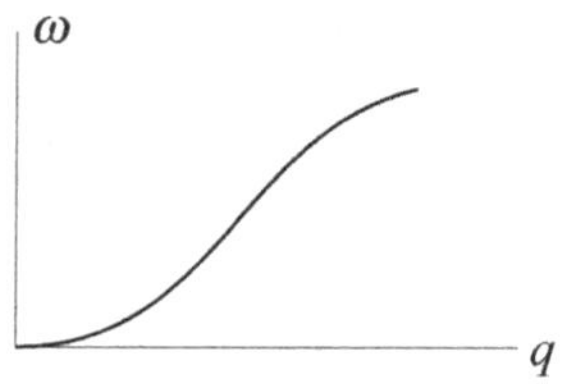


Fig. 6.19

is already a quadratic form, similar to the phonon case, cf. Chapter 4. We can diagonalize it by the Fourier transform:

$$\mathcal{H} = \text{const.} + \sum_{q}\Big(2S\sum_{ij}\tilde{J}_{ij}\big(1 - e^{-i\boldsymbol{q}\cdot(\boldsymbol{R}_i-\boldsymbol{R}_j)}\big)\Big)\, a^{\dagger}_{\boldsymbol{q}} a_{\boldsymbol{q}} \tag{6.73}$$

or

$$\mathcal{H} = \text{const.} + \sum_{q}\omega_{\boldsymbol{q}} a^{\dagger}_{\boldsymbol{q}} a_{\boldsymbol{q}} \ , \tag{6.74}$$

where the spin-wave spectrum is

$$\omega_{\boldsymbol{q}} = +2S\sum \tilde{J}(\boldsymbol{R}_i - \boldsymbol{R}_j)\big(1 - e^{-i\boldsymbol{q}\cdot(\boldsymbol{R}_i-\boldsymbol{R}_j)}\big) = 2S\big[\tilde{J}(\boldsymbol{q}=\boldsymbol{0}) - \tilde{J}(\boldsymbol{q})\big] \ . \tag{6.75}$$

As $q \to 0$ the spectrum is quadratic in $q$; for the nearest-neighbour interaction $\tilde{J}_{ij} = \tilde{J}\delta_{i,j\pm1}$ it is

$$\omega_{\boldsymbol{q}} \simeq 2S\tilde{J}q^2a^2 \ , \tag{6.76}$$

see Fig. 6.19. Here $S$ is a nominal spin (e.g. for spin $\frac{1}{2}$, $2S = 1$). (By the way, from (6.75) we see again that the ferromagnetic ordering becomes unstable if $J(\boldsymbol{q})$ has a minimum not at $\boldsymbol{q} = \boldsymbol{0}$ but at a certain nonzero $\boldsymbol{q}_0$: then the spin-wave energy $\omega_{\boldsymbol{q}}$ becomes negative for certain values of $\boldsymbol{q}$ close to $\boldsymbol{q}_0$, signalling an instability of the initial ferromagnetic state and a transition to another type of magnetic ordering, e.g. the helicoidal one with momentum $\boldsymbol{q}_0$.)

Spin waves (magnons) $a^{\dagger}_{\boldsymbol{q}}$, $a_{\boldsymbol{q}}$ are approximately bosons (at low temperatures, when there are only a few of them and they are, on average, far from each other). Then their occupation number is

$$n_{\boldsymbol{q}} = \frac{1}{e^{\omega_{\boldsymbol{q}}/T} - 1} \ . \tag{6.77}$$

(Their number is not conserved, thus, according to the treatment of Chapter 3, we take their chemical potential $\mu$ in (3.3) as zero. But the external magnetic field in (6.21) is linearly coupled to the magnetization, i.e. to the number of magnons, and it can play the role of their chemical potential.)

The total number of magnons at temperature $T$ is (in the 3d case)

$$n = \frac{1}{(2\pi)^3} \int d^3\boldsymbol{q}\, n_{\boldsymbol{q}} = \frac{N}{4\pi^2} \left(\frac{T}{2S\tilde{J}}\right)^{\frac{3}{2}} \int_0^\infty \frac{\sqrt{x}\, dx}{e^x - 1} . \tag{6.78}$$

Each magnon (spin reversal) reduces the magnetization, i.e. the magnetization is

$$M(T) = M(0) \left[ 1 - \text{const.} \left(\frac{T}{2S\tilde{J}}\right)^{3/2} \right] . \tag{6.79}$$

This dependence is indeed observed experimentally in isotropic ferromagnets. Note that the behaviour of the magnetization close to $T = 0$, given by the spin-wave theory, is quite different from that of the mean field treatment of Section 6.2: according to the latter, the magnetization would approach the saturation value as $T \to 0$ exponentially.

How can we visualize a spin wave? Suppose we have created a magnon with wavevector $Q$, $a_Q^\dagger\, |0\rangle \simeq S_Q^-\, |0\rangle$. We have to use a (quasi)classical description if we want to visualize what is going on and want to be able to draw pictures, etc. Thus we assume that there is a macroscopic occupation of the spin wave with the wavevector $Q$. Then the average is

$$\langle Q|S_q^-|Q\rangle = a\delta(q - Q) , \tag{6.80}$$

where $a$ is some (in general complex, $a = |a|\, e^{-i\varphi}$) constant, describing the amplitude and phase of the spin wave. (One can make this treatment rigorous by introducing the so-called *coherent states*, equivalent to the classical description of a coherent electromagnetic wave instead of the description in terms of photons; actually this state $|Q\rangle$ with the macroscopic occupation is such a coherent state.) Then the average projection of the spin on the $x$-axis is

$$\begin{aligned} \langle S_n^x\rangle = \langle Q|\, S_n^x\, |Q\rangle = \langle Q|\, \frac{S_n^+ + S_n^-}{2}\, |Q\rangle &= \langle Q|\, \tfrac{1}{2} \left\{ \int dq \left[ e^{iqn} S_q^+ + e^{-iqn} S_q^- \right] \right\} |Q\rangle \\ &= |a| \left\{ \frac{e^{iQn+i\varphi} + e^{-iQn-i\varphi}}{2} \right\} = |a| \cos(Qn + \varphi) . \end{aligned} \tag{6.81}$$

Similarly,

$$\langle S_n^y\rangle = \langle Q|\, S_n^y\, |Q\rangle = \langle Q|\, \frac{S_n^+ - S_n^-}{2i}\, |Q\rangle = |a| \sin(Qn + \varphi) . \tag{6.82}$$

And the $z$-projection of the spins is the same for each site and just decreases slightly,

$$\langle S^z\rangle = \tfrac{1}{2} - \langle a_n^\dagger a_n\rangle = \tfrac{1}{2} - |a|^2 , \tag{6.83}$$

cf. (6.70); i.e. the spins are actually only slightly tilted from the original direction. Actually the phase is $\varphi = \omega t$, and at each site one has a picture of *precession*

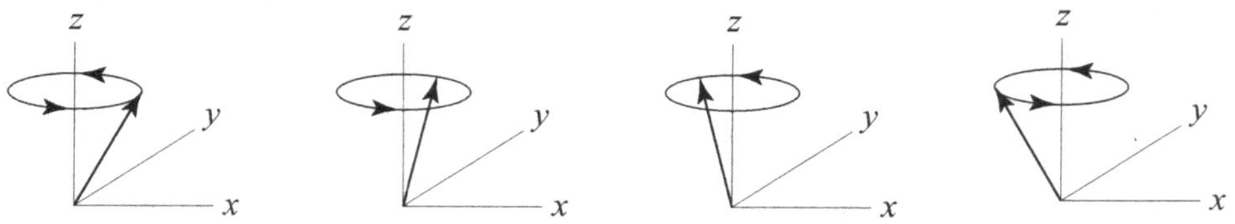


Fig. 6.20

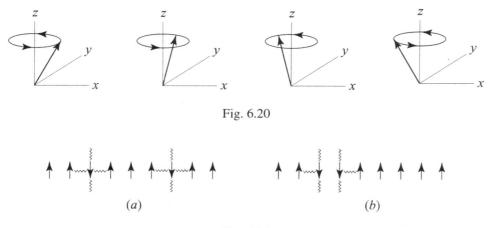

Fig. 6.21

around the average (molecular) field, Fig. 6.20: the spin at each site rotates on a cone. But the 'instantaneous' picture (snapshot) corresponds to a wave $\sim \cos Qn$ in $\langle S^x \rangle$ and $\sim \sin Qn$ in $\langle S^y \rangle$.

The terms omitted in (6.72) include, e.g. terms of the type $a^\dagger a^\dagger aa$, i.e. they describe the *interaction* between magnons. Because actually magnons are not exactly bosons, when two of them come together, there is a 'Pauli-like' on-site repulsion. But there also exists a certain attraction between spin deviations, illustrated in Fig. 6.21.

In Fig. 6.21($a$) we show two independent magnons (reversed spins) far from each other. Each of them 'spoils' $z$ bonds (here, in Fig. 6.21($a$), four bonds, denoted by the wavy lines). When two reversed spins come close together and become nearest neighbours (Fig. 6.21($b$)), they cost *less* energy, as they destroy fewer bonds (here in total six wrong bonds instead of eight in the case of Fig. 6.21($a$)). This decrease of energy means an attraction between spin deviations; it can in principle lead to the formation of *bound states* of two magnons. Such bound states always exist for the Heisenberg interaction in the 1d and 2d cases, and under certain conditions in three-dimensional systems too. If, instead of a reversed spin, we simply remove an electron with a spin, i.e. make a hole-doping, similar arguments tell us that these two holes may prefer to stay together forming a bound state. This is, in a nutshell, one of the mechanisms of the formation of Cooper pairs suggested, e.g. for the high-temperature superconductivity cuprates.

### *Another way to treat spin waves*

There exists another, very powerful and convenient method to study excitation spectra, which is applicable to many situations – the *method of equations of motion*.

The general scheme of this method is the following. The Schrödinger equation for a wavefunction $|\Psi\rangle$ is

$$\hat{\mathcal{H}}\,|\Psi\rangle = E|\Psi\rangle\;. \tag{6.84}$$

Let $|\Psi\rangle = \hat{A}\,|0\rangle$, where $|0\rangle$ is the ground state (vacuum), and the (yet unknown) operator $\hat{A}$ creates a given state $|\Psi\rangle$ out of $|0\rangle$. Equation (6.84) can then be rewritten as $\hat{\mathcal{H}}\hat{A}\,|0\rangle = E\hat{A}\,|0\rangle$, or, adding and subtracting $\hat{A}\hat{\mathcal{H}}$, as $(\hat{\mathcal{H}}\hat{A} - \hat{A}\hat{\mathcal{H}} + \hat{A}\hat{\mathcal{H}})\,|0\rangle = E\hat{A}\,|0\rangle$. In other words, the commutator satisfies the equation

$$[\hat{\mathcal{H}}, \hat{A}]\,|0\rangle = E\hat{A}\,|0\rangle - \hat{A}\hat{\mathcal{H}}\,|0\rangle\;. \tag{6.85}$$

We assume that the ground state $|0\rangle$ is an eigenstate of the Hamiltonian, $\hat{\mathcal{H}}\,|0\rangle = E_0\,|0\rangle$; choosing the beginning of the energy, we can put $E_0 = 0$, which we will do below. Then (6.85) takes the form

$$[\hat{\mathcal{H}}, \hat{A}]\,|0\rangle = E\hat{A}\,|0\rangle\;. \tag{6.86}$$

We can now treat equation (6.86) not as a Schrödinger equation for the wavefunction, but can say that the *operator* $\hat{A}$ obeys the equation

$$E\hat{A} = [\hat{\mathcal{H}}, \hat{A}]\;. \tag{6.87}$$

If this equation is satisfied, then also the original Schrödinger equation (6.84) will be satisfied. (Until now we have not made any approximations and used only identity transformations in going from (6.84) to (6.87).) The approach in which, instead of the Schrödinger equation for the wavefunction (6.84), we consider the equivalent equation (6.87) for operators, is called the *Heisenberg representation* (in this method time- or frequency-dependence is transferred from the *wavefunctions* to *operators*; see the more detailed discussion in Chapter 8).

Actually these two approaches go back to the early days of the formulation of quantum mechanics. In the 1920s two approaches had been suggested. One was the method of Schrödinger, which dealt with the wavefunctions and with equations for them. But slightly earlier Heisenberg suggested the matrix formulation of quantum mechanics, which dealt with *matrices*, which are actually *operators* (or, vice versa, one can represent operators as matrices in a given basis). The terminology 'Schrödinger representation' and 'Heisenberg representation' reflects the origin of these two methods.

Consider now a ferromagnet with spin $S = \frac{1}{2}$ with the ground state $|0\rangle = |\uparrow\uparrow\uparrow\uparrow\cdots\rangle$. Let us write down the equation of motion (6.87) for the operator $S_l^-$, which, as we have seen above, creates spin deviations and consequently spin waves. With the exchange Hamiltonian $\mathcal{H}$ (6.68) the equation (6.87) becomes, using commutation relations (6.65),

$$ES_l^- = [\mathcal{H}, S_l^-] = 2\sum_j J_{lj}\left\{-S_l^- S_j^z + S_l^z S_j^-\right\} . \tag{6.88}$$

(The factor of 2 in (6.88) comes from the terms with $l = i$ and $l = j$ in (6.68); interchanging the summation indices $i \leftrightarrow j$ in the second term we obtain (6.88).)

Now, we have started from the operator $S_i^-$, and obtained in the right-hand side more complicated operators, the products $S_i^- S_j^z$, $S_i^z S_j^-$. Generally speaking, one now has to write down equations similar to (6.87), (6.88) for these new operators. One gets, in this way, a set of equations for more and more complicated operators which one can solve by interrupting, or truncating, these equations at a certain stage, making certain assumptions about the nature of the ground state. Here we do it by making a *decoupling* similar to a mean field: we replace $S_i^z$ by the average magnetization $\langle S_i^z\rangle = \langle S_j^z\rangle = \langle S\rangle$. Then we obtain

$$ES_j^- = -2\langle S\rangle S_j^- \sum_i J_{ij} + 2\langle S\rangle \sum_i J_{ij} S_j^- . \tag{6.89}$$

In the *ferromagnetic* case $J_{ij} < 0$; denote again $\tilde{J}_{ij} = -J_{ij}$. After Fourier transforming we get

$$ES^-(\boldsymbol{q}) = 2\langle S\rangle \left\{\tilde{J}(\boldsymbol{0}) - \tilde{J}(\boldsymbol{q})\right\} S^-(\boldsymbol{q}) , \tag{6.90}$$

which gives the excitation spectrum

$$E(\boldsymbol{q}) = 2\langle S\rangle \left\{\tilde{J}(\boldsymbol{0}) - \tilde{J}(\boldsymbol{q})\right\} . \tag{6.91}$$

The spectrum (6.91) coincides with the magnon spectrum $\omega_{\boldsymbol{q}}$, (6.75), with one important modification:

$$E(\boldsymbol{q}) = \frac{\langle S\rangle}{S}\,\omega_{\boldsymbol{q}} , \tag{6.92}$$

i.e. it is 'self-consistent', and in this approximation the energy of the spin wave changes with the average magnetization $\langle S\rangle$ and with the temperature: the spectrum becomes *softer* (the energy of the excitation decreases) when $T \to T_c$. This is indeed very reasonable: as each spin is kept in its direction by the combined action (molecular field) of its neighbours, the spin stiffness (the 'cost' of reversing a given spin) should depend on the average spin of the surrounding ions and should decrease with decreasing $\langle S\rangle$. This is what equation (6.92) corresponds to.

If one now calculates the magnetization $M(T)$ using this spectrum, one obtains the standard spin-wave law (6.79) as $T \to 0$, and one reproduces the mean field results $M \sim \sqrt{T_c - T}$ close to $T_c$, i.e. it is a very good interpolation from $T = 0$ to $T_c$.

### 6.3.2 Antiferromagnetic magnons. Zero-point oscillations and their role

Let us consider now the antiferromagnetic case, $J > 0$. As the ground state we take here the state with two sublattices $|\uparrow\downarrow\uparrow\downarrow\uparrow\downarrow\rangle \equiv |0\rangle$ (the Néel state). The spin operators act on different sublattices as follows:

$$\begin{aligned} &\text{Sublattice } j\text{: } \uparrow \qquad S_j^z\,|0\rangle = +S\,|0\rangle\,, \quad S_j^+\,|0\rangle = 0\,, \\ &\text{Sublattice } l\text{: } \downarrow \qquad S_l^z\,|0\rangle = -S\,|0\rangle\,, \quad S_l^-\,|0\rangle = 0\,. \end{aligned} \tag{6.93}$$

Again we introduce spin deviation operators, as in (6.69), but in this case for each sublattice separately:

$$\begin{aligned} S_j^z &= \phantom{-}S - a_j^\dagger a_j\,, \qquad S_j^+ \simeq \sqrt{2S}\,a_j\,, \qquad S_j^- \simeq \sqrt{2S}\,a_j^\dagger\,, \\ S_l^z &= -S + b_l^\dagger b_l\,, \qquad S_l^- \simeq \sqrt{2S}\,b_l\,, \qquad S_l^+ \simeq \sqrt{2S}\,b_l^\dagger\,. \end{aligned} \tag{6.94}$$

Because of the antiferromagnetic (Néel) structure of the ground state, in this case the creation operators for spin waves are $S^-$ for one sublattice and $S^+$ for the other.

Substituting (6.94) into the Hamiltonian (6.68) and making a Fourier transform, we get (again omitting higher order terms):

$$\mathcal{H} = E_0 + 2S\sum_q \Big\{ J(\mathbf{0})\big(a_q^\dagger a_q + b_q^\dagger b_q\big) + J(\boldsymbol{q})\,\big(a_q^\dagger b_q^\dagger + b_q a_q\big)\Big\}\,. \tag{6.95}$$

In contrast to the ferromagnetic case, the expression (6.95), albeit a quadratic form, is still nondiagonal (it contains terms $a^\dagger b^\dagger$, $ab$). But we know what to do with such terms: we can diagonalize this expression using the Bogolyubov canonical transformation (5.23), as we have done for the Bose gas:

$$a_q = u_q\alpha_q + v_q\beta_q^\dagger\,, \qquad b_q = u_q\beta_q + v_q\alpha_q^\dagger\,. \tag{6.96}$$

As $u_q^2 - v_q^2 = 1$, we can represent $u_q$ and $v_q$, for example, as

$$u_q = \cosh\theta_q\,, \qquad v_q = \sinh\theta_q\,. \tag{6.97}$$

As in Chapter 5, the requirement of cancellation of 'dangerous' terms $\alpha\alpha$, $\alpha^\dagger\alpha^\dagger$, $\beta\beta$, $\beta^\dagger\beta^\dagger$, or equivalently the minimization of the energy $E = \langle\mathcal{H}\rangle$, gives

$$\tanh 2\theta_q = J(\boldsymbol{q})/J(\mathbf{0})\,, \tag{6.98}$$

and in effect we get

$$\begin{aligned} \mathcal{H} &= E_0 + \sum_q\big(\omega_q - 2SJ(\mathbf{0})\big) + \sum_q \omega_q(\alpha_q^\dagger\alpha_q + \beta_q^\dagger\beta_q) \\ &= \text{const.} + \sum_q \omega_q\left[\left(\alpha_q^\dagger\alpha_q + \tfrac{1}{2}\right) + \left(\beta_q^\dagger\beta_q + \tfrac{1}{2}\right)\right] \end{aligned} \tag{6.99}$$

with the spin-wave spectrum

$$\omega_q = 2S\sqrt{J^2(\mathbf{0}) - J^2(\mathbf{q})} \,. \tag{6.100}$$

For $\mathbf{q} \to \mathbf{0}$ $\omega_q \sim |\mathbf{q}|$, i.e. we obtain here a *linear* spectrum (vs. quadratic in the ferromagnetic case). There are two degenerate modes, $\alpha$ and $\beta$. The last two terms in (6.99) combine and give the term $\sim 2\sum_q \omega_q(n_q + \frac{1}{2})$, cf. the case of phonons, Chapter 4.

The presence of $\frac{1}{2}$ in $(n_q + \frac{1}{2})$ in (6.99) tells us that, in analogy with phonons in crystals, Chapter 4, there exist *zero-point fluctuations* in quantum antiferromagnets, which reduce sublattice magnetization even at $T = 0$, and which can even suppress long-range order completely. This is connected with the presence of nondiagonal terms in (6.95): the initial Néel ground state $|0\rangle$ is *not an eigenstate* of $\mathcal{H}$, and *pairs* of spin excitations at neighbouring sites (in different sublattices) can be created. (One can easily see that the fully ordered *ferromagnetic state* $|\uparrow\uparrow\uparrow\uparrow\uparrow\cdots\rangle$ is an eigenstate of the exchange Hamiltonian.) In an antiferromagnet the term $S_j^- S_l^+$ in (6.68) gives $S_j^- S_l^+ |\uparrow\underset{j}{\downarrow}\underset{l}{\uparrow}\downarrow\uparrow\rangle = |\uparrow\underset{j}{\downarrow}\underset{l}{\downarrow}\uparrow\uparrow\downarrow\cdots\rangle$, i.e. it reverses two spins and creates two excitations, leading to quantum fluctuations. This means that the actual ground state of the antiferromagnetic Heisenberg model (6.21), (6.68) is *not* just the state $|\uparrow\downarrow\uparrow\downarrow\uparrow\downarrow\cdots\rangle \equiv |0\rangle$ (the Néel state), but contains an admixture of states with different number of spin flips. In effect zero-point fluctuations are present even at $T = 0$, and the sublattice magnetization $\langle S\rangle$ is less than $S$. Indeed, the deviation of the average spin, e.g. in the $b$-sublattice from the nominal value $S$, is

$$\langle \delta S_l^z\rangle = S - \langle S_l^z\rangle = \langle b_l^\dagger b_l\rangle = \frac{2}{N}\sum_q \langle b_q^\dagger b_q\rangle \,. \tag{6.101}$$

Using (6.96)–(6.98) we can express it in the following way (see the details in Ziman (1979) and Kittel (1987)):

$$\langle \delta S^z\rangle = \text{const.} + 2\sum_q \frac{J(\mathbf{0})}{\omega_q}\left(n_q + \tfrac{1}{2}\right) \,, \tag{6.102}$$

where $\omega_q$ is the spectrum (6.100).

**Check this**, using the relations (6.96), expressing the coefficients $u_q$, $v_q$ through $\tanh 2\theta_q$ (6.98), using the bosonic commutation relations for the operators $\alpha_q$, $\beta_q$, and taking into account that the nondiagonal terms of the type $\alpha^\dagger\beta^\dagger$, $\alpha\beta$ should drop out.

The expression for spin deviations (6.102) is exactly analogous to the expression for the vibration amplitude of the lattice (4.74). Therefore we should expect that all the consequences thereof, discussed in Section 4.4.4, would also apply to antiferromagnets. Thus the deviations from the maximum sublattice magnetization at $T = 0$ in $d$-dimensional space are $\sim \int d^d\boldsymbol{q}/q$, i.e. they are finite for 3d ($d^3\boldsymbol{q} \simeq q^2\,dq$) or 2d ($d^2\boldsymbol{q} \simeq q\,dq$) cases, but for the 1d case the integral diverges logarithmically, $\langle b_l^\dagger b_l \rangle \sim \int dq/q$, which means that *there is no long-range order in a 1d antiferromagnet even at* $T = 0$! (cf. the case of crystals, (4.74)–(4.77)). This is due to quantum fluctuations, driven by the terms $S^+S^-$ in the Hamiltonian. A model without these terms, $\mathcal{H} = \sum J_{ij} S_i^z S_j^z$, is different in this respect. This *Ising model* is in a sense classical (vs. the Heisenberg model which is intrinsically quantum).

It is instructive at this point to elaborate somewhat on the parallel with ordinary lattice phonons. We have seen that many features of antiferromagnets remind us of a harmonic lattice: the spectrum of collective excitations (antiferromagnetic magnons vs. acoustic phonons) is linear as $q \to 0$; there exist zero-point vibrations, cf. (4.65), (4.74) and (6.102); because of that, one-dimensional order becomes unstable even at $T = 0$, etc. Physically this is connected with the fact that, in contrast to a ferromagnet, for which a perfectly ordered state is an exact eigenstate of the Hamiltonian, both in an antiferromagnet and in a lattice the ordered state we usually start with is not an eigenstate; there are terms in the Hamiltonian which lead to the creation of pairs of excitations.

Mathematically, as we have seen above, both the linear spectrum of antiferromagnetic magnons and the presence of zero-point deviations (6.102) follow from the Bogolyubov canonical transformation (6.96) (giving the term $\sum(n_{\boldsymbol{q}} + \frac{1}{2})$ in (6.102)). We have seen that the same term $\frac{1}{2}$ in $(n + \frac{1}{2})$ which is responsible for zero-point oscillations at $T = 0$ (where $\langle n_{\boldsymbol{q}} \rangle = 0$) is also present in the case of phonons. However we did not do any Bogolyubov transformation in Chapter 4! What is really the connection here, then?

The answer is the following. When considering lattice vibrations, one usually proceeds classically, treating classical equations of motion for atomic coordinates $x_i$, or for small deviations $u_i$, see Chapter 4. Only at a later stage did we go to the second quantization formalism and introduced phonon operators $b^\dagger$, $b$. The corresponding connection has the form (see (4.10)) $u_i \sim (b_i^\dagger + b_i)$. If we had used this representation from the very beginning, and written the initial Hamiltonian through the phonon operators, then, e.g. the term $\frac{1}{2}B(u_i - u_j)^2$ in the lattice Hamiltonian (4.16) (elastic energy) would contain both the terms $b_i^\dagger b_j$ and $b_i^\dagger b_j^\dagger$, $b_i b_j$, exactly as in the case of weakly interacting bosons (5.21) or antiferromagnets (6.95). And then we should have used again the same Bogolyubov canonical transformation to get rid of these terms, even in ordinary crystals! Thus it is indeed not accidental that the behaviour of antiferromagnets (antiferromagnetic magnons) resembles that of lattice (phonons).

On the other hand, one can proceed in the opposite way and try to obtain the magnon spectrum classically, going to the second quantization description at a later stage (as we have done with the lattice dynamics). This is indeed possible; the corresponding classical description can be formulated and is actually widely used in various problems of magnetism. The corresponding equations are the Landau–Lifshits equations mentioned in Section 2.4. They are especially useful for discussions of problems such as the dynamics of domain walls, etc.

Whereas at $T = 0$ the long-range antiferromagnetic order is unstable in one-dimensional systems, at finite temperatures these deviations may diverge also in two-dimensional antiferromagnets or ferromagnets.

**Problem:** For an isotropic *ferromagnet* show that the magnetization is *zero at finite temperatures* in the 1d and 2d cases.

**Solution:** From equation (6.78), with $n_{\boldsymbol{q}} = 1/(e^{\omega_{\boldsymbol{q}}/T} - 1)$ (6.77), and with $\omega_{\boldsymbol{q}}$ given by (6.75), (6.76), we obtain that the deviations from $\langle S\rangle$ at $T \neq 0$ are

$$\langle \delta S\rangle \sim \langle n\rangle \sim \int \frac{d^d\boldsymbol{q}}{e^{cq^2/T} - 1} \underset{(q\to 0)}{\sim} \int_0 \frac{q^{d-1}dq}{q^2/T}\,. \tag{6.103}$$

Thus for the 3d case $\langle \delta S\rangle \sim T \int q^2 dq/q^2$ which is OK; spin deviations are finite (and given by (6.79)).

But already for the 2d case $\langle \delta S\rangle \sim \int q\, dq/q^2$ is logarithmically divergent! And in the 1d case this divergence is even stronger.

**Problem:** Do the same for an isotropic *antiferromagnet* at $T \neq 0$. Remember that the antiferromagnetic spectrum is $\omega_{\boldsymbol{q}} \sim q$ instead of $\sim q^2$ for a ferromagnet. But, on the other hand, the expression for spin deviations in this case is given by equation (6.102), i.e. it contains an extra $\omega_{\boldsymbol{q}}$ in the denominator (and also the term $\frac{1}{2}$ in the integrand, describing zero-point fluctuations).

**Solution:** All the calculations here are exactly the same as in the case of phonons, Section 4.4.4. Thus the conclusions are the same: there is no long-range order in 1d either at $T = 0$ or $T \neq 0$, and in 2d at $T \neq 0$.

Thus for the isotropic (Heisenberg) interaction there is no long-range order at $T \neq 0$ in 1d and 2d systems. For 1d antiferromagnets there is no long-range order even at $T = 0$. This is again connected with the fact that elementary excitations (magnons or spin waves) are *gapless*, their spectrum obeys $\omega_q \to 0$ as $q \to 0$ (and sufficiently fast).

We want to repeat here this important point, already discussed in Sections 2.6 and 4.4. The general statement (the Goldstone theorem) is: if there exists a continuous broken symmetry, there should exist gapless excitations (if the interactions are not long-range ones).

Once again, here are examples of this general statement in specific cases:

*Magnets*

The continuous symmetry that is broken in the low-temperature ordered state is the continuous spin rotation: arbitrary directions, $\uparrow$, or $\nearrow$, or $\searrow$, etc. are allowed and are equivalent for the Heisenberg interaction (6.21). As a result there exist gapless spin waves. Spin waves with $q \to 0$ correspond to a rotation of magnetization (or sublattice magnetization) of the whole sample. As all spin orientations are equivalent, all such states are degenerate, it costs no energy to go from one to another, and consequently $\omega(q=0)=0$. But if the interaction is anisotropic (e.g. $JS_i^z S_j^z$ – this is called the Ising interaction), then the spin excitations *have a gap*, and the conclusions would be different.

*Crystallization*

In liquids there exists a continuous symmetry: continuous translations $\boldsymbol{r} \to \boldsymbol{r} + \delta\boldsymbol{r}$ (and continuous rotations). In crystals, however, we have a periodic structure, and the allowed translations are discrete, $\boldsymbol{r} \to \boldsymbol{r} + \boldsymbol{an}$. Broken continuous symmetry again leads to a gapless collective mode – acoustic phonons, with the spectrum $\omega_q = sq$. The mode with $q \to 0$ corresponds to a shift of the crystal as a whole, which explains why these phonons are gapless.

*Bose condensation*

The order parameter here is $\langle a_0^\dagger \rangle$, or $\langle a_0 \rangle$. What is the broken symmetry in this case? As explained in Chapter 5, there exists a gauge transformation of operators, an arbitrary change of phase: $a_q^\dagger \to e^{-i\varphi} a_q^\dagger$, $a_q \to e^{+i\varphi} a_q$. Usually all physical observables contain all operators always *in pairs*, $a^\dagger a$, $a^\dagger a^\dagger a a$, etc. and nothing depends on $\varphi$. But in the Bose-condensed state the average, not of a pair, but of only *one* operator $\langle a_0^\dagger \rangle$, is nonzero and plays the role of an order parameter, which means a *broken gauge symmetry* (again it is a continuous symmetry which is broken: all values of the phase $0 \le \varphi < 2\pi$ were possible). An ordered phase is characterized by *fixed* $\varphi$ (exactly as in ferromagnets, where below $T_c$ we chose some fixed orientation of $\boldsymbol{M}$). Consequently in the Bose condensed state there exists a mode $\omega_q$ *without a gap* – the Bogolyubov sound. One may thus connect the Bogolyubov sound with the oscillations of the phase of the order parameter.

Thus we again repeat: the conclusion is that, because of the presence of such gapless excitations, there is no long-range order in models or systems with broken

Table 6.1

| Ferromagnetic | Antiferromagnetic |
|---|---|
| Ground state: $\lvert 0\rangle = \lvert\uparrow\uparrow\uparrow\uparrow\uparrow\cdots\rangle$ | Crudely: the long-range ordered state is the Néel state $\lvert 0\rangle = \lvert\uparrow\downarrow\uparrow\downarrow\uparrow\cdots\rangle$ |
| It is the *exact ground state* of the Heisenberg exchange Hamiltonian $\mathcal{H} = -\tilde{J}\sum_{\langle i,j\rangle} \boldsymbol{S}_i \cdot \boldsymbol{S}_j$ | It is *not* an exact eigenstate of $\mathcal{H}$ (mathematically, one has to make a Bogolyubov transformation to diagonalize the Hamiltonian and to obtain the spectrum of elementary excitations) |
| Spin waves: $\omega_{\boldsymbol{k}} = ck^2$ | Spin waves: $\omega_{\boldsymbol{k}} = ck$ |
| Spin deviations: $\sim \int d^d\boldsymbol{k}\, n_{\boldsymbol{k}} \simeq T\int \frac{d^d\boldsymbol{k}}{\omega_{\boldsymbol{k}}} \sim T\int \frac{d^d\boldsymbol{k}}{k^2}$ | Spin deviations: $\sim \int \frac{d^d\boldsymbol{k}}{\omega_{\boldsymbol{k}}}\left(n_{\boldsymbol{k}} + \frac{1}{2}\right)$ |
| $T = 0$ : LRO, *no spin deviations* in 1d, 2d and 3d systems | $T = 0$ : 1d – divergent, *no LRO* even at $T = 0$ ! 2d – OK 3d – OK |
| $T \neq 0$ : 1d – divergent, no LRO 2d – divergent, no LRO 3d – OK | $T \neq 0$ : 1d – divergent, no LRO 2d – divergent, no LRO 3d – OK |

continuous symmetry in 1d and 2d cases at $T \neq 0$; this is the Mermin–Wagner theorem discussed in Chapter 4.

We summarize the behaviour of ferro- and antiferromagnetic systems (from the point of view of the role of fluctuations) in Table 6.1 (LRO stands for long-range order).

The behaviour at $T \neq 0$ is a consequence of the Mermin–Wagner theorem.

**Problem** (a real problem).
The excitation spectrum (gapless) may be, e.g. $\omega_{\boldsymbol{k}} \sim k$ or $\omega_{\boldsymbol{k}} \sim k^2$. In crystals (sound waves) or in antiferromagnets, for which the simple ordered state *is not the exact ground state* and *there exist zero-point oscillations*, the spectrum is linear, $\omega = sk$, so that $\int d^d\boldsymbol{k}/\omega_{\boldsymbol{k}}$ is divergent for $d = 1$ (1d systems) but not in 2d (at $T = 0$). (At $T \neq 0$ the linear spectrum is sufficient to destroy ordering also in the 2d case.)

For a *ferromagnet* the ground state of the type $\uparrow\uparrow\uparrow\uparrow$ is an *exact eigenstate* of the Hamiltonian, and there are no zero-point fluctuations; as a result there exists long-range order at $T = 0$, $\langle S\rangle = S$ (e.g. $\langle S\rangle_{T=0} = +\frac{1}{2}$). At the same time the spin-wave spectrum is quadratic, $\omega_{\boldsymbol{k}} = ck^2$.

*Is it accidental that the situation with zero-point motion goes together with a linear spectrum, and an exact state without zero-point oscillations gives quadratic dispersion?*

For instance if there would have been zero-point oscillations, but with excitations with the spectrum $\omega_{\mathbf{k}} \sim k^2$, then the fluctuations (e.g. spin deviations in magnets), given in this case by an expression of the type of (4.74) or (6.102), at $T = 0$ would behave as $\int d^d\mathbf{k}/\omega_{\mathbf{k}} = \int d^d\mathbf{k}/k^2$, i.e. they would diverge also at $d = 2$ (2d systems) – and there would be *no ordering* at $T = 0$ in both 1d and 2d cases.

But worse than that: then at $T \neq 0$ we would have had, for example, for a crystal (cf. (4.78))

$$\langle u^2 \rangle \sim \int \frac{d^d\mathbf{k}}{\omega_{\mathbf{k}} \frac{\omega_{\mathbf{k}}}{T}} = T \int \frac{d^d\mathbf{k}}{\omega_{\mathbf{k}}^2} \sim T \int \frac{d^d\mathbf{k}}{c^2 k^4}\,, \tag{6.104}$$

and $\langle u^2 \rangle$ would diverge, i.e. the crystal with such a spectrum would be unstable at any temperature even in our real 3d world! (And even in the 4d case.)

Maybe just because of that, this is not the case? In other words, is there a theorem that states:

In the presence of quantum fluctuations at $T = 0$ (or of the zero-point motion) the excitation spectrum (spectrum of the Goldstone mode) is linear (or at least $\omega_{\mathbf{k}} = ck^{\alpha}$ with $\alpha < \frac{3}{2}$, so as to make a 3d system stable at $T \neq 0$); and when we have the exact ground state (an ordered state which is an eigenstate of the Hamiltonian) and there are no quantum fluctuations at $T = 0$, then the spectrum of the Goldstone mode may be (should be?) $\omega \sim ck^2$?

Note also: the spectrum of surface modes (vibrations of the surface of a liquid, for example) is $\omega_{\mathbf{k}} \sim k^{3/2}$, just enough to make the surface stable at $T = 0$ (the amplitude of zero-point oscillations is $\int d^2\mathbf{k}/k^{3/2} \sim \int dk/\sqrt{k}$ which is convergent).

## 6.4 Some magnetic models

The study of different versions of magnetic models is now a rather well-developed field. We often have a good feeling of what to expect; many mathematical methods can be used for these systems and were actually developed in this context. It is also very useful and instructive because it is often possible to map other physical problems onto magnetic ones and then use the magnetic 'know-how' to solve or at least to better understand these problems.

### 6.4.1 One-dimensional models

These models are often *exactly soluble* (not all of them of course). The results of the exact solutions are often different from the simple mean field results.

*(1a) 1d Ising model with nearest-neighbour interaction*

The Hamiltonian of the Ising model in one dimension can be written as

$$\mathcal{H} = J \sum_{\langle ij \rangle} S_i^z S_j^z = 2J \sum_i S_i^z S_{i+1}^z \tag{6.105}$$

($\langle ij \rangle$ means summation over nearest neighbours – '*nn*'). It is easy to solve this model exactly – to find the partition function $Z(T, H)$ at arbitrary temperature and magnetic field $H \parallel z$ (described by an extra term in the Hamiltonian (6.105) of the type $-H \sum_i S_i^z$), see, e.g. Ziman (1979). Note that for the Ising model the situation for ferromagnetic and antiferromagnetic couplings is practically the same: in both cases there exists long-range ordering at $T = 0$, ferromagnetic in one case and a simple two-sublattice Néel state in the other. One can formally go from one case to the other by changing the spin quantization axes for every second site to the opposite.

There exists an elegant method for solving this problem – the so-called transfer matrix method; it can also be used for treating some other 1d models. One can write down the partition function as ($\beta = 1/T$)

$$\begin{aligned} Z &= \operatorname{Tr} e^{-\beta\mathcal{H}} = \operatorname{Tr} \prod_i \exp(-2\beta J S_i S_{i+1}) = \prod_i \operatorname{Tr} \exp(-2\beta J S_i S_{i+1}) \\ &= \sum_{S_1} \sum_{S_2} \cdots \sum_{S_N} \exp\Big\{-2\beta J \sum_{i=1}^{N} S_i S_{i+1}\Big\} . \end{aligned} \tag{6.106}$$

Taking into account the fact that every Ising spin $S_i$ can take only two values, $\pm\frac{1}{2}$, we can write down the partition function (6.106) as the trace of a product of matrices $\mathbf{T}_{SS'}$ such that $\langle S|\mathbf{T}|S'\rangle = \exp(-2\beta J S S')$ ($\langle +1|\mathbf{T}| + 1\rangle = \langle -1|\mathbf{T}| - 1\rangle = e^{-\beta J/2}$, $\langle -1|\mathbf{T}| + 1\rangle = \langle +1|\mathbf{T}| - 1\rangle = e^{+\beta J/2}$):

$$\begin{aligned} Z &= \sum_{S_1} \sum_{S_2} \cdots \sum_{S_N} \langle S_1|\mathbf{T}|S_2\rangle \langle S_2|\mathbf{T}|S_3\rangle \cdots \langle S_N|\mathbf{T}|S_1\rangle \\ &= \sum_{S_1} \langle S_1|\mathbf{T}^N|S_1\rangle = \operatorname{Tr}(\mathbf{T}^N) = \lambda_+^N + \lambda_-^N , \end{aligned} \tag{6.107}$$

where $\lambda_+$ and $\lambda_-$ are two eigenvalues of the matrix $\mathbf{T}$, and we have taken the periodic boundary condition $S_{N+1} = S_1$. One can easily generalize this treatment to the Ising model in a parallel magnetic field (with the extra term $-H_{\text{ext}} \sum_i S_i$ added to (6.105)). Thus the solution of the problem is reduced to finding the

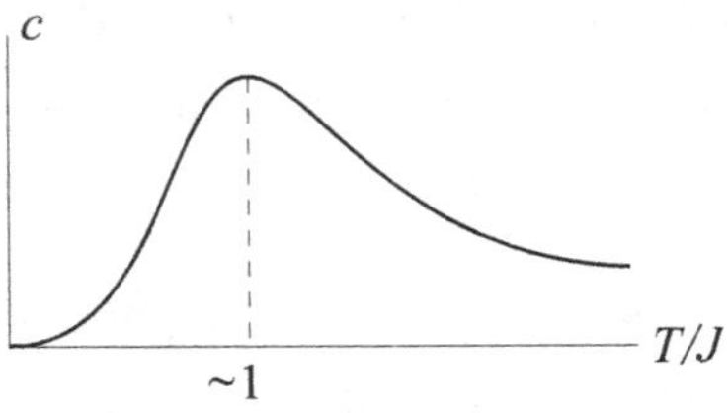


Fig. 6.22

(maximum) eigenvalue of the transfer matrix **T**, which in this case amounts to solving a quadratic equation.

Using this method we can easily obtain all the thermodynamic properties of the 1d Ising model. Thus, for example, the specific heat is (exact result)

$$c_V = \left(\frac{J}{T}\right)^2 \operatorname{sech}^2\left(\frac{J}{T}\right), \tag{6.108}$$

see Fig 6.22.

The maximum of $c_V(T)$ can be easily understood. The energy spectrum of the Ising model consists of discrete levels. The energy of a given bond is $-\frac{1}{2}|J|$ for the ground state (parallel spins for the ferromagnetic case) and $+\frac{1}{2}|J|$ for the wrong bond (antiparallel spins). Thus the creation of such a wrong bond costs us $|J|$; it is the 'cheapest' elementary excitation here. (One can easily see that in fact this means the creation of a 'domain wall' – the spins are, say, $\uparrow\uparrow\uparrow\uparrow\uparrow$ to the left of this defect, but $\downarrow\downarrow\downarrow\downarrow\downarrow$ to the right of it; this character of elementary excitation is very common in 1d systems, as we will see below. Of course, in a general excited state we can create many such wrong bonds.) The specific heat (6.108) in this essentially two-level-like situation has a broad maximum at the temperature of the order of level splitting – in this case at $T \sim J$; this is called the *Schottky anomaly*.

Note once again that for the Ising model the ferromagnetic ($J < 0$) and antiferromagnetic ($J > 0$) cases behave in the same way (e.g. $c_V$ is an even function of $J$). The ground states of the kind $\uparrow\uparrow\uparrow\uparrow$ for $J < 0$ (ferromagnetic case) and $\uparrow\downarrow\uparrow\downarrow$ for $J > 0$ (antiferromagnetic case) are both *exact* ground states of the Ising Hamiltonian (6.105), in contrast to the antiferromagnetic Heisenberg model (6.21). In this sense the Ising model is classical, it does not have significant quantum effects. The broken symmetry here is *discrete*, it is the reversal of spin $\uparrow \Longleftrightarrow \downarrow$; the spectrum of elementary excitations contains a gap equal to $|J|$ for both cases, i.e. there is no Goldstone mode, and correspondingly the Mermin–Wagner theorem does not apply. However this system is ordered only at $T = 0$, and there is no long-range order at any finite $T > 0$.

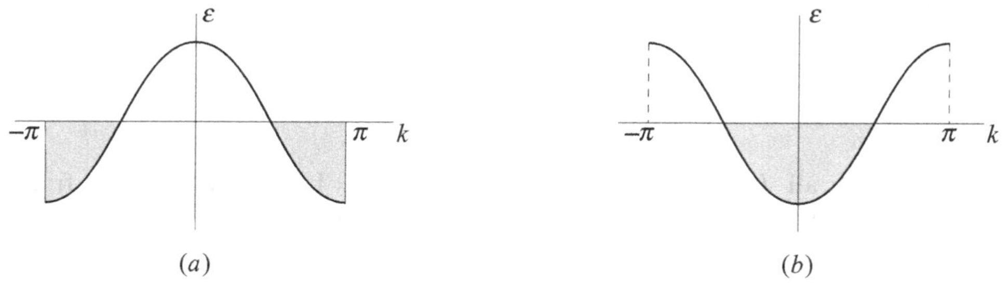


Fig. 6.23

*(1b) 1d xy model, spins 1/2*

In this model the spins are allowed to point in any direction in the $xy$-plane. The Hamiltonian of the $xy$ model is

$$\mathcal{H} = J \sum_{\langle ij \rangle} (S_i^x S_j^x + S_i^y S_j^y) = \tfrac{1}{2} J \sum_{\langle ij \rangle} (S_i^+ S_j^- + S_i^- S_j^+) \,. \tag{6.109}$$

Remembering (6.69), (6.71), one may hope to map this model onto a model like $J \sum_{\langle ij \rangle} a_i^\dagger a_j$, with bosonic operators $a$, $a^\dagger$. But in the standard Holstein–Primakoff representation (6.69) the operators $a$, $a^\dagger$ are only *approximate bosons*. In the 1d case, and for $S = \frac{1}{2}$, quantum effects are especially important, in particular the *on-site anticommutativity* of $S_i^+$, $S_i^-$, see (6.67). Thus one should expect that the conventional Holstein–Primakoff approximation would be a rather poor approximation in this case.

However, there exists another possibility: it is possible to map the model (6.109) *exactly* onto a model of noninteracting *fermions* by the so-called Jordan–Wigner transformation. In effect $S_i^+$ is transformed into $c_i^\dagger$, and $S_i^-$ goes over to $c_i$ (with some complicated phase factors), where $c_i^\dagger$, $c_i$ are spinless *fermions*; the Hamiltonian (6.109) is transformed into

$$\mathcal{H} = J \sum_{\langle ij \rangle} c_i^\dagger c_j \,, \tag{6.110}$$

which can be easily diagonalized by the Fourier transform (this is the tight-binding model for spinless fermions):

$$\mathcal{H} = \sum \varepsilon_k c_k^\dagger c_k \,, \qquad \varepsilon_k = 2J \cos k \,. \tag{6.111}$$

The ground state (the state with the minimum energy) is reached when all fermion states with negative energy are filled. The wavefunction of the ground state is thus the Fermi sphere of these spinless fermions $c_k$, with $k_F = \pi$, see Fig. 6.23($a$) for the case $J > 0$, Fig. 6.23($b$) for $J < 0$.

One can show that the projection of the total spin is

$$\mathcal{S}^z = \tfrac{1}{2}\left(N_\uparrow - N_\downarrow\right) = \left\langle \sum_{k=-\pi}^{\pi} \left(c_k^\dagger c_k - \tfrac{1}{2}\right)\right\rangle . \tag{6.112}$$

Thus, e.g. if all $k$-states were filled, $c_k^\dagger c_k = n_k = 1$ for all $k$, then from equation (6.112) $\mathcal{S}^z = N - \frac{1}{2}N = \frac{1}{2}N$, i.e. this would correspond to all spins being $\uparrow$. If all $k$-states are empty, then $\mathcal{S}^z = -\frac{1}{2}N$, which corresponds to all spins being $\downarrow$. The ground state (half of the states are filled) corresponds to the total spin $\mathcal{S}^z = 0$. But it is not an ordered (LRO) antiferromagnet, although there are antiferromagnetic short-range correlations, or short-range order (SRO).

Note also that, similar to the Ising model, for the one-dimensional $xy$ model the sign of the exchange interaction $J$ in (6.109) does not matter: it is clear from Fig. 6.23 that the change of sign of $J$ corresponds simply to the transformation $k \longrightarrow k + \pi$, and the properties of the solution remain the same.

This mapping of the one-dimensional $xy$ model into spinless fermions is used nowadays to discuss many problems, e.g. the physics of elementary excitations in polymers, etc.

### *(1c) 1d Heisenberg model for $S = 1/2$*

The Hamiltonian of the 1d Heisenberg model with nearest-neighbour interaction is

$$\mathcal{H} = J' \sum_i \mathbf{S}_i \cdot \mathbf{S}_{i+1} = J' \sum_i \left[ S_i^z S_{i+1}^z + \tfrac{1}{2}(S_i^+ S_{i+1}^- + S_i^- S_{i+1}^+)\right] . \tag{6.113}$$

We use here the standard way to write down the Hamiltonian of the 1d Heisenberg model, with the summation over one site index $i$ (cf. (6.21)). Note that the exchange integral $J'$ introduced thus differs by a factor of 2 from our previous definition, $J' = 2J$. This definition here is convenient, because then the energies of the singlet and triplet states of the pair of spins $(S_1, S_2)$ would be $-\frac{3}{4}J'$ for a singlet and $+\frac{1}{4}J'$ for a triplet, see also below, so that their difference is $J'$.

The ferromagnetic case is simple: the ordered state with all spins, e.g. up is the exact ground state at $T = 0$. The antiferromagnetic model is much more interesting because of the presence of spin deviations in the ground state, as discussed above. An exact solution of this model was obtained in the 1930s by H. Bethe, who found an expression for the wavefunction of the ground state and calculated certain properties. Thus the exact value of the ground state energy is $E/N = -J'(\ln 2 - \frac{1}{4})$, i.e. the energy per bond is between the value $-\frac{3}{4}J'$ which one would get for an isolated singlet dimer, and the mean field value $-\frac{1}{4}J'$. The method proposed by Bethe (which is called the *Bethe Ansatz*) is rather complicated; still not all properties are calculated, and many results were obtained only relatively recently.

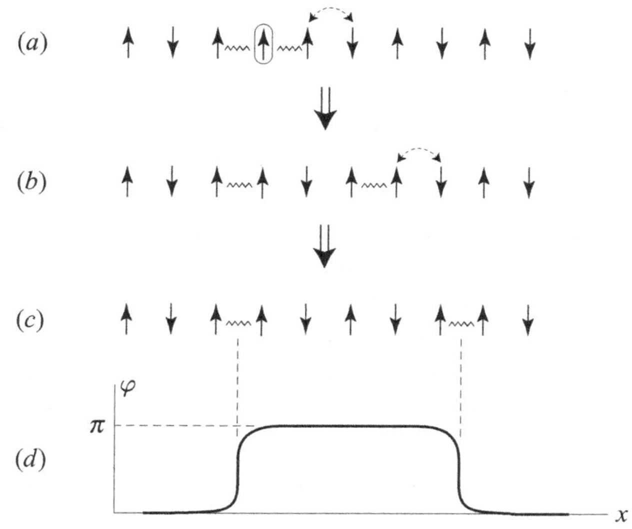


Fig. 6.24

It has a number of interesting features, some of which will be discussed in the next section.

### *6.4.2 Resonating valence bonds, spinons and holons*

The usual *semiclassical* picture of an antiferromagnetic state of a 1d chain would be the following:

- The ground state would be of the Néel type: ↑ ↓ ↑ ↓ ↑ ↓ ↑ ↓.
- The simplest excitation would correspond to one reversed spin: ↑ ↓ ↑ [↑] ↑ ↓ ↑ ↓.

In our previous treatment this approach gave spin waves, or magnons. Such an excitation changes the total spin by 1, i.e. the quantum number of a magnon is $S = 1$. This is actually why magnons could be treated as approximate *bosons*: the general rule is that *integer* spins correspond to bosons, and *half-integer* spins to fermions (although just in the 1d case this rule, strictly speaking, is no longer true, and the situation may be more complicated).

However, as already discussed above, in the Heisenberg antiferromagnet due to quantum fluctuations (caused by terms $S_i^+ S_j^- + S_i^- S_j^+$ in (6.113)) the following process is possible. We start from the state with one reversed spin, Fig. 6.24($a$), and interchange, using $S_i^+ S_j^-$, two spins in the vicinity of the reversed one, as shown by the dashed arrow in Fig. 6.24($a$). As a result we would have the situation shown in Fig. 6.24(*b*). We can continue this process further, and go to the state of Fig. 6.24(*c*), etc. In effect *one* spin excitation (a reversed spin) is split into *two* excitations, each of which is actually a *domain wall*. Between them we again have good antiferromagnetic ordering, but with even and odd sublattices interchanged. In other words the sign of the order parameter changes to the opposite, or its phase $\varphi$ changes by $\pi$, see Fig. 6.24(*d*).

Such *domain walls*, or *solitons*, are the actual elementary excitations in the 1d Heisenberg antiferromagnet (although the picture presented above is strictly speaking not valid, because there is no long-range Néel order in the ground state to start with). As the initial excitation (reversed spin) had $S = 1$ and it is now split into two equivalent 'defects', this means that each of these new excitations carries spin $\frac{1}{2}$, i.e. they are more like *fermions*. In the previous section we saw that the $xy$ model can be transformed into free fermions. These are related features, but the isotropic Heisenberg case is much more complicated – it rather gives *interacting fermions*. (As mentioned above, strictly speaking, in 1d systems the standard connection between spin and statistics is not valid, i.e. the fact that these excitations carry spin $\frac{1}{2}$ does not yet automatically imply that they are fermions.) But in any case these solitons are really the elementary excitations in one-dimensional Heisenberg antiferromagnets, and the treatment of the properties of such systems in terms of these excitations is definitely much better than the one using ordinary magnons. The fact that such objects are the actual elementary excitations does not only follow from our approximate (and in fact not very realistic, see below) treatment, but has been rigorously proven using the Bethe Ansatz method (Faddeev and Tachtajan).

Now, the same quantum effects ($S^+S^-$ terms) which split the spin-wave excitation into two, lead to the fact that the Néel state $\uparrow\downarrow\uparrow\downarrow\uparrow$ is not really the ground state: there appears an admixture of configurations with reversed pairs of spins. We know from quantum mechanics that the ground state of an isolated *pair* of spins with an antiferromagnetic interaction is not simply $\uparrow\downarrow$, but rather an antisymmetric combination – a real singlet. Therefore it may be worthwhile to study an alternative trial state, *making real singlets in the ground state*.

The Néel state is $\uparrow\downarrow\uparrow\downarrow\uparrow\downarrow$. Its energy is (using the Hamiltonian (6.113))

$$E_{\text{Néel}} = -\frac{1}{4}NJ' \; . \tag{6.114}$$

Let us now form real singlets. The singlet state for a pair of spins 1 and 2 is

$$|\text{Singl.}\rangle = \frac{1}{\sqrt{2}}\,|1\!\uparrow 2\!\downarrow - 1\!\downarrow 2\!\uparrow\rangle \; . \tag{6.115}$$

Fig. 6.25

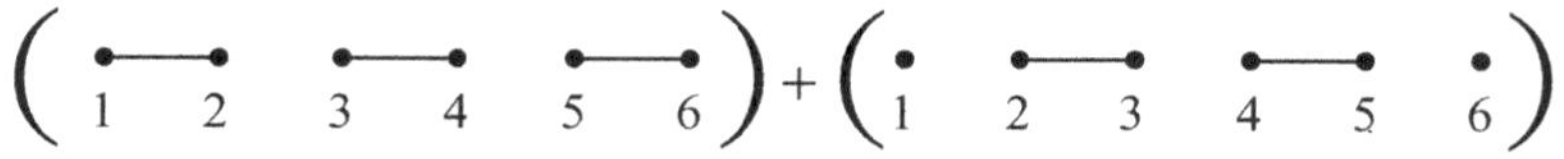

Fig. 6.26

The energy of a singlet is

$$\langle \text{Singl.}|\, J' \boldsymbol{S}_1 \cdot \boldsymbol{S}_2 \,|\text{Singl.}\rangle = -J' \cdot \frac{3}{4} \,. \tag{6.116}$$

**Check:** this directly and by using the expression for the total spin $(\boldsymbol{S}_1 + \boldsymbol{S}_2)^2$ and the fact that $\boldsymbol{S}^2 = S(S+1)$.

**Solution:** $\boldsymbol{S}_{\text{tot}}^2 = (\boldsymbol{S}_1 + \boldsymbol{S}_2)^2 = \boldsymbol{S}_1^2 + \boldsymbol{S}_2^2 + 2\boldsymbol{S}_1 \cdot \boldsymbol{S}_2$, from which $\boldsymbol{S}_1 \cdot \boldsymbol{S}_2 = \frac{1}{2}(\boldsymbol{S}_{\text{tot}}^2 - \boldsymbol{S}_1^2 - \boldsymbol{S}_2^2)$. For a singlet state, $S_{\text{tot}} = 0$ and for $S_1 = S_2 = \frac{1}{2}$, with $\boldsymbol{S}^2 = S(S+1)$, this gives $\langle \boldsymbol{S}_1 \cdot \boldsymbol{S}_2 \rangle = -\frac{3}{4}$, from which we obtain the result (6.116).

Why does the factor $\frac{3}{4}$ appear in (6.116)? The term $S_1^z S_2^z$ gives $-\frac{1}{4}$. In the Néel state only this term contributes. But in a real singlet state (6.115) not only $S_1^z S_2^z$ contributes, but also $S_1^x S_2^x$ and $S_1^y S_2^y$ in $\boldsymbol{S}_1 \cdot \boldsymbol{S}_2$ (the singlet state is spherically symmetric). Each of them gives the same contribution $-\frac{1}{4}$, that is why the total energy is $-\frac{3}{4}J'$.

Now, let us take instead of the Néel state with the energy (6.114) another trial state, the one made of singlets (actually we form *valence bonds*, as in the $H_2$ molecule, see Fig. 6.25):

$$|\Psi\rangle_{\text{VB}} = |\text{Singl.}\rangle_{12}\, |\text{Singl.}\rangle_{34}\, |\text{Singl.}\rangle_{56} \cdots \,. \tag{6.117}$$

The energy of this state is

$$E_{\text{VB}} = -\frac{3}{4} J' \cdot \frac{N}{2} = -\frac{3}{8} J' N \,. \tag{6.118}$$

In comparison with the Néel state (6.114) we have lost *half* of the bonds, but *gained a factor of 3* at each of the remaining singlet bonds. As a result the energy of this *valence bond state*, or *valence bond crystal*, $E_{\text{VB}}$ is less than $E_{\text{Néel}}$. Actually we can decrease the energy still further by using an exchange between singlets (the so-called *resonance*, see Fig. 6.26), i.e. interchanging 'singlet' and 'empty' bonds. It

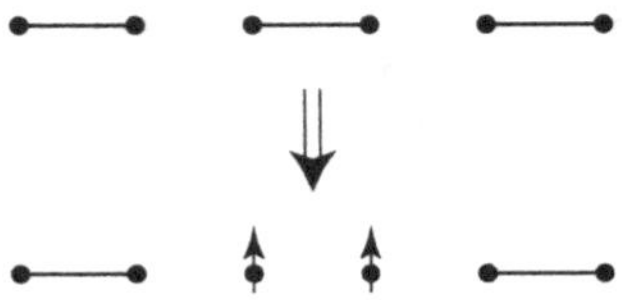

Fig. 6.27

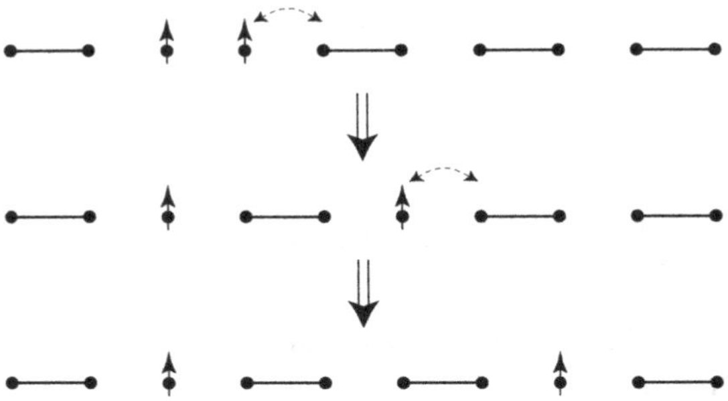

Fig. 6.28

can be shown that this mixing indeed decreases the energy still further. This mechanism of extra stabilization of such a state due to resonance between different configurations is known to work, e.g. in the *benzene* molecule, 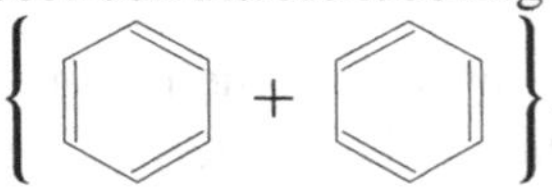.

The very concept of resonance appeared in this context in chemistry and is largely due to L. Pauling. In our problem this is what is now called the *resonating valence bond* (RVB) state (the concept introduced to solid state physics by P. W. Anderson in 1973); it is at present a very popular notion, in particular in application to high-$T_c$ superconductors, but not only there.

Let us now consider elementary excitations in the (R)VB state. Similar to the Néel configuration, we can first create a spin 1 excitation by 'breaking the pair' – exciting one pair from the ground state singlet into a triplet state, Fig. 6.27. Again, as in Fig. 6.24, we can 'recommute' the singlet bond several times, and in effect we will get two separated spin $\frac{1}{2}$ objects, between which there will still be a paired state, but with the interchanged sublattices of singlet and 'empty' bonds, Fig. 6.28. We thus again have created two excitations with spin $\frac{1}{2}$ – solitons, or kinks (domain walls between two different domains). From Fig. 6.28 it is clear that these excitations are indeed $S = \frac{1}{2}$ objects. These excitations are now called *spinons*.

One can show that excitations of this kind – spinons – are indeed the actual elementary excitations in the exact treatment of 1d Heisenberg antiferromagnets. They have a gapless spectrum. Note, however, that this is *not* a consequence of the Goldstone theorem: although we have here a continuous symmetry, this symmetry

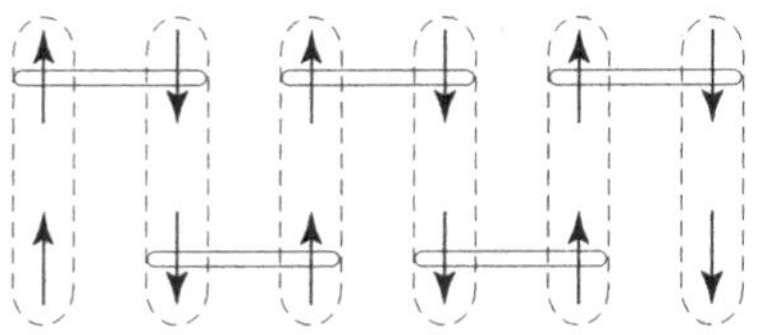

Fig. 6.29

is *not broken* in the ground state; there is no long-range order in the ground state in this case. The singlet pairs in the RVB state are spherically symmetric objects, there is no preferred orientation in the spin direction, and the continuous spin-rotation symmetry is not broken.

This remark is important when we compare the properties of antiferromagnetic spin chains with spin $\frac{1}{2}$ and those with other spins, e.g. spin 1. For a long time it was believed that all the main features in these cases are the same. However, relatively recently D. Haldane has shown that this is not the case, and that there exists a gap in the spectrum of elementary excitations in chains with integer spins $S = 1, 2, \ldots$, whereas no gap exists for half-integer spins $\frac{1}{2}, \frac{3}{2}, \ldots$. This *Haldane gap* is now indeed observed in several quasi-one-dimensional antiferromagnets with spin 1.

A simple qualitative picture which can illustrate the origin of this difference is the following. One can treat spin 1 at a site as composed of two parallel spins $\frac{1}{2}$. Then one can visualize the following situation: one can form singlet pairs of these $S = \frac{1}{2}$ 'subspins' with neighbours both to the left and to the right, see Fig. 6.29. In contrast to the case of spin $\frac{1}{2}$ considered earlier, here all bonds are equivalent, and such a state is nondegenerate, whereas it was doubly degenerate for spin $\frac{1}{2}$ (either (12)(34)(56)..., or (01)(23)(45)(67)... singlets). Consequently, we cannot create here the usual 'domain walls', as there are no different domains. This simplified picture explains the appearance of the Haldane gap, although the real mathematical proof of that is quite elaborate.

Let us now return to spin $\frac{1}{2}$ antiferromagnets in higher dimensions and let us see whether the concept of the RVB state can be used in these cases as well.

**Problem:** Compare, similarly to the 1d-case, the Néel and VB (or RVB) states in (*a*) a 2d square lattice, (*b*) a 2d triangular lattice.

**Solution:** The case of a square lattice is simple: the energy of the Néel state shown in Fig. 6.30 is

$$E_{\mathrm{N}} = -J' \cdot \tfrac{1}{4} \cdot 2N = -\tfrac{1}{2} J' N \,, \tag{6.119}$$

where $N$ is the number of sites, each has four nearest neighbours, or four antiferromagnetic bonds per site, but each such bond belongs to two sites.

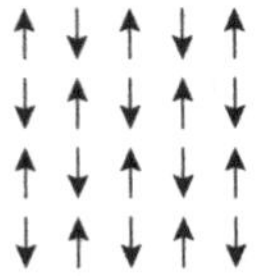

Fig. 6.30

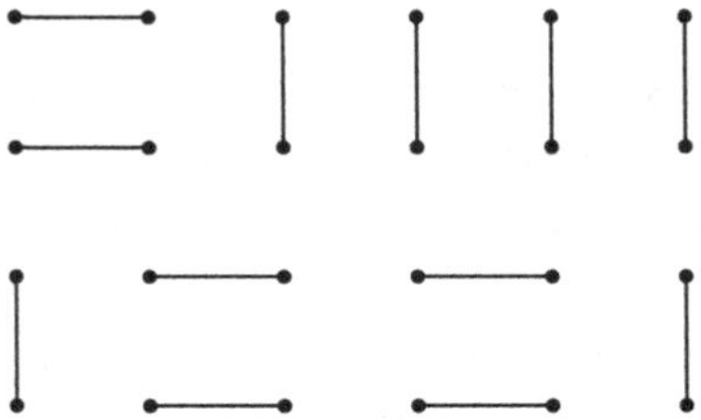

Fig. 6.31

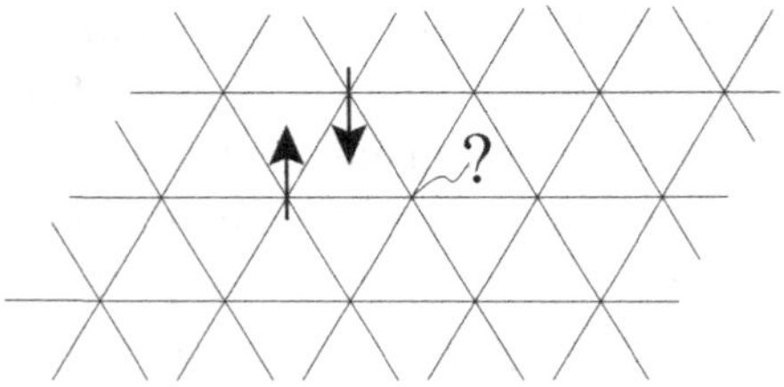


Fig. 6.32

A typical valence bond state is shown in Fig. 6.31. There exist actually many possibilities; we can connect lattice sites by bonds in many different ways which will give rise to RVB. The energy of each of these states is

$$E_{\mathrm{VB}} = -\frac{3}{4}J' \cdot \frac{N}{2} = -\frac{3}{8}J'N \; . \tag{6.120}$$

(the energy is $-\frac{3}{4}J'$ per singlet bond, but there is one such bond per two sites). Thus the simple *valence bond* state (without resonance switching of bonds) here is *worse* than the Néel configuration. What will be the situation with resonance? Numerical calculations show that the Néel state of Fig. 6.30 (with LRO, but with a lot of fluctuations, of course) is still the ground state of the 2d Heisenberg model on a square lattice.

The triangular lattice is a bit more tricky. Let us try to make the Néel state; we immediately see that we are in trouble already at the first step, see Fig. 6.32. When we consider the basic block of such a lattice – a triangle with antiferromagnetic interactions – the situation is not trivial: if we only consider ordering of spins

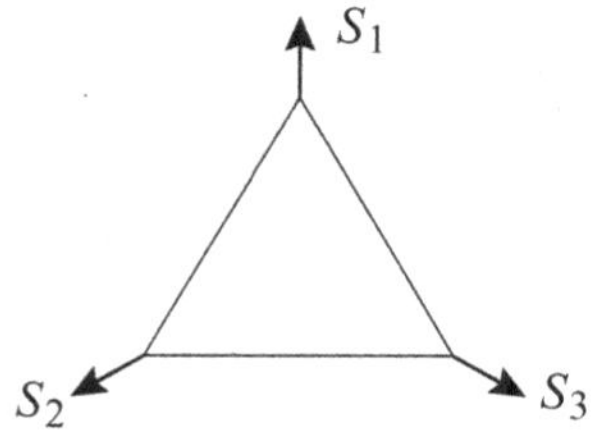


Fig. 6.33

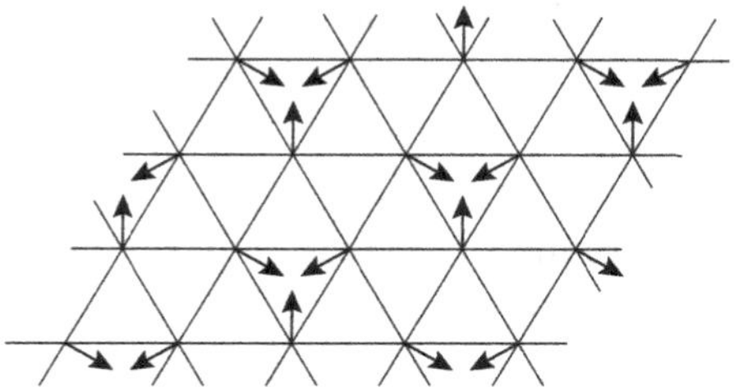

Fig. 6.34

up and down, then whatever we do here, one bond is always *wrong*; this is called *frustration*. One cannot subdivide a triangular lattice into *two* sublattices with spins up and down so that every site of one is surrounded by the sites of the other (i.e. when all nearest neighbours of one sublattice belong to the other). Thus the simple Néel solution with collinear spins is not good, and we have to think of something else. The best classical state is the one shown in Figs. 6.33 and 6.34. In this state the spins on the triangle are directed at the angle $2\pi/3$ to each other. For such a state the whole lattice is subdivided into three sublattices, and at each triangle we have the same situation, all the angles between neighbouring spins are the same, $\pm 2\pi/3$. The energy of such state is

$$E_{2\pi/3} = J' \cdot \underbrace{\cos\frac{2\pi}{3} \cdot \frac{1}{4}}_{(S_i \cdot S_j)} \cdot \underbrace{\frac{Nz}{2}}_{(z=6)} = -\frac{3}{8}J'N \; , \tag{6.121}$$

where $z$ is the number of nearest neighbours.

The valence bond state shown in Fig. 6.35 has the energy

$$E_{\mathrm{VB}} = -\frac{3}{4}J' \cdot \frac{N}{2} = -\frac{3}{8}J'N \; . \tag{6.122}$$

Thus in a 2d triangular lattice the energy of the valence bond state $E_{\mathrm{VB}}$ is equal to the energy of the best classical long-range ordered state $E_{2\pi/3}$, i.e. the valence bond state (random covering of the lattice by singlets), even without their resonance

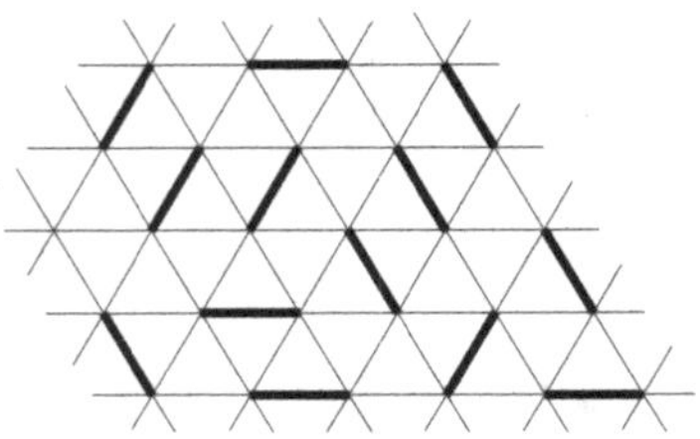

Fig. 6.35

(which would decrease the VB energy still further) is degenerate with the best 'classical, Néel-like' state, and the resonance could have made the RVB state the ground state.

These considerations are of course only suggestive; they do not yet prove that the RVB state is indeed the ground state of the Heisenberg model on a 2d triangular lattice. The point is that both the RVB state and the state with some long-range order would be modified (and their energies would be reduced) by quantum fluctuations. Thus a priori we cannot say which state would have the lowest energy. The most detailed numerical calculations carried out until now seem to show that there still exists in this case a long-range order of the '120 degrees' type. However, for example, already a small doping can make corresponding systems RVB-like.

The RVB state is a *spin liquid* (it is made of singlets, but without any long-range order). The notion of RVB is at present an important concept in different fields, e.g. in the theory of high-$T_c$ superconductivity and in the physics of frustrated magnets such as, e.g. the *kagome lattice*, a 2d lattice of corner-sharing triangles, or the *pyrochlore lattice*, a 3d lattice of corner-sharing tetrahedrons. What is important is the fact that in such spin-liquid states with short-range RVB (short-range antiferromagnetic correlations) the excitation spectrum is *gapful*; this is the typical situation for such quantum liquid states (although there may exist also spin liquids with a gapless spectrum). This is also important for many situations with quantum critical points, see Section 2.6 and Section 10.2 below.

One extra interesting and important feature of the RVB state is the unusual character of elementary excitations in it which we already encountered, when considering the 1d case. The creation of such excitations is illustrated in Fig. 6.36: starting from a particular VB state, Fig. 6.36($a$), we first break one singlet bond, making it a triplet, Fig. 6.36($b$), and then 'recommute' singlets, moving $\uparrow$ spins apart, Fig. 6.36($c$). Thus, similar to the 1d Heisenberg model we get *two* independent excitations, each with spin $\frac{1}{2}$, which are in fact very similar to the spinons introduced above for the 1d case. Note that they have no charge (at each site we have a positive nucleus with charge $+e$, and a localized electron with charge $-e$). Let us now *remove an electron*, i.e. create *a hole* (e.g. by doping). It is much easier to remove

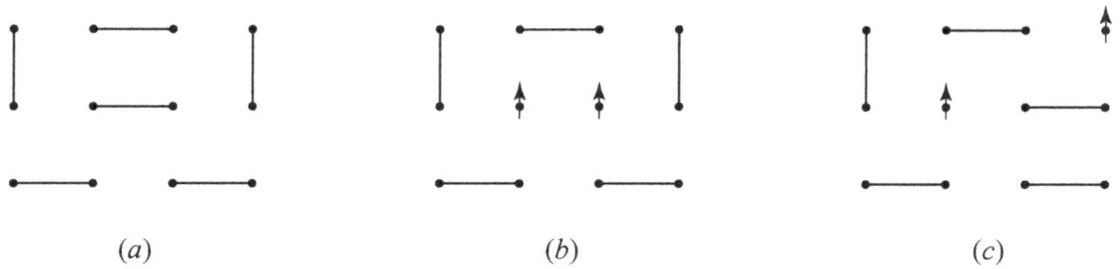

Fig. 6.36

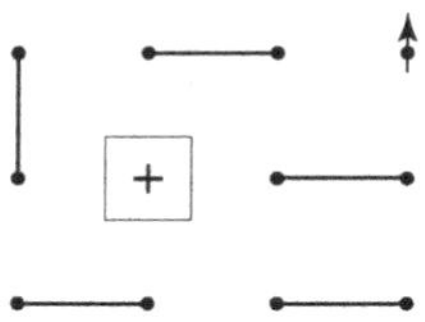

Fig. 6.37

an electron from a *broken bond* (e.g. from the site ↑ in Fig. 6.36(*c*)) – we do not have to spend energy to break a singlet. Then we will have the situation shown in Fig. 6.37, that is we will create a site (a hole) $\boxed{+}$ *with charge* $+e$ and *without spin* – a charged but spinless excitation, which is now called a *holon*.

This kind of decoupling of spin and charge degrees of freedom, or spin–charge separation, is definitely valid in 1d models. As to the 2d-case, this is still questionable and not proven. But this is a very appealing picture; it is widely used in describing systems with strong electron correlations, such as cuprate superconductors (the main proponent of this idea is P. W. Anderson).

### 6.4.3 Two-dimensional models

Now, we have already gradually switched from 1d to 2d models. What about them?

*(3a) 2d Ising model*

The two-dimensional Ising model is exactly soluble on a square lattice and in some other cases; it was the first example of an exact solution of a nontrivial 2d problem (the famous solution of Onsager). For the Ising model on a square lattice there is a real phase transition, and the long-range order appears at

$$T_c = \frac{|J'|}{\operatorname{arc\,tanh}(\sqrt{2}-1)} . \tag{6.123}$$

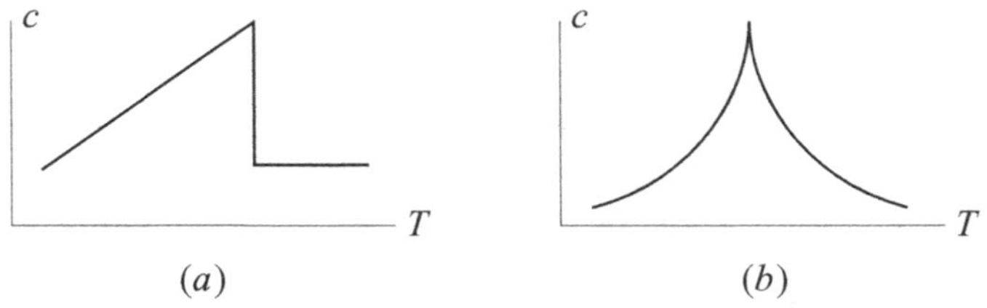


Fig. 6.38

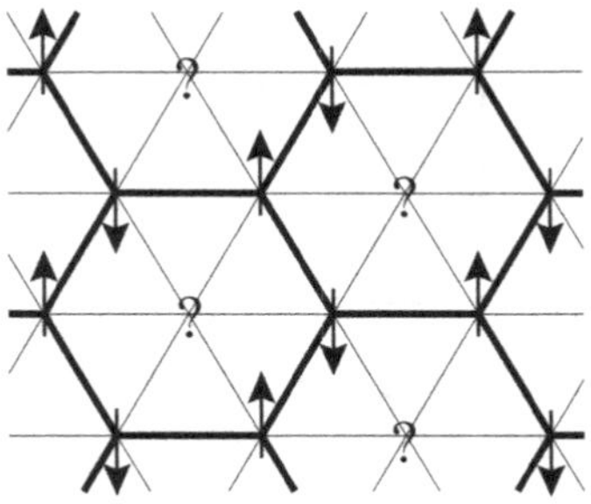


Fig. 6.39

It is a second-order phase transition, but not mean-field-like; thus, e.g. the specific heat has not a jump, Fig. 6.38(*a*), but a logarithmic divergence, Fig. 6.38(*b*), i.e. close to $T_c$ we have $c(T) \sim -\ln|T - T_c|$.

One can also calculate exactly some other properties for this case. For example the order parameter (magnetization or sublattice magnetization) behaves close to $T_c$ as $M \sim (T_c - T)^{1/8}$. Note that these exact results for the specific heat and for the order parameter satisfy the scaling relations of Section 2.5, as they should (the logarithmic dependence of the specific heat in scaling relations means that the corresponding exponent should be taken as zero).

An interesting situation exists in the two-dimensional Ising model on a triangular lattice. As discussed above, see, e.g. Fig. 6.32, in such a case we have frustration, and many states have the same energy (the three-sublattice solution with 120° spins, discussed in the previous section, cannot be formed in the Ising model, where all spins must be either up or down). And indeed the exact solution of this model (G. Wannier) shows that there exists a macroscopic number of degenerate ground states, so that this system has finite entropy at $T = 0$.

The simplest way to understand this is to form first a two-sublattice antiferromagnetic ordering on a honeycomb sublattice of a triangular lattice, Fig. 6.39. Then the spins at the centres of each hexagon are in zero molecular field, and we can put their spins at random. All such states will have the same energy. The number of

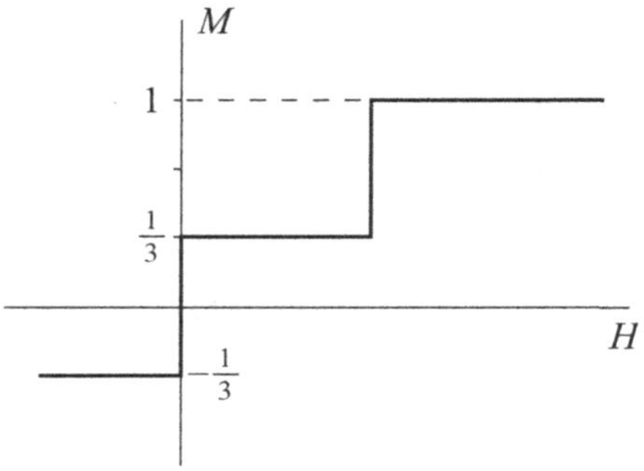


Fig. 6.40

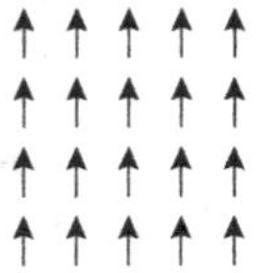

Fig. 6.41

such states is $2^{N/3}$, i.e. the ground state entropy can be estimated as $S \sim \frac{1}{3}N \ln 2$. The exact value is not far from this simple estimate.

Yet another interesting consequence of this simple picture is that, as one can easily see, we expect in this case that in an arbitrary small external field, say up, all these 'free' spins immediately orient along the field, so that there appears a net magnetization of the sample equal to $\frac{1}{3}\mu_B$ per site. Similarly, at any small $H < 0$ the magnetization will be negative, $-\frac{1}{3}\mu_B$. Thus there will be a jump of magnetization at $H = 0$, and the second jump to full magnetization $1\mu_B$ will occur at still higher critical field (at $T = 0$). As a result the magnetization at $T = 0$ will change with (parallel) field as shown in Fig. 6.40. Such *magnetization plateaux* are very typical for frustrated magnetic systems.

### *(3b) 2d xy model. Topological excitations (vortices)*

Let us now consider the two-dimensional $xy$ model with the Hamiltonian of the type (6.109) on a square lattice. In contrast to the 1d case, here one does not yet know the exact solution. An interesting feature of this model is that it possesses interesting types of excitations – topological excitations. Thus let us take the simple ferromagnetic state shown in Fig. 6.41. One possible type of elementary excitation is ordinary spin waves. But we can also create configurations, e.g. like those shown in Figs. 6.42 and 6.43.

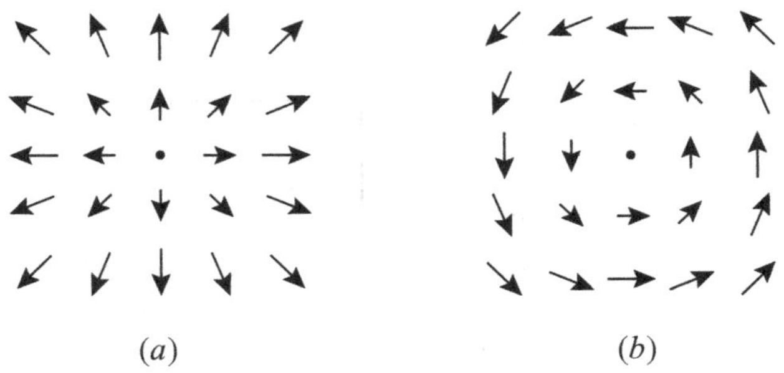


Fig. 6.42

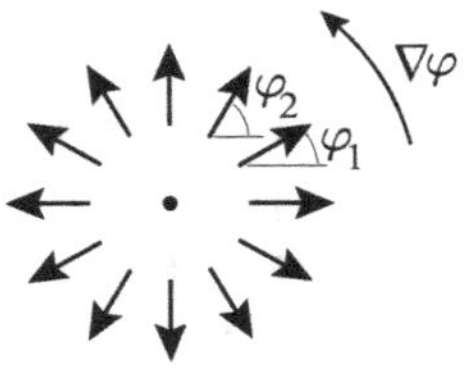


Fig. 6.43

Far from the centres of such excitations the spins are locally parallel, as they should be in a ferromagnet, but when one approaches the centres this rule is of course violated. We cannot destroy such excitations (defects) by continuously rotating the spins – they are *topological* defects. They are very similar to vortices in liquid He. (Actually, starting from the ground state of Fig. 6.41, we cannot create *one* such excitation, because it has a different topological quantum number than the ground state and will necessarily lead to a distortion of the spin structure at infinity. But we can create a *pair* of such excitations, similar to a vortex–antivortex pair, so that the corresponding 'field' (distribution of arrows) will look like the field of a dipole, and the corresponding distortion will decrease with distance rapidly enough.)

In a Bose gas the broken symmetry is the gauge symmetry, the *phase* of the order parameter, $a \to ae^{i\varphi}, 0 < \varphi < 2\pi$. The order parameter there is $\langle a_0 \rangle = |a_0|e^{i\varphi}$. In the $xy$ model there exists the same symmetry: the order parameter is here the two-dimensional vector $\boldsymbol{S} = |S|e^{i\varphi}$, with the phase showing its orientation in the $xy$-plane being $0 \le \varphi < 2\pi$. Thus there exists an exact mapping of the Bose gas onto the $xy$ model, and not only in the 2d case! The connection between these two models is the following:

Bose condensation $\longleftrightarrow$ ordering in the $xy$ model;
Bogolyubov sound $\longleftrightarrow$ spin waves (magnons);
Vortices $\longleftrightarrow$ topological excitations in the $xy$ model.

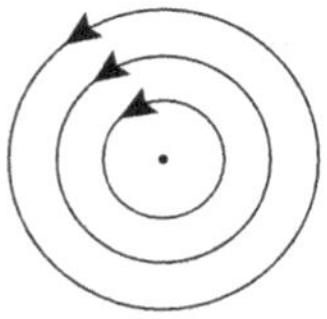

Fig. 6.44

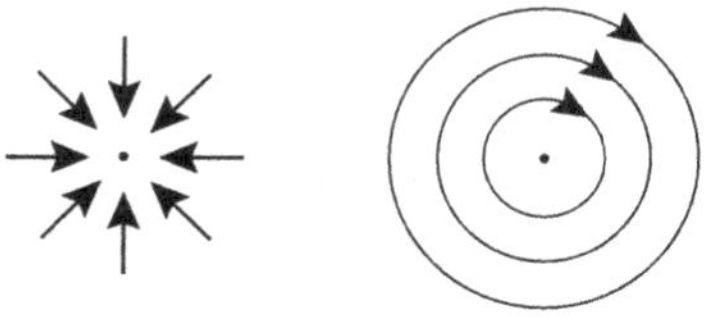

Fig. 6.45

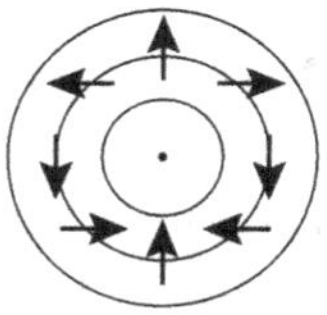

Fig. 6.46

In a superfluid the current (or velocity of superfluid motion) is $\boldsymbol{v}_{sf} \sim \nabla\varphi$, and in a vortex the current circulates around the core, see Fig. 6.44. In the spin model this would correspond to a spin configuration shown, e.g. in Figs. 6.42 and 6.43, i.e. this spin configuration is characterized by nonzero gradient of $\varphi$ such that the integral along a contour surrounding the centre is

$$\oint \nabla\varphi \cdot d\boldsymbol{l} = 2\pi \ . \qquad (6.124)$$

A defect or excitation of the opposite sign, Fig. 6.45, would correspond to an *antivortex* (opposite circulation equal to $-2\pi$). The circulation around the vortex is, in general, $2\pi n$, where $n$ is called the *winding number*. In previous figures we had $n = \pm 1$. One can draw spin configurations for example corresponding to $n = 2$, Fig. 6.46, etc.

There is no usual phase transition in the 2d $xy$ model at nonzero temperatures, at which real long-range order would appear (this also follows from the Mermin–Wagner theorem, see Sections 4.4.4 and 6.3.2), but there is a point $T^*$ at which the correlation between spins $\langle \boldsymbol{S}_0 \cdot \boldsymbol{S}_r \rangle$, which above $T^*$ behaves as $\sim r^{-\zeta(T)}$ with $\zeta = 2(T/T^*)$, below $T^*$, still decaying with distance, becomes so strong that the susceptibility $\chi \sim \sum_r \langle \boldsymbol{S}_0 \cdot \boldsymbol{S}_r \rangle$ diverges. This is also a special kind of phase transition, which is called a Beresinskii–Kosterlitz–Thouless (BKT, or often simply KT) transition. Although strictly speaking there is no long-range order at $0 < T < T^*$ (the average $\langle \boldsymbol{S} \rangle = 0$, in accordance with the Mermin–Wagner theorem), the fact that the susceptibility diverges means that the magnetic response of such a system is in many respects similar to that of an ordered state. For example, as we mentioned already, the Bose condensation phenomenon may be mapped onto the $xy$ model. It turns out that below the BKT transition the two-dimensional Bose system (e.g. thin films of liquid $^4$He) would be superfluid, despite the formal absence of long-range order.

Actually the Kosterlitz–Thouless transition is connected with the topological excitations – vortices – which we have just discussed. One can show that the energy of an isolated vortex with one flux (winding number $\pm 1$) is equal to

$$E_{\mathrm{v}} = \pi J \ln \left( \frac{R}{\xi} \right) , \tag{6.125}$$

where $\xi$ is the coherence length (crudely speaking it is the radius of the inner part, the core of the vortex, where the modulus of the order parameter goes to zero – the so-called normal core in superfluids or superconductors), and $R$ is the dimension of the sample. At the same time each core can be at any point of the sample with the area $\sim R^2$, i.e. the entropy of such a system is

$$S = \ln \frac{R^2}{\xi^2} = 2 \ln \frac{R}{\xi} . \tag{6.126}$$

The total free energy of the system is then

$$F = E - TS = (\pi J - 2T) \ln \frac{R}{\xi} . \tag{6.127}$$

When $T < T^* = \pi J/2$, the free energy is minimized when there are no vortices in the system; such a state would have magnetic stiffness, or, in a 2d Bose system, superfluid properties. For $T > T^*$ it is favourable to create vortices, and these properties would disappear.

One can also give a somewhat different interpretation of this phenomenon. One can show that the interaction between vortices separated by distance $r$ is also logarithmic, $\sim \ln(r/\xi)$. This interaction is repulsive for the vortices with the same circulation and attractive for vortex and antivortex. If there exist in a system vortices and antivortices (e.g. thermally excited), at $T > T^*$ they will be unbound, but for $T < T^*$ they will be bound in pairs (the existence of free vortices is unfavourable at $T < T^*$ by the same arguments as follow from (6.127)). Unbound vortices destroy superfluidity, whereas the bound ones do not.

One more remark is relevant here. The topological character of 'vortices' in the 2d $xy$ model means that we cannot destroy such excitations by a continuous deformation of the spin structure. It turns out that despite the apparent similarity, in this respect the Heisenberg model is principally different from the $xy$ model. For the Heisenberg model we can also draw spin configurations similar to those shown in Fig. 6.42. However in this case we can 'annihilate' this defect by continuously deforming the spin structure, moving spins out of the plane, so that in the end, for example, all spins would point up, restoring perfect ferromagnetic ordering. This remark actually has a rather deep meaning: what is important here is the *space of the order parameter*. Whereas in the $xy$ model (or, to this end, for the Bose condensation) the order parameter is a complex scalar, and its possible values 'live *on a circle*' (are characterized by an arbitrary phase $0 \leq \varphi \leq 2\pi$), the order parameter $\boldsymbol{S}$ of the Heisenberg model is a vector characterized by its modulus and two Euler angles, i.e. it can take any value *on a sphere* with the radius $|\boldsymbol{S}|$. The 'vortex' of Fig. 6.42 is a state represented by a circle on this sphere, e.g. its 'equator'. But we can continuously deform this circle, moving it e.g. to the 'North Pole' (all spins pointing up), where it would reduce to a point and finally disappear. It is clear that this process is forbidden if the space of the order parameter is not a sphere, but a 2d circle, as for the $xy$ model. Different spaces of possible order parameters, which can exist in different ordered states in condensed matter systems, determine possible types of excitations in them, and finally their thermodynamic and transport properties.

## 6.5 Defects and localized states in magnetic and other systems

An interesting problem arises when there exist *external defects* in an otherwise regular material (e.g. impurities with different spin $S'$ in a system with spin $S$, or a site with different exchange interaction $J' \neq J$). A similar (mathematically) problem arises in lattice theory if there exists an impurity atom with mass $M'$ different from the mass $M$ of the regular lattice, or in a system of tight-binding electrons with an impurity.

Often there appear in this case *localized* states, or *impurity* states: localized phonons, or localized magnons, or localized (impurity) electron states.

Mathematically the problem is the following: using the Hamiltonian of the type of (4.26), or (6.68), or the standard tight-binding description of electronic systems, see, e.g. the term $H'$ in (12.4), and adding an impurity at the origin of the coordinates, we can write down the equations for the corresponding operators (phonon operators, or the spin ones, or the electron creation and annihilation operators) in the form

$$\omega c_i = \varepsilon_0 c_i + \sum_j t_{ij} c_j + \lambda \delta_{i,0} c_0 \, , \tag{6.128}$$

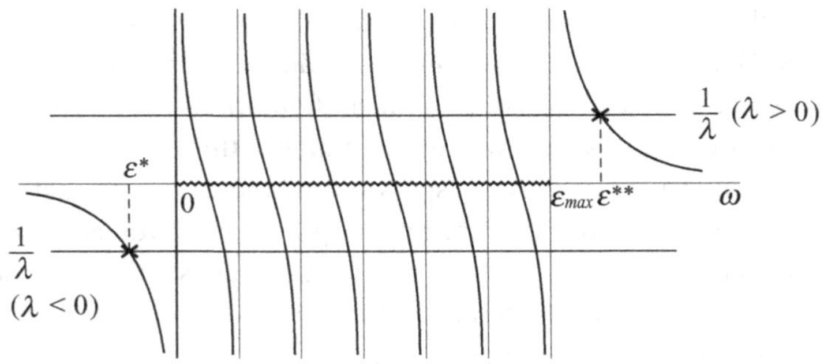


Fig. 6.47

cf. (6.89), see also the standard description of the tight-bound model for electrons. Here the last term describes a perturbation $\lambda$ at the site $i = 0$ (e.g. different spin, or different mass, or different potential for electrons).

The standard solution of this problem is the following. Let us make a Fourier transform $c_i = \sum_q e^{i\boldsymbol{q}\cdot\boldsymbol{R}_i} c_{\boldsymbol{q}}$. Then equation (6.128) takes the form

$$[\omega - \varepsilon_0 - t(\boldsymbol{q})]\, c_{\boldsymbol{q}} = \lambda \delta_{i,0} c_0 = \lambda \sum_{\boldsymbol{k}} c_{\boldsymbol{k}} \,. \tag{6.129}$$

From equation (6.129) we find

$$c_{\boldsymbol{q}} = \frac{\lambda \sum_{\boldsymbol{k}} c_{\boldsymbol{k}}}{\omega - \varepsilon_0 - t(\boldsymbol{q})} \equiv \frac{\lambda \sum_{\boldsymbol{k}} c_{\boldsymbol{k}}}{\omega - \varepsilon_{\boldsymbol{q}}} \,, \tag{6.130}$$

where $\varepsilon_{\boldsymbol{q}} = \varepsilon_0 + t(\boldsymbol{q})$ is the spectrum of the system without impurities. Now, take $\sum_{\boldsymbol{q}}$ of both the left- and right-hand sides of (6.130):

$$\sum_{\boldsymbol{q}} c_{\boldsymbol{q}} = \lambda \sum_{\boldsymbol{k}} c_{\boldsymbol{k}} \cdot \sum_{\boldsymbol{q}} \frac{1}{\omega - \varepsilon_{\boldsymbol{q}}} \,, \tag{6.131}$$

or, denoting $A = \sum_{\boldsymbol{q}} c_{\boldsymbol{q}}$, we see that $A = \lambda A \sum_{\boldsymbol{q}} \frac{1}{\omega - \varepsilon_{\boldsymbol{q}}}$, or

$$\frac{1}{\lambda} = \sum_{\boldsymbol{q}} \frac{1}{\omega - \varepsilon_{\boldsymbol{q}}} \,. \tag{6.132}$$

Let us consider a typical situation in which the bare spectrum is confined, $0 < \varepsilon_{\boldsymbol{q}} < \varepsilon_{\max}$. If we initially consider the system in a box so that the values of $\boldsymbol{q}$ are discrete (and $\varepsilon_{\boldsymbol{q}}$ too), the graphic solution of (6.132) looks as shown in Fig. 6.47. The solutions are given by the crossing of the straight line $1/\lambda$ with the curves representing the right-hand side of equation (6.132). Thus we see that there exist solutions inside the former band of excitations (each energy is slightly shifted, but for the volume $V \to \infty$ these solutions fill the entire initial band $0 < \omega < \varepsilon_{\max}$,

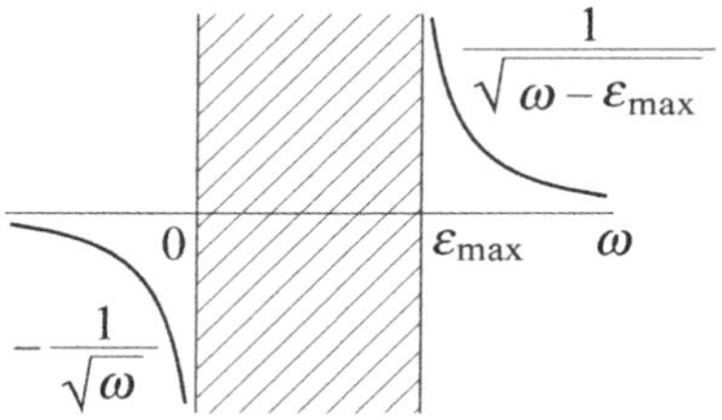


Fig. 6.48

the wavy line in Fig. 6.47). But besides that there may appear solutions *outside* the band – the localized modes.

For the situation described by equation (6.129) such states, if they do exist, lie below the band (at $\varepsilon^* < 0$) for $\lambda < 0$ (attraction), or at $\varepsilon^{**} > \varepsilon_{max}$, i.e. above the continuum of the band states for $\lambda > 0$ (repulsion).

Now, in which cases do such split-off localized states exist, and how does their energy depend on the strength of the perturbation $\lambda$? The answer is different for 1d, 2d and 3d systems.

Let us take a continuous limit of (6.132) (i.e. take $\int d\boldsymbol{q}$ instead of $\sum_q$),

$$\frac{1}{\lambda} = \int d\boldsymbol{q}\, \frac{1}{\omega - \varepsilon_{\boldsymbol{q}}} = \int_0^{\varepsilon_{max}} \frac{\rho(\varepsilon)\, d\varepsilon}{\omega - \varepsilon} . \tag{6.133}$$

The result depends on the behaviour of the density of states $\rho(\varepsilon)$ at the edge of the band ($\varepsilon \to 0$, $\varepsilon \to \varepsilon_{max}$).

(1) One-dimensional systems. Here $\rho(\varepsilon) \sim 1/\sqrt{\varepsilon}$ as $\varepsilon \to 0$ (if the spectrum is quadratic, $\varepsilon_q \sim q^2$, as for electrons or for ferromagnons). The integral in equation (6.133) is divergent, and we would have the situation shown in Fig. 6.48. Thus there is a bound (or antibound) state for any $\lambda$, however small. For $\lambda = -|\lambda| \to 0$ the energy of the localized state behaves as

$$\varepsilon^* \sim -\lambda^2 . \tag{6.134}$$

The result depends also on the initial spectrum $\varepsilon(q)$, of course.

(2) Two-dimensional systems. In the 2d case the typical density of states at the edge of the band is $\rho(\varepsilon \to 0) = \text{const.}$ (for electrons). Then the integral in (6.133) is logarithmically divergent. The bound state then exists also at any $\lambda$, and for small $\lambda$ its energy is

$$\varepsilon^* \sim -e^{-1/\lambda} . \tag{6.135}$$

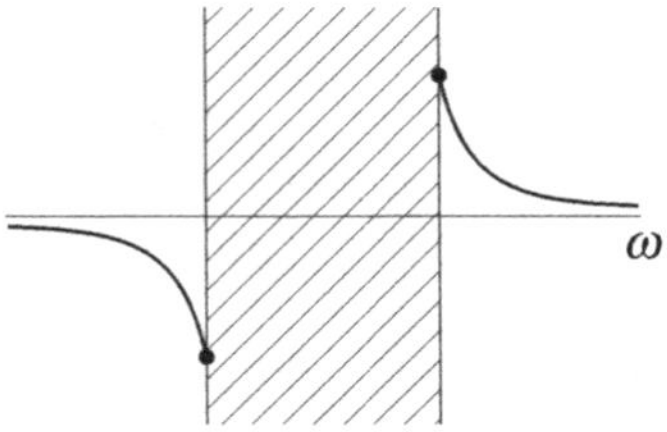


Fig. 6.49

(3) Three-dimensional case. Here, for electrons, $\rho(\varepsilon) \sim \sqrt{\epsilon}$ and the integral is *finite*. This means that Fig. 6.47 actually looks like that shown in Fig. 6.49, and the bound state exists only for $\lambda$ exceeding a certain critical value, $|\lambda| > \lambda_c$.

For different types of spectrum (e.g. if $\varepsilon_{\boldsymbol{q}} \sim q$, and not $q^2$) the criteria for the existence, and the energies of the bound states may be different (and the form of the perturbation may also be different from that introduced in (6.128)), but the general scheme is always like the one described above (it is often called the Slater–Koster method). One can also use the same technique for considering possible bound state formation due to interaction between excitations (e.g. the formation of the bound state of magnons, cf. the qualitative discussion in Section 6.3.1, Fig. 6.21, or the famous solution of the Cooper pair problem which led finally to the Bardeen–Cooper–Schrieffer theory of superconductivity).

# 7
# Electrons in metals

## 7.1 General properties of Fermi systems

This very short chapter serves as a reminder of the main properties of Fermi systems and a useful collection of respective formulae. We consider here noninteracting electrons with the spectrum $\varepsilon(\boldsymbol{p}) = p^2/2m$. In the degenerate Fermi gas the number of states in the interval $(p, p + dp)$ is (per spin)

$$V\,\frac{d^3\boldsymbol{p}}{(2\pi\hbar)^3} = V\,\frac{4\pi p^2\,dp}{(2\pi\hbar)^3} = V\,\frac{p^2\,dp}{2\pi^2\hbar^3}\,. \tag{7.1}$$

Fermions obey the Pauli principle, and at $T = 0$ the electrons in metals occupy all states up to a certain momentum – the Fermi momentum $p_{\mathrm{F}}$, so that the electron density (for both spins) is

$$n = \frac{N}{V} = \frac{1}{\pi^2\hbar^3}\int_0^{p_{\mathrm{F}}} p^2\,dp = \frac{p_{\mathrm{F}}^3}{3\pi^2\hbar^3}\,, \tag{7.2}$$

or the Fermi momentum

$$p_{\mathrm{F}} = (3\pi^2)^{1/3}\hbar n^{1/3} \tag{7.3}$$

and the Fermi energy

$$\varepsilon_{\mathrm{F}} = \frac{p_{\mathrm{F}}^2}{2m} = (3\pi^2)^{2/3}\frac{\hbar^2}{2m}n^{2/3}\,. \tag{7.4}$$

The distribution function of electrons $n(p)$ has the well-known form shown in Fig. 7.1. At $T = 0$, $n(p) = 1$ for $|p| < p_{\mathrm{F}}$ and $n(p) = 0$ outside this region; $|p| = p_{\mathrm{F}}$ determines the Fermi surface. At finite temperatures some of the electrons are excited from the states below the Fermi surface to empty states above it, i.e. there appear electron–hole pairs. Consequently the distribution function $n(p)$ is broadened, see Fig. 7.1. The width of the region around $\varepsilon_{\mathrm{F}}$ in which $n(p)$ significantly changes is $\sim T$.

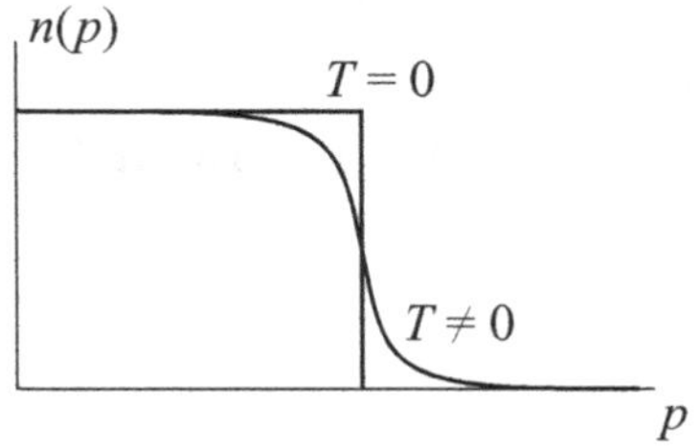


Fig. 7.1

Often one introduces the quantity $\tilde{r}_s$ (the average distance between electrons) by the relation

$$\frac{V}{N} = \frac{1}{n} = \frac{4\pi \tilde{r}_s^3}{3} . \tag{7.5}$$

Then one can express different quantities through $\tilde{r}_s$:

$$p_F = \left(\frac{9\pi}{4}\right)^{1/3} \frac{1}{\hbar\, \tilde{r}_s} = \frac{1.92}{\hbar\, \tilde{r}_s} = \frac{3.63}{\tilde{r}_s/a_0}\, \text{Å}^{-1} , \tag{7.6}$$

where

$$a_0 = \frac{\hbar^2}{me^2} \tag{7.7}$$

is the Bohr radius. Usually one uses the dimensionless parameter $r_s = \tilde{r}_s/a_0$, measuring $\tilde{r}_s$ in units of the Bohr radius $a_0$, and writes all formulae in terms of $r_s$, omitting $a_0$. Then the situation with $r_s \ll 1$ corresponds to a high-density electron gas, and that with $r_s \gg 1$ to the low-density limit. Typically in metals $r_s \sim 2$–$3$ (see Table 2.1 in Ashcroft and Mermin (1976)). From (7.6) the Fermi velocity is

$$v_F = \frac{p_F}{m} = \frac{4.2}{r_s} \cdot 10^8 \, \frac{\text{cm}}{\text{sec}} , \tag{7.8}$$

and the Fermi energy is

$$\varepsilon_F = \frac{50.1\,\text{eV}}{r_s^2} , \tag{7.9}$$

or the degeneracy temperature $T_F$ is

$$T_F = \frac{\varepsilon_F}{k_B} = \frac{58.2}{r_s^2} \cdot 10^4\,\text{K} . \tag{7.10}$$

Note that the parameter $r_s$ is actually a measure of the relative importance of the Coulomb interaction in a metal. Omitting numerical factors, we see that the ratio of the typical value of the Coulomb interaction $\tilde{V}_C = e^2/\tilde{r}_s$ (the Coulomb interaction at the average distance between electrons) to the characteristic value of the kinetic energy, which is of the order of $\varepsilon_F$, is

$$\tilde{V}_C/\varepsilon_F = \frac{e^2}{\tilde{r}_s}\bigg/\frac{\hbar^2}{m\tilde{r}_s^2} = \frac{me^2}{\hbar^2}\tilde{r}_s = \frac{\tilde{r}_s}{a_0} = r_s \,. \tag{7.11}$$

(We have used here equations (7.4), (7.5) and (7.7).) Thus the small values of $r_s$ (dense electron systems) correspond to weak interactions, $\tilde{V}_C/\varepsilon_F \ll 1$, whereas large values of $r_s$ (low-density electron gas) correspond to systems with strong interactions. These considerations will be very important for us later on, in Chapters 10–12. Note also that for the typical values of $r_s$ for ordinary metals, $r_s \sim 2$–$3$, the electron–electron interaction is of the same order as the kinetic energy or the bandwidth, which is actually the most difficult case for theoretical treatment. Luckily, in most normal metallic systems one can still use the description very much resembling that of a Fermi gas; the justification of this possibility is given by the Landau Fermi liquid theory, see Chapter 10.

The density of states of electrons which, in particular, is used in the transformation from $\int dp$ to $\int \rho(\varepsilon)\,d\varepsilon$, for free electrons in the three-dimensional case has the following form:

$$\rho(\varepsilon) = \frac{m}{\pi^2\hbar^2}\sqrt{\frac{2m\varepsilon}{\hbar^2}} = \frac{3}{2}\frac{n}{\varepsilon_F}\sqrt{\frac{\varepsilon}{\varepsilon_F}} \,. \tag{7.12}$$

At the Fermi level

$$\rho_F = \rho(\varepsilon_F) = \frac{mp_F}{\pi^2\hbar^3} = \frac{3}{2}\frac{n}{\varepsilon_F} \,. \tag{7.13}$$

### *7.1.1 Specific heat and susceptibility of free electrons in metals*

The total energy of the electron system at temperature $T$ and the density of electrons are given by the following general expressions:

$$E = \int d\varepsilon\, \varepsilon\rho(\varepsilon)\, f(\varepsilon) \tag{7.14}$$

$$n = \int d\varepsilon\, \rho(\varepsilon)\, f(\varepsilon) \,. \tag{7.15}$$

Here $\rho(\varepsilon)$ is given by (7.12), and $f(\varepsilon)$ is the Fermi distribution function (3.5),

$$f(\varepsilon) = \frac{1}{e^{(\varepsilon-\mu)/T} + 1} \,. \tag{7.16}$$

The chemical potential $\mu(T)$ should be determined from (7.15) and put into (7.14), and then the specific heat, entropy and other characteristics can be calculated.

The integrals in (7.14), (7.15) – the so-called Sommerfeld integrals – are calculated as follows (cf. Landau and Lifshits (1980)). The general form is

$$I = \int_0^\infty \frac{\varphi(\varepsilon)\, d\varepsilon}{e^{(\varepsilon-\mu)/T} + 1}\,, \tag{7.17}$$

where $\varphi$ is a function such that the integral is convergent. Let us make a change of variables: $\varepsilon - \mu = Tz$,

$$I = \int_{-\mu/T}^\infty \frac{\varphi(\mu + Tz)}{e^z + 1}\, T\, dz = T \int_0^{\mu/T} dz\, \frac{\varphi(\mu - Tz)}{e^{-z} + 1} + T \int_0^\infty dz\, \frac{\varphi(\mu + Tz)}{e^z + 1}\,. \tag{7.18}$$

In the first integral we write $\frac{1}{e^{-z}+1} = 1 - \frac{1}{e^z+1}$; then

$$I = \int_0^\mu \varphi(\varepsilon)\, d\varepsilon + T \int_0^\infty \frac{\varphi(\mu + Tz) - \varphi(\mu - Tz)}{e^z + 1}\, dz\,. \tag{7.19}$$

For low temperatures $T \ll \varepsilon_{\mathrm{F}}$ we can make an expansion in $Tz$ in the second integral and integrate by parts:

$$\begin{aligned} I &= \int_0^\mu \varphi(\varepsilon)\, d\varepsilon + 2T^2 \varphi'(\mu) \int_0^\infty \frac{z\, dz}{e^z + 1} + \tfrac{1}{2} T^4 \varphi'''(\mu) \int \frac{z^3\, dz}{e^z + 1} + \cdots \\ &= \int_0^\mu \varphi(\varepsilon)\, d\varepsilon + \frac{\pi^2}{6} T^2 \varphi'(\mu) + \frac{7\pi^4}{360} T^4 \varphi'''(\mu)\,. \end{aligned} \tag{7.20}$$

$$\left( \begin{aligned} &\int_0^\infty \frac{z^{x-1}\, dz}{e^z + 1} = (1 - 2^{1-x})\, \Gamma(x)\, \zeta(x)\,, \quad \zeta(x) = \sum_{n=1}^\infty \frac{1}{n^x} \text{ is the Riemann } \zeta \text{ function} \\ &\zeta(\tfrac{3}{2}) = 2.61,\ \zeta(\tfrac{5}{2}) = 1.34,\ \ldots;\ \Gamma(\tfrac{3}{2}) = \tfrac{1}{2}\sqrt{\pi},\ \Gamma(\tfrac{5}{2}) = \tfrac{3}{4}\sqrt{\pi},\ \ldots \end{aligned} \right) \tag{7.21}$$

As a result, from (7.14)–(7.20) we obtain:

$$n = \int_0^{\varepsilon_{\mathrm{F}}} \rho(\varepsilon)\, d\varepsilon + \left\{ (\mu - \varepsilon_{\mathrm{F}})\, \rho(\varepsilon_{\mathrm{F}}) + \frac{\pi^2}{6} T^2 \rho'(\varepsilon_{\mathrm{F}}) \right\} \tag{7.22}$$

(with $\varepsilon_{\mathrm{F}} = \mu(T = 0)$). From the condition that $n(T) = n(0)$ we find

$$\mu = \varepsilon_{\mathrm{F}} - \frac{\pi^2}{6} T^2 \frac{\rho'(\varepsilon_{\mathrm{F}})}{\rho(\varepsilon_{\mathrm{F}})} = \varepsilon_{\mathrm{F}} \left[ 1 - \frac{1}{3} \left( \frac{\pi T}{2\varepsilon_{\mathrm{F}}} \right)^2 \right]\,. \tag{7.23}$$

For the energy, finally we get

$$E = E_0 + \frac{\pi^2}{6} T^2 \rho(\varepsilon_{\mathrm{F}}) \tag{7.24}$$

and the specific heat of free electrons is

$$c_V = \left(\frac{\partial E}{\partial T}\right)_V = \frac{\pi^2}{3} T \, \rho(\varepsilon_F) = \frac{\pi^2}{2} n \left(\frac{T}{\varepsilon_F}\right) = \left(\frac{\pi}{3}\right)^{2/3} \frac{mT}{\hbar^2} n^{1/3} \equiv \gamma T \, . \tag{7.25}$$

Here we have used (7.13), (7.3), and introduced the standard notation $\gamma = c(T)/T$.

**Problem:** Find the entropy of a Fermi gas at low temperatures.

**Solution:**

$$c = T\frac{dS}{dT} \quad \Longrightarrow \quad S = \int_0^T \frac{1}{T'} c(T')\,dT' = \int_0^T \frac{1}{T'} \gamma T'\,dT' = \gamma T \, , \tag{7.26}$$

i.e. the entropy behaves exactly as the specific heat itself.

**Problem:** Estimate $c_P - c_V$ for $T \to 0$.

**Solution:**

$$c_P - c_V = +T \frac{\left(\frac{\partial S}{\partial P}\right)_T \left(\frac{\partial V}{\partial T}\right)}{\left(\frac{\partial V}{\partial P}\right)_T} \, , \qquad \text{and} \qquad \left(\frac{\partial V}{\partial T}\right)_P = -\left(\frac{\partial S}{\partial P}\right)_T \, . \tag{7.27}$$

If as $T \to 0$, $S \sim T^n$ (the entropy should go to zero as $T \to 0$, according to the Nernst theorem, or the third law of thermodynamics), then

$$c_P - c_V \sim T \frac{\left(\frac{\partial S}{\partial P}\right)_T^2}{\left(\frac{\partial V}{\partial P}\right)_T} \sim T^{2n+1} \, . \tag{7.28}$$

Thus for electrons, for which $c = \gamma T$ and $S = \gamma T$, i.e. the coefficient $n$ in equation (7.28) is equal to 1, $c_P - c_V \sim T^3$, i.e. it is much smaller than the specific heat $c$ itself. Therefore we can ignore the difference between $c_P$ and $c_V$ for electrons at low temperatures.

Yet another conclusion can be obtained from the treatment presented above. According to general thermodynamic relations, see e.g. (7.27), the thermal expansion satisfies

$$\frac{1}{V}\left(\frac{dV}{dT}\right) = -\frac{1}{V}\frac{\partial S}{\partial P} \, ,$$

thus the electronic contribution to the thermal expansion is $\sim T$, i.e. in metals the thermal expansion $\alpha(T)$ at low temperatures is linear in temperature,

$$\alpha(T) \sim \text{const} \cdot T \, . \tag{7.29}$$

This result is very similar to the Grüneisen equation for phonons (4.53): here also the thermal expansion is proportional to the specific heat (and both are $\sim T$). (Do not confuse the Grüneisen constant $\gamma$ in (4.53) with the coefficient $\gamma$ in the electronic specific heat (7.25)!) Sometimes one speaks about the Grüneisen constant for electrons $\gamma_e$, determined by the same relation as (4.53), with $\alpha$ being the electronic contribution to the thermal expansion, and $c_V$ being the electronic specific heat. Note also that as the phonon specific heat at low temperatures is $\sim T^3$, see equation (4.39), the phonon contribution to the thermal expansion, according to (4.53), is also $\sim T^3$, thus indeed at the lowest temperatures both the specific heat and thermal expansion in metals are determined by the electron contribution.

Similarly (even simpler) one can calculate the Pauli (spin) susceptibility of free electrons:

$$\chi_p = \mu_{\mathrm{B}}^2\, \rho(\varepsilon_{\mathrm{F}}) \; . \tag{7.30}$$

Thus both the specific heat $c$ and magnetic susceptibility $\chi$ are proportional to $\rho(\varepsilon_{\mathrm{F}})$, and we have the ratio (the so-called *Wilson ratio*)

$$R_{\mathrm{W}} = \frac{\pi^2 \chi}{3\mu_{\mathrm{B}}^2 \gamma} = 1 \; , \tag{7.31}$$

i.e. the Wilson ratio for free electrons is 1.

When we include an interaction between electrons, both $c_V$ and $\chi$ are modified, in principle differently; and the Wilson ratio tells us a lot about what is going on. For example, for metals close to ferromagnetism (e.g. Pd, Pt) $\chi$ is enhanced (exchange enhancement). Thus the Wilson ratio is $\sim$1–2 for normal metals (Fermi liquids), and may be $\gg$1 in nearly ferromagnetic metals (see Chapter 10). The notion of the Wilson ratio is widely used, in particular, in treating heavy-fermion systems, see Chapter 13.

# 8

# Interacting electrons. Green functions and Feynman diagrams (methods of field theory in many-particle physics)

In this chapter I will give a short summary of the Green function method and Feynman diagram techniques in application to condensed matter physics. This in itself is quite a big field, and the full description of it, with all derivations and all details, requires a quite lengthy discussion. I will only present the main ideas of this method and give 'recipes' which can be used in practical calculations. The detailed discussion of these points, with all the derivations and proofs, can be found in many books specially devoted to these problems, such as the books by Mattuck (1992) and Fetter and Walecka (2003). This method is also discussed in detail in the books by Abrikosov, Gor'kov and Dzyaloshinskii (1975) and by Mahan (2000), mentioned in the Introduction, where also many applications can be found. A very good summary of this method is also contained in the book by Schrieffer (1999).

## 8.1 Introduction to field-theoretical methods in condensed matter physics

In general, electrons can interact with external forces or potentials, e.g. impurities, with phonons, with magnons, and also interact between themselves. Postponing the discussion of interaction with other excitations till later and concentrating on the electron–electron interaction, we can write the electron Hamiltonian in ordinary quantum mechanics as

$$\mathcal{H} = \sum \mathcal{H}_i + \tfrac{1}{2} \sum_{ij} V(\boldsymbol{r}_i - \boldsymbol{r}_j) \tag{8.1}$$

$$\mathcal{H}_i = \frac{\hbar^2}{2m} \nabla_i^2 + \tilde{v}(\boldsymbol{r}) \,. \tag{8.2}$$

Here $\tilde{v}(\boldsymbol{r}_i)$ is the external potential and $V(\boldsymbol{r} - \boldsymbol{r}')$ is the electron–electron interaction. Thus, e.g. for electrons this is the Coulomb interaction,

$$V(\boldsymbol{r} - \boldsymbol{r}') = \frac{e^2}{|\boldsymbol{r} - \boldsymbol{r}'|} \,. \tag{8.3}$$

In the second quantization the interaction with the external potential has the form

$$\tilde{v}(\boldsymbol{r})\,\rho(\boldsymbol{r}) = \tilde{v}(\boldsymbol{r}) \sum_{\sigma} \hat{\Psi}^{\dagger}_{\sigma}(\boldsymbol{r})\,\hat{\Psi}_{\sigma}(\boldsymbol{r})\,, \tag{8.4}$$

where $\hat{\Psi}^{\dagger}_{\sigma}(\boldsymbol{r})$, $\hat{\Psi}_{\sigma}(\boldsymbol{r})$ are creation and annihilation *operators* for electron with spin $\sigma$ at position $\boldsymbol{r}$. The interaction between electrons is

$$V(\boldsymbol{r}-\boldsymbol{r}')\,\rho(\boldsymbol{r})\,\rho(\boldsymbol{r}') = \sum_{\sigma\sigma'} V(\boldsymbol{r}-\boldsymbol{r}')\,\hat{\Psi}^{\dagger}_{\sigma}(\boldsymbol{r})\,\hat{\Psi}_{\sigma}(\boldsymbol{r})\,\hat{\Psi}^{\dagger}_{\sigma'}(\boldsymbol{r}')\,\hat{\Psi}_{\sigma'}(\boldsymbol{r}')\,. \tag{8.5}$$

Here $\sigma, \sigma' = \pm\frac{1}{2}$ are spin indices.

It is usually convenient to work in the momentum representation. Making the Fourier transform $\hat{\Psi}(\boldsymbol{r}) = \frac{1}{\sqrt{\Omega}} \sum_{\boldsymbol{p}} e^{i\boldsymbol{p}\cdot\boldsymbol{r}} c_{\boldsymbol{p}}$, we get finally for the total Hamiltonian

$$\begin{aligned}\mathcal{H} = {} & \sum_{\boldsymbol{p},\sigma} \varepsilon_{\boldsymbol{p}} c^{\dagger}_{\boldsymbol{p},\sigma} c_{\boldsymbol{p},\sigma} + \sum_{\boldsymbol{p}\boldsymbol{q},\sigma} \tilde{v}(\boldsymbol{q})\, c^{\dagger}_{\boldsymbol{p}+\boldsymbol{q},\sigma}\, c_{\boldsymbol{p},\sigma} \\ & + \frac{1}{2\Omega} \sum_{\boldsymbol{p}\boldsymbol{p}'\boldsymbol{q},\sigma\sigma'} V(\boldsymbol{q})\, c^{\dagger}_{\boldsymbol{p}+\boldsymbol{q},\sigma}\, c^{\dagger}_{\boldsymbol{p}'-\boldsymbol{q},\sigma'}\, c_{\boldsymbol{p}',\sigma'}\, c_{\boldsymbol{p},\sigma}\,.\end{aligned} \tag{8.6}$$

Here $\Omega$ is the total volume (needed for normalization in the Fourier transform; we use here the notation $\Omega$ to distinguish it from the interaction $V$).

The electron density operator $\rho(\boldsymbol{r}) = \Psi^{\dagger}(\boldsymbol{r})\,\Psi(\boldsymbol{r})$ is transformed into

$$\rho(\boldsymbol{q}) = \sum_{\boldsymbol{p},\sigma} c^{\dagger}_{\boldsymbol{p},\sigma}\, c_{\boldsymbol{p}+\boldsymbol{q},\sigma}\,. \tag{8.7}$$

One can conveniently describe the interaction terms in (8.6) using pictures – diagrams. Thus, for example, the interaction of an electron with an external (impurity) potential $v$ will look like

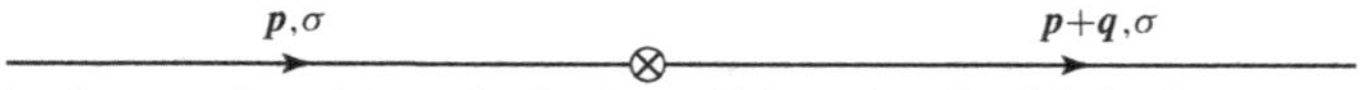


(8.8)

Particle in the state $(\boldsymbol{p}, \sigma)$ is annihilated (operator $c_{\boldsymbol{p},\sigma}$ in the second term in (8.6))

In the 'vertex' (cross) stands $\tilde{v}(\boldsymbol{q})$

Particle in the state $(\boldsymbol{p}+\boldsymbol{q}, \sigma)$ is created (operator $c^{\dagger}_{\boldsymbol{p}+\boldsymbol{q},\sigma}$ in the second term in (8.6))

In this process the momentum is not conserved, because here we describe scattering by an external potential, e.g. by an impurity which we assume to be infinitely heavy and which has no recoil. The line $\xrightarrow{\boldsymbol{p},\sigma}$ thus describes the electron with momentum $\boldsymbol{p}$ and spin $\sigma$, and the cross corresponds to the interaction with the external potential.

The last term in (8.6) (the interaction between electrons) may be depicted as

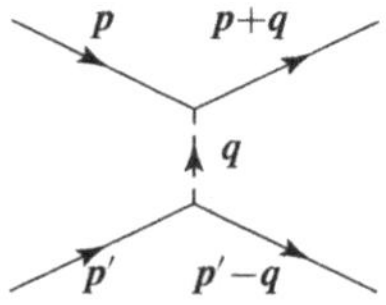


(8.9)

Here the solid line $\xrightarrow{p}$ again describes the electron, and the dashed line $\overset{q}{\dashrightarrow}$ represents the interaction $V(\boldsymbol{q})$. Thus in fact here we describe this process (Coulomb interaction) as an exchange of a virtual photon: one electron emits a photon $\overset{q}{\dashrightarrow}$, and another absorbs it. There exists another, alternative form to depict the electron–electron interaction, which we will also use sometimes:

(8.10)

In this form the interaction corresponds to the vertex with two electron lines going in and two out.

In the process (8.9) the total momentum is conserved (this is why in (8.6) we have these particular momenta, $\boldsymbol{p}$, $\boldsymbol{p}'$, $\boldsymbol{p}+\boldsymbol{q}$, $\boldsymbol{p}'-\boldsymbol{q}$: the total sum of momenta of *incoming* and *outgoing* particles should be the same). Sometimes one writes the interaction term as

$$\frac{1}{2\Omega} \sum_{\substack{\boldsymbol{p}_1,\boldsymbol{p}_2,\boldsymbol{p}_3,\boldsymbol{p}_4 \\ (\boldsymbol{p}_1+\boldsymbol{p}_2=\boldsymbol{p}_3+\boldsymbol{p}_4), \\ \sigma\sigma'}} V(\boldsymbol{p}_1-\boldsymbol{p}_4)\, c^{\dagger}_{\boldsymbol{p}_1,\sigma}\, c^{\dagger}_{\boldsymbol{p}_2,\sigma'}\, c_{\boldsymbol{p}_3,\sigma'}\, c_{\boldsymbol{p}_4,\sigma}\,, \tag{8.11}$$

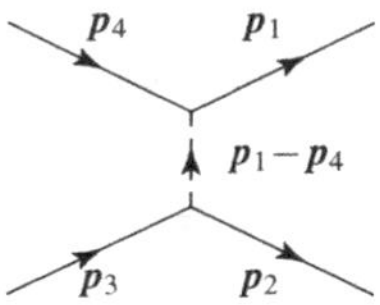


(8.12)

(electrons with momenta $\boldsymbol{p}_3$, $\boldsymbol{p}_4$ which were present in the initial state are annihilated, and two other electrons, with momenta $\boldsymbol{p}_1$ and $\boldsymbol{p}_2$, are created instead). Similar pictures can also be drawn for other types of interaction (e.g. for the electron–phonon interaction, see below).

The most important fact is that such diagrams, 'pictures', are not just a way to depict different terms of the Hamiltonian and different processes; they may be really used *to calculate* many properties of the material. This is one of the virtues of the *Green function* method; the corresponding technique is the *Feynman diagram technique*.

## 8.2 Representations in quantum mechanics

The following section is of a more technical character and can be omitted at first reading, although it is necessary if one wants to understand better how the techniques used below can be really derived.

To explain the method of Green functions and Feynman diagrams we have to go back for a while and discuss different formulations of quantum mechanics – so-called different *representations*.

1. Standard quantum mechanics is usually described in the so-called *Schrödinger representation* in which the *wavefunction* $\Psi(r,t)$ is time dependent, and operators are constant in time:

$$i\frac{\partial}{\partial t}\Psi_{\mathrm{S}}(t) = \mathcal{H}\Psi_{\mathrm{S}}(t) \tag{8.13}$$

(in the future we often put $\hbar = 1$).

The formal operator solution of (8.13) is

$$\Psi_{\mathrm{S}}(t) = e^{-i\mathcal{H}t}\Psi_{\mathrm{S}}(0)\ . \tag{8.14}$$

2. An alternative is the *Heisenberg representation*, in which we take wavefunctions as time independent and ascribe all time dependence to operators, with the dependence

$$\hat{A}_{\mathrm{H}}(t) = e^{i\mathcal{H}t}\hat{A}(0)\,e^{-i\mathcal{H}t}\ , \tag{8.15}$$

and the wavefunction

$$\Psi_{\mathrm{H}} = \Psi_{\mathrm{S}}(0) = e^{i\mathcal{H}t}\Psi_{\mathrm{S}}(t)\ . \tag{8.16}$$

Then the physically measurable quantities (averages) do not change:

$$\langle\hat{A}\rangle = \langle\Psi_{\mathrm{S}}(t)|\hat{A}|\Psi_{\mathrm{S}}(t)\rangle \quad \left(\equiv \int \Psi_{\mathrm{S}}^{*}(t)\,\hat{A}\,\Psi_{\mathrm{S}}(t)\,dr\right)$$

$$\stackrel{\text{from (8.14)}}{=} \langle\Psi_{\mathrm{S}}(0)|e^{i\mathcal{H}t}\,\hat{A}\,e^{-i\mathcal{H}t}|\Psi_{\mathrm{S}}(0)\rangle = \langle\Psi_{\mathrm{H}}|\hat{A}_{\mathrm{H}}(t)|\Psi_{\mathrm{H}}\rangle\ . \tag{8.17}$$

Now the $\Psi$ function is taken as constant, but the operator $\hat{A}_{\mathrm{H}}(t)$ obeys the equation (cf. equation (6.87))

$$i\frac{\partial}{\partial t}\hat{A}_{\mathrm{H}}(t) = \left[\hat{A}_{\mathrm{H}}(t), H\right]\ . \tag{8.18}$$

3. Yet another convenient form is the 'mixed' representation – the so-called *interaction representation*, which is actually the basis of the Feynman diagram technique.

If the Hamiltonian consists of two parts,

$$\mathcal{H} = \mathcal{H}_0 + \mathcal{H}' , \tag{8.19}$$

where $\mathcal{H}_0$ is the 'bare' Hamiltonian of noninteracting particles (or anything we can solve simply), and $\mathcal{H}'$ is an interaction, then we can introduce the representation where *operators* depend on time as in the case of free particles, i.e. as in (8.15), *but with* $\mathcal{H} = \mathcal{H}_0$,

$$\hat{A}_{\mathrm{I}}(t) = e^{i\mathcal{H}_0 t}\,\hat{A}(0)\,e^{-i\mathcal{H}_0 t} , \tag{8.20}$$

and the *wavefunctions* also are time dependent, but only because of the presence of the interaction $\mathcal{H}'$:

$$\Psi_{\mathrm{I}}(t) = e^{i\mathcal{H}_0 t}\,\Psi_{\mathrm{S}}(t) = e^{i\mathcal{H}_0 t} e^{-i\mathcal{H}t}\,\Psi_{\mathrm{S}}(0) . \tag{8.21}$$

If $\mathcal{H}' = 0$, then indeed in (8.21) we have $e^{i\mathcal{H}_0 t} e^{-i\mathcal{H}_0 t} = 1$, and $\Psi_{\mathrm{I}}(t)$ (8.21) becomes identical with $\Psi_{\mathrm{H}}$ (8.16), i.e. time independent. In general, the operators $\mathcal{H}_0$ and $\mathcal{H}$ in (8.21) *do not commute*, and $e^{i\mathcal{H}_0 t} e^{-i(\mathcal{H}_0+\mathcal{H}')t} \neq e^{-i\mathcal{H}'t}$ !

**Problem:** Try to find the expression for $e^{\hat{A}} e^{\hat{B}}$ when the operators $\hat{A}$ and $\hat{B}$ do not commute.

**Solution:**

$$e^{\hat{A}} e^{\hat{B}} = e^{\hat{A}+\hat{B}} e^{\frac{1}{2}[\hat{A},\hat{B}]} , \tag{8.22}$$

if only one commutator $[\hat{A}, \hat{B}]$ is nonzero; if the next commutators, e.g. $\left[\hat{A}, [\hat{A}, \hat{B}]\right]$, are nonzero, then the corresponding formula looks more complicated.

From the previous definitions one can show that

$$i\frac{\partial}{\partial t}\Psi_{\mathrm{I}}(t) = \mathcal{H}'_{\mathrm{I}}(t)\,\Psi_{\mathrm{I}}(t) , \tag{8.23}$$

where

$$\mathcal{H}'_{\mathrm{I}}(t) = e^{i\mathcal{H}_0 t}\,\hat{\mathcal{H}}'_{\mathrm{S}}\,e^{-i\mathcal{H}_0 t} \tag{8.24}$$

according to the general rule (8.20).

The interaction representation is convenient because we can usually choose $\mathcal{H}_0$ such that the time dependence, or $\omega$-dependence in a Fourier transform, is simple, for example starting from the noninteracting electrons (the ideal Fermi gas, see Chapter 7). Then we can solve equation (8.23) or an equivalent one by perturbation theory, which in turn can be conveniently represented by Feynman diagrams. This is actually the main reason for introducing the interaction representation.

Let us try formally to solve equation (8.23) with $\mathcal{H}'_{\mathrm{I}}$ given by (8.24). We *cannot* write the solution as

$$\Psi_{\mathrm{I}}(t) = \mathrm{const} \cdot e^{-i\int^t \mathcal{H}'_{\mathrm{I}}(t')\,dt'} \tag{$*$}$$

because now the operators $\mathcal{H}'_{\mathrm{I}}(t)$ do not commute with themselves at different moments $(t_1, t_2)$. But we can try to solve it by perturbation theory. For that we write down an integral equation for $\Psi_{\mathrm{I}}(t)$, equivalent to (8.23):

$$\Psi_{\mathrm{I}}(t) = \Psi_{\mathrm{I}}(t_0) - i\int_{t_0}^{t} \mathcal{H}'_{\mathrm{I}}(t')\,\Psi_{\mathrm{I}}(t')\,dt' \tag{8.25}$$

and seek the solution as $\Psi_{\mathrm{I}}(t) = \Psi_{\mathrm{I}}^{(0)}(t) + \Psi_{\mathrm{I}}^{(1)}(t) + \cdots$. In lowest order $\Psi_{\mathrm{I}}^{(0)} = \Psi_{\mathrm{I}}(t_0)$ (as though $\mathcal{H}'$ in (8.25) were absent). Then

$$\begin{aligned}
\Psi_{\mathrm{I}}^{(1)}(t) &= -i\int_{t_0}^{t} \mathcal{H}'_{\mathrm{I}}(t_1)\,dt_1 \cdot \Psi_{\mathrm{I}}(t_0)\,,\\
\Psi_{\mathrm{I}}^{(2)}(t) &= -\int_{t_0}^{t} \mathcal{H}'_{\mathrm{I}}(t_1)\,dt_1 \int_{t_0}^{t_1} \mathcal{H}'_{\mathrm{I}}(t_2)\,dt_2 \cdot \Psi_{\mathrm{I}}(t_0)\,, \quad \text{etc.}
\end{aligned} \tag{8.26}$$

One can write the general solution in terms of the so-called $S$-matrix:

$$\Psi_{\mathrm{I}}(t) = S(t, t_0)\,\Psi_{\mathrm{I}}(t_0)\,, \tag{8.27}$$

and for the $S$-matrix we have the series

$$\begin{aligned}
S(t, t_0) = 1 &- i\int_{t_0}^{t_1} \mathcal{H}'_{\mathrm{I}}(t_1)\,dt + \cdots\\
&+ (-i)^n \int_{t_0}^{t} \mathcal{H}'_{\mathrm{I}}(t_1)\,dt_1 \int_{t_0}^{t_1} \mathcal{H}'_{\mathrm{I}}(t_2)\,dt_2 \cdots \int_{t_0}^{t_{n-1}} \mathcal{H}'_{\mathrm{I}}(t_n)\,dt_n\\
&+ \cdots\,.
\end{aligned} \tag{8.28}$$

It is important that here we have the sequence of times such that $t > t_1 > t_2 \ldots t_n > t_0$, or, in a general term of the type $\int\int\int \mathcal{H}'_{\mathrm{I}}(t_{m_1})\,\mathcal{H}'_{\mathrm{I}}(t_{m_2}) \ldots \mathcal{H}'_{\mathrm{I}}(t_{m_n})$ all operators $\mathcal{H}'$ should be *time-ordered* (time increases from right to left, $t_{m_1} > t_{m_2} > \ldots > t_{m_n}$). Thus the formal solution for the $S$-matrix should be written as

$$S(t, t_0) = \mathrm{T}\exp\left\{-i\int_{t_0}^{t} \mathcal{H}'_{\mathrm{I}}(t')\,dt'\right\}\,, \tag{8.29}$$

where by T we mean time-ordering under the integral. Thus in general $\mathrm{T}[\hat{A}(t_1)\hat{A}(t_2)\hat{A}(t_3)]$ is $\hat{A}(t_1)\hat{A}(t_2)\hat{A}(t_3)$ if $t_1 > t_2 > t_3$, or $\pm\hat{A}(t_2)\hat{A}(t_1)\hat{A}(t_3)$ for $t_2 > t_1 > t_3$, etc. (Do not confuse the symbol T here with the temperature!) Note also the signs: for bosons we always have the plus sign, but for fermions one interchange of fermion operators gives a minus, two interchanges give plus, etc., which is the usual rule for fermions.

The connection between operators in the Heisenberg and in the interaction representation can now be written as

$$\hat{A}_{\mathrm{H}}(t) = S^{-1}(t)\,\hat{A}_{\mathrm{I}}(t)\,S(t)\,. \tag{8.30}$$

## 8.3 Green functions

In this section we give a brief introduction to the method of Green functions. For a more complete treatment, see e.g. the books by Abrikosov *et al.* (1975), Doniach and Sondheimer (1974), Fetter and Walecka (2003), Mahan (2000), Mattuck (1992) and Schrieffer (1999).

Instead of working with the full many-body wavefunctions, let us introduce certain combinations thereof – *Green functions*, which of course contain not the full information, but nevertheless describe a lot, and through which one can directly express many different measurable quantities.

The definition of the most common type of Green functions (they are sometimes called *causal* Green functions, to discriminate them from the *retarded* and *advanced* Green functions) is:

$$G_{\sigma\sigma'}(\boldsymbol{r}_1, t_1; \boldsymbol{r}_2, t_2) = -i\,\langle 0|\mathrm{T}\big\{\hat{\Psi}_\sigma(\boldsymbol{r}_1, t_1)\,\hat{\Psi}^\dagger_{\sigma'}(\boldsymbol{r}_2, t_2)\big\}|0\rangle\,. \tag{8.31}$$

Here $\hat{\Psi}_\sigma(\boldsymbol{r}, t)$ is the electron *operator* in second quantization form in the Heisenberg representation (do not confuse it with the wavefunction in (8.13)–(8.16)), $\langle 0|\ldots|0\rangle$ is the average over the ground state (*in principle unknown!*), and T stands for the so-called T-product – the time-ordered sequence of operators standing after this symbol (cf. (8.29)):

$$G_{\sigma\sigma'}(\boldsymbol{r}_1, t_1; \boldsymbol{r}_2, t_2) = \begin{cases} -i\,\langle 0|\hat{\Psi}_\sigma(\boldsymbol{r}_1, t_1)\,\hat{\Psi}^\dagger_{\sigma'}(\boldsymbol{r}_2, t_2)|0\rangle & \text{for } t_1 > t_2 \\ \;\;\, i\,\langle 0|\hat{\Psi}^\dagger_{\sigma'}(\boldsymbol{r}_2, t_2)\,\hat{\Psi}_\sigma(\boldsymbol{r}_1, t_1)|0\rangle & \text{for } t_1 < t_2\,. \end{cases} \tag{8.32}$$

For nonmagnetic systems Green functions are diagonal in the spin indices, $G_{\sigma\sigma'} = G\,\delta_{\sigma\sigma'}$; in future we will mostly omit the indices $\sigma, \sigma'$. Different signs for $t_1 > t_2$ and $t_1 < t_2$ in (8.32) are needed because of Fermi statistics; for bosons (e.g. phonons) the sign will be the same.

How can we understand qualitatively why we have introduced such a strange object, with the T-product? What is the physical meaning of these Green functions? 'The proof of the pudding is in the eating', according to a British proverb. We will see soon that Green functions are indeed very convenient objects. One can express a lot of physical quantities through them. Green functions also give information about the spectrum of elementary excitations in the system. But here we will first present some qualitative arguments:

Consider the initial system in its ground state $|0\rangle$, and let us create an extra electron at the moment $t_2$ at point $\boldsymbol{r}_2$: $|2\rangle = \hat{\Psi}^\dagger(\boldsymbol{r}_2, t_2)|0\rangle$. This electron will propagate in time, and we want to know 'the fate' of this electron, i.e. to look at the probability that at time $t_1$ it will be at point $\boldsymbol{r}_1$ (note that in the process it will interact with other electrons, the system may be excited, etc.). The corresponding wavefunction is $|1\rangle = \hat{\Psi}^\dagger(\boldsymbol{r}_1, t_1)|0\rangle$. The amplitude of the probability we are interested in is $\langle 1|2\rangle = \langle 0|\hat{\Psi}(\boldsymbol{r}_1, t_1)\hat{\Psi}^\dagger(\boldsymbol{r}_2, t_2)|0\rangle$.

Similarly, we can consider the process in which we initially create a hole, e.g. $|\tilde{2}\rangle = \hat{\Psi}(\boldsymbol{r}_2, t_2)|0\rangle$, and consider its propagation to the state $|\tilde{1}\rangle = \hat{\Psi}(\boldsymbol{r}_1, t_1)|0\rangle$. Then we want to know $\langle\tilde{1}|\tilde{2}\rangle = \langle 0|\hat{\Psi}^\dagger(\boldsymbol{r}_1, t_1)\hat{\Psi}(\boldsymbol{r}_2, t_2)|0\rangle$.

Note that it is reasonable to consider such processes if $t_1 > t_2$ (first we create a particle, and then it propagates to a new state).

It turns out that it is convenient to combine these two objects, $\langle 1|2\rangle$ and $\langle\tilde{1}|\tilde{2}\rangle$ (which are both functions of $(\boldsymbol{r}_1, t_1, \boldsymbol{r}_2, t_2)$), into one, ordering the times as defined in (8.31), (8.32). The Green function thus defined describes the motion of an added electron from $(\boldsymbol{r}_2, t_2)$ to $(\boldsymbol{r}_1, t_1)$, with $t_1 > t_2$, or the motion of an added hole from $(\boldsymbol{r}_1, t_1)$ to $(\boldsymbol{r}_2, t_2)$ (and here $t_2 > t_1$; to combine the description of both these processes in one function we interchanged $(\boldsymbol{r}_1, t_1) \leftrightarrow (\boldsymbol{r}_2, t_2)$ for the hole). Actually this definition goes back to Feynman who described positrons (in our case holes) as electrons moving backwards in time.

From the definition (8.31) the particle density $n(\boldsymbol{r})$ (also denoted sometimes $\rho(\boldsymbol{r})$) is

$$n(\boldsymbol{r}) = -2i \lim_{\substack{\boldsymbol{r}=\boldsymbol{r}' \\ t' \to t+0}} G(\boldsymbol{r}, t; \boldsymbol{r}', t') \tag{8.33}$$

(the factor of 2 comes from summation over spins). For spatially homogeneous systems $G(\boldsymbol{r}, t; \boldsymbol{r}', t') = G(\boldsymbol{r} - \boldsymbol{r}', t - t')$. Its Fourier transform (the Green function in the momentum representation) $G(\boldsymbol{p}, \omega)$ is given by

$$G(\boldsymbol{r} - \boldsymbol{r}', t - t') = \int \frac{d^3 p\, d\omega}{(2\pi)^4} G(\boldsymbol{p}, \omega)\, e^{i[\boldsymbol{p}\cdot(\boldsymbol{r}-\boldsymbol{r}') - \omega(t-t')]} \,. \tag{8.34}$$

One can show that the momentum distribution function for electrons $n(\boldsymbol{p})$ is expressed through the Green function as

$$n(\boldsymbol{p}) = -i \lim_{t \to -0} \int G(\boldsymbol{p}, \omega)\, e^{-i\omega t} \frac{d\omega}{2\pi} \,, \tag{8.35}$$

cf. (8.33).

One can also express through the electron Green function $G(\boldsymbol{p}, \omega)$ all thermodynamic properties of the system. There are several ways to do that. One of them is to use the general expression for the thermodynamic potential $\tilde{\Omega}$ (we denote it here by $\tilde{\Omega}$ so as not to confuse it with the volume $\Omega$) (1.29), (1.30). At $T = 0$ the entropy is $S = 0$, and

$$d\tilde{\Omega} = -\mu\, dN\,. \tag{8.36}$$

Using the relation

$$n = -2i \lim_{t\to -0} \int G(\boldsymbol{p}, \omega)\, e^{-i\omega t}\, \frac{d\omega\, d^3\boldsymbol{p}}{(2\pi)^4}\,, \tag{8.37}$$

following from (8.35) (the factor of 2 again comes from summation over spins) and integrating (8.36) over $\mu$, we obtain

$$\tilde{\Omega}(\mu) = 2i\Omega \int_0^{\mu} d\mu' \lim_{t\to -\infty} \int G(\boldsymbol{p}, \omega)\, e^{-i\omega t}\, \frac{d\omega\, d^3\boldsymbol{p}}{(2\pi^4)}\,. \tag{8.38}$$

Thus we see that many important properties of the interacting system may be expressed through the Green function.

By analogy with the one-particle Green function $G(x; x')$ or $G(\boldsymbol{p}, \omega)$, we can introduce also two-particle and higher-order Green functions. In particular, the two-particle Green function is important for treating the response of the system to an external perturbation such as an electromagnetic field, for the discussion of eventual instabilities of the system and for treating transport properties of the system, such as resistivity or thermal conductivity.

## 8.4 Green functions of free (noninteracting) electrons

For free electrons the one-electron operators $\Psi^{\dagger}(\boldsymbol{r}, t)$, $\Psi(\boldsymbol{r}, t)$ in (8.31), (8.32) are connected with the creation and annihilation operators $c_{\boldsymbol{p}}^{\dagger}$, $c_{\boldsymbol{p}}$ by the usual relation

$$\Psi(\boldsymbol{r}, t) = \frac{1}{\sqrt{\Omega}} \sum_{\boldsymbol{p}} c_{\boldsymbol{p}} e^{i[\boldsymbol{p}\cdot\boldsymbol{r} - \varepsilon_0(\boldsymbol{p})t]}\,. \tag{8.39}$$

Let us put this in (8.32): for Green functions of free fermions $G_0(\boldsymbol{r}, t)$ we then get

$$G_0(\boldsymbol{r}, t) = -\frac{i}{\Omega} \sum_{\boldsymbol{p}} e^{i[\boldsymbol{p}\cdot\boldsymbol{r} - \varepsilon_{\boldsymbol{p}} t]} \times \begin{cases} 1 - f_{\boldsymbol{p}} & (t > 0) \\ -f_{\boldsymbol{p}} & (t < 0)\,, \end{cases} \tag{8.40}$$

where by $\varepsilon_{\boldsymbol{p}}$ we denote the bare spectrum of noninteracting electrons $\varepsilon_0(\boldsymbol{p})$, and $f_{\boldsymbol{p}} = \langle n_{\boldsymbol{p}} \rangle = \langle c_{\boldsymbol{p}}^{\dagger} c_{\boldsymbol{p}} \rangle$ is the Fermi distribution function at $T = 0$,

$$f_{\boldsymbol{p}} = \begin{cases} 1 & \text{for } |\boldsymbol{p}| < p_{\mathrm{F}} \\ 0 & \text{for } |\boldsymbol{p}| > p_{\mathrm{F}}\,. \end{cases}$$

Using this, one can show that

$$\boxed{G_0(\boldsymbol{p},\omega) = \frac{1}{\omega - \varepsilon_{\boldsymbol{p}} + i\delta\,\mathrm{sign}(|\boldsymbol{p}| - p_{\mathrm{F}})}} \tag{8.41}$$

where $\delta$ is an infinitesimally small positive number. (Alternatively we can write in the denominator $i\delta\,\mathrm{sign}(\varepsilon_{\boldsymbol{p}} - \mu)$.)

**Problem:** Check (8.41) using (8.34)–(8.40).

Hint: Work back from (8.34) using (8.41).

The bare Green function $G_0$ (8.41) obeys the equation

$$\left(i\frac{\partial}{\partial t} - \varepsilon_{\boldsymbol{p}}\right) G_0(\boldsymbol{p}, t) = \delta(t)\,, \tag{8.42}$$

i.e. it really is a 'Green function' as introduced in mathematics (the solution of the corresponding differential equation with a point source on the right-hand side). But it will no longer be so simple for interacting systems.

Often one counts the energy from the chemical potential $\mu$:

$$\xi_{\boldsymbol{p}} = \varepsilon_{\boldsymbol{p}} - \mu\,. \tag{8.43}$$

Then

$$G_0(\boldsymbol{p},\omega) = \frac{1}{\omega - \xi_{\boldsymbol{p}} + i\delta_{\boldsymbol{p}}}\,, \qquad \delta_{\boldsymbol{p}} = \delta\,\mathrm{sign}\,\xi_{\boldsymbol{p}}\,. \tag{8.44}$$

From equations (8.41) and (8.44) we immediately see that the poles of the Green function correspond to energies of (quasi)particles. This is true not only for the Green functions of noninteracting electrons, but also in general; this is one of the reasons why the notion of Green functions is actually very useful. One can show (see below) that the real part of the poles gives the energy, and the imaginary part describes the damping (finite lifetime) of quasiparticles. For free electrons this is evident: Im $G_0$ is proportional to $\delta \to 0$, and the lifetime is infinite, as it should be for a noninteracting system. For interacting systems the excitations in general have finite lifetimes. However one can still speak about quasiparticles as well-defined objects if their damping is not too strong, i.e. if the real part of the pole $\xi_{\boldsymbol{p}}$ is $\mathrm{Re}\,\xi_{\boldsymbol{p}} > \mathrm{Im}\,\xi_{\boldsymbol{p}}$. And in this case the poles of $G(\boldsymbol{p},\omega)$ describe such excitations.

## 8.5 Spectral representation of Green functions

One can show that the Green function $G(\boldsymbol{p}, \omega)$ is an analytical function in the complex $\omega$-plane. In this case one can represent it as

$$G(\boldsymbol{p}, \omega) = \int_{-\infty}^{\infty} \frac{A(\boldsymbol{p}, \omega')\, d\omega'}{\omega - \omega' + i\omega'\delta} \tag{8.45}$$

($\delta = +0$). This is called the *spectral representation*; the function $A(\boldsymbol{p}, \omega)$ is the *spectral function*. For an ideal Fermi gas (with the Green function $G_0(\boldsymbol{p}, \omega)$ (8.44)) the function $A(\boldsymbol{p}, \omega')$ is

$$A(\boldsymbol{p}, \omega) = \delta\big(\omega - [\varepsilon_{\boldsymbol{p}} - \mu]\big) = \delta(\omega - \xi_{\boldsymbol{p}})\ . \tag{8.46}$$

Thus for free noninteracting electrons the spectral function is the delta function at the position of the pole – at the energy of the elementary excitation (here at $\xi_{\boldsymbol{p}} = \varepsilon_{\boldsymbol{p}} - \mu$). In general $A(\boldsymbol{p}, \omega)$ may strongly differ from this simple form, but if it still contains a relatively narrow peak, it can be interpreted as a quasiparticle (with a finite lifetime determined by the width of the peak). But in general the spectral function can also contain broad features which describe the incoherent part of the spectrum, see Section 8.5.2 below.

One can obtain an important expression for the Green function $G$ and for its spectral function $A$. From the definition of the Green function one can show that

$$\begin{aligned} G(\boldsymbol{p}, t) &= -i\,\langle 0|\mathrm{T}\{c_{\boldsymbol{p}}(t)\, c^{\dagger}_{\boldsymbol{p}}(0)\}|0\rangle \\ &= \begin{cases} -i\,\langle 0|c_{\boldsymbol{p}}(0) e^{-i\mathcal{H}t} c^{\dagger}_{\boldsymbol{p}}(0)|0\rangle\, e^{iE_0^N t} & \text{for } t > 0 \\ i\,\langle 0|c^{\dagger}_{\boldsymbol{p}}(0) e^{i\mathcal{H}t} c_{\boldsymbol{p}}(0)|0\rangle\, e^{-iE_0^N t} & \text{for } t < 0\,, \end{cases} \end{aligned} \tag{8.47}$$

where we have used the definition of the Green function (8.31), (8.32), with the Fourier transforms of operators $c^{\dagger}_{\boldsymbol{p}}$, $c_{\boldsymbol{p}}$, and we took into account (8.17); $E_0^N$ is the energy of the ground state of the system with $N$ electrons. Putting between the $c$ and $c^{\dagger}$ operators in (8.47) the complete system of functions $\sum_m |\Psi_m^{N\pm1}\rangle\langle\Psi_m^{N\pm1}| = 1$ (here $m$ is the index of the quantum state, $N \pm 1$ denotes that we are dealing with a system with $N \pm 1$ particles, i.e. one electron added to or removed from our system), we get

$$G(\boldsymbol{p}, t) = \begin{cases} -i \sum_m |(c^{\dagger}_{\boldsymbol{p}})_{m,0}|^2\, e^{-i(\omega_m^{N+1}+\mu)t} & \text{for } t > 0 \\ +i \sum_m |(c_{\boldsymbol{p}})_{m,0}|^2\, e^{-i(\omega_m^{N-1}+\mu)t} & \text{for } t < 0\,. \end{cases} \tag{8.48}$$

For the spectral function $A(\boldsymbol{p}, \omega)$ we then have:

$$A(\boldsymbol{p}, \omega) = \sum_m |(c^{\dagger}_{\boldsymbol{p}})_{m,0}|^2 \delta(\omega - \omega_m^{N+1}) + \sum_m |(c_{\boldsymbol{p}})_{m,0}|^2\, \delta(\omega - \omega_m^{N-1})\,, \tag{8.49}$$

i.e. the positive-frequency part of $G$ and $A$ ($\omega > 0$) describes the creation (addition) of one electron to our system, whereas the negative-frequency part ($\omega < 0$) describes annihilation (removal) of an electron; for more details see Section 8.5.2.

In general one can express $A(\boldsymbol{p}, \omega)$ through the imaginary part of $G(\boldsymbol{p}, \omega)$ itself:

$$\operatorname{Im} G(\boldsymbol{p}, \omega + \mu) = \begin{cases} -\pi\, A(\boldsymbol{p}, \omega) & \text{for } \omega > 0 \\ \pi\, A(\boldsymbol{p}, \omega) & \text{for } \omega < 0\,. \end{cases} \tag{8.50}$$

Here we have used the identity

$$\left.\frac{1}{x \pm i\varepsilon}\right|_{\varepsilon \to 0} = \frac{\overline{P}}{x} \mp i\pi\delta(x)\,, \tag{8.51}$$

where $\overline{P}$ is the symbol for the principal part of an integral, $\overline{P}\int g(x) \equiv ⨍ g(x)$. With equation (8.50), the spectral representation (8.45) gives the Green function through the integral of $\operatorname{Im} G$ – it is the so-called *dispersion relation* (similar to the Kramers–Kronig relation in optics).

The general properties of the spectral function $A(\boldsymbol{p}, \omega)$ are the following: it is real, positive

$$A(\boldsymbol{p}, \omega) = A^*(\boldsymbol{p}, \omega) > 0\,, \tag{8.52}$$

and obeys the sum rule

$$\int_{-\infty}^{\infty} A(\boldsymbol{p}, \omega)\, d\omega = 1\,. \tag{8.53}$$

### 8.5.1 Physical meaning of the poles of $G(\boldsymbol{p}, \omega)$

We have already mentioned that the poles of the Green function describe excitations (quasiparticles). In general, close to a pole (with some finite (small) imaginary part) one can write

$$G(\boldsymbol{p}, \omega) \simeq \frac{Z_p}{\omega - \tilde{\varepsilon}_p + i\gamma}\,. \tag{8.54}$$

(More accurately one has to use here the so-called retarded and/or advanced Green functions.) Here $\tilde{\varepsilon}_p$ is the *energy* of the quasiparticle, $\gamma$ is its damping, and the residue $Z_p \leq 1$ gives the weight of the *real electron* in the quasiparticle (in an interacting system the excitations – quasiparticles – are not *bare* electrons, but are renormalized, 'dressed' by the cloud of other electronic excitations). The factor $Z_p$ is called wavefunction renormalization, or simply the $Z$-factor.

One can show that there exists an important relation: there exists in general a *jump* in the distribution function of electrons $n(\boldsymbol{p})$ at $p_{\mathrm{F}}$ (the same Fermi momentum

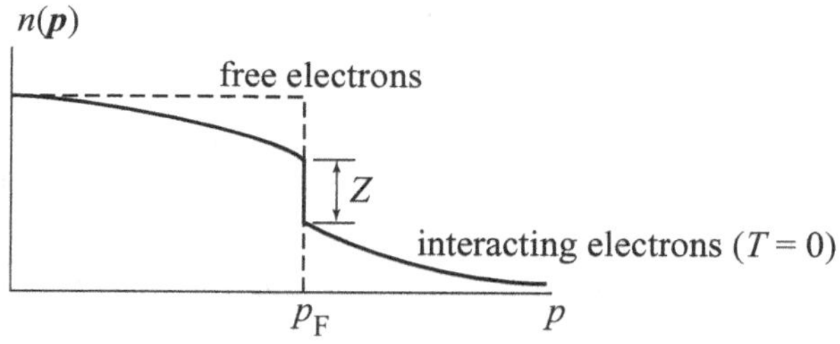


Fig. 8.1

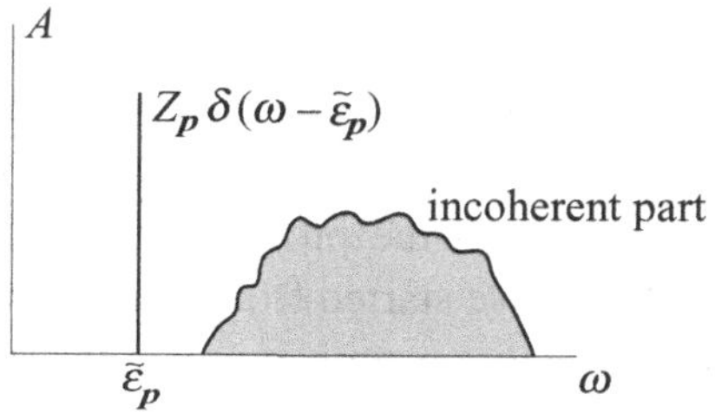


Fig. 8.2

$p_F$ (7.3) as in noninteracting systems), and this jump is given by $Z_p$:

$$n_\sigma(p_F - 0) - n_\sigma(p_F + 0) = Z \tag{8.55}$$

(for electrons with isotropic spectrum $Z_{p_F} = Z$ is a constant at the Fermi surface; in general, for a complicated Fermi surface, it may depend on the direction of $\boldsymbol{p}$). Thus the distribution function of electrons $n(\boldsymbol{p})$ looks as shown in Fig. 8.1. In systems in which $Z \neq 0$ the nature of single-particle excitations is similar to those of real electrons or holes. However, there may be situations in which $Z = 0$ – these are the *non-Fermi liquid* systems (see next chapter).

**Problem:** Using the properties of the spectral function $A(\boldsymbol{p}, \omega)$, show that $0 \leq Z \leq 1$.

**Solution:** The pole contribution (8.54) corresponds to the spectral function $A(\boldsymbol{p}, \omega) \simeq Z_{\boldsymbol{p}}\,\delta(\omega - \varepsilon_{\boldsymbol{p}})$. But according to the general properties of the spectral function $A(\boldsymbol{p}, \omega)$ it is positive, and the sum rule (8.53) gives that $Z \leq 1$ ($Z = 1$ for an ideal Fermi gas, and it is less than 1 if there exist, besides the pole, other (positive) incoherent contributions to $A(\boldsymbol{p}, \omega)$).

Actually what happens is that due to interactions a part of the spectral weight $A(\boldsymbol{p}, \omega)$, which for the noninteracting system was all contained in the $\delta$-function $\delta(\omega - \varepsilon_{\boldsymbol{p}})$, is now transferred into the *incoherent* part, Fig. 8.2, and the weight or intensity of the remaining pole at $\tilde{\varepsilon}_{\boldsymbol{p}}$ is decreased. As mentioned above, $Z_{p_F}$ also gives the jump of the electron distribution function $n(\boldsymbol{p})$ at the Fermi surface; it is also reduced by interaction.

There is one more very important general result in this field: one can prove that when one starts from the noninteracting case and then adds interactions, the volume of the Fermi surface, or in other words the value of the Fermi momentum $p_{\mathrm{F}}$, does not change. This is known as the *Luttinger theorem* (see Chapter 10 for more details).

### 8.5.2 Physical meaning of the spectral function $A(\boldsymbol{p}, \omega)$

As discussed above, from the definition of the Green function (8.31), (8.32) and from equations (8.45), (8.49), one can show (see, e.g. Schrieffer (1999)) that the positive-frequency part of $A(\boldsymbol{p}, \omega)$ (for $\omega > 0$) describes the process when we *add* one electron to our system. This is the process studied for example by inverse photoemission (IPES or BIS). If we started from the ground state of an $N$-particle system $|\Psi_0^N\rangle$ and created an electron with momentum $\boldsymbol{p}$, $c_{\boldsymbol{p}}^{\dagger}|\Psi_0^N\rangle = |\Phi_{\boldsymbol{p}}\rangle$, then the state $|\Phi_{\boldsymbol{p}}\rangle$ is not an eigenstate of the system of $N+1$ electrons. One can decompose $|\Phi_{\boldsymbol{p}}\rangle$ into eigenstates $|\Psi_m^{N+1}\rangle$. Then the probability of finding the system (after one electron was added) in the state with energy $\omega$ is

$$P_{\omega}(\boldsymbol{p})\, d\omega = \int_{\omega}^{\omega+d\omega} A(\boldsymbol{p}, \omega')\, d\omega' \,. \tag{8.56}$$

Thus for $\omega > 0$ the spectral function $A(\boldsymbol{p}, \omega)$ gives directly the intensity of the spectrum of angle-resolved inverse photoemission (IPES) – the probability of finding the system in the state with energy $\omega$ and momentum $\boldsymbol{p}$ after adding one electron. Similarly, for $\omega < 0$, $A(\boldsymbol{p}, \omega)$ describes the probability of *extracting*, *removing* an electron, leaving the system in the state with energy $\omega$, which is the process of (angle-resolved) photoemission (ARPES). For the ideal gas there is a one-to-one correspondence of $\omega$ and $\boldsymbol{p}$, i.e. given $\boldsymbol{p}$ we know the energy $\omega = \varepsilon_{\boldsymbol{p}}$; this is described by (8.46). In this case the state $c_{\boldsymbol{p}}^{\dagger}|\Psi_0^N\rangle$ is an eigenstate with energy $\varepsilon_{\boldsymbol{p}}$ (or $\varepsilon_{\boldsymbol{p}} + \mu$), and corresponding PES or IPES spectra would formally contain $\delta$-function peaks. But in a system with an interaction this is no longer the case, and there is in general an *incoherent* part in the spectral function $A(\boldsymbol{p}, \omega)$ besides the $\delta$-functions (or Lorentzians with small width) describing quasiparticles. Thus the schematic form of spectral function shown in Fig. 8.2 (with corresponding broadening) is actually a typical form of photoemission (or inverse photoemission) spectra.

Thus, an important short summary:

The one-electron Green function describes quasiparticles (if they exist!) – these are the poles of $G(\boldsymbol{p}, \omega)$. The real part of the pole gives the energy of the quasiparticle, and the imaginary part describes its decay (it is the inverse lifetime of the excitation). The residue of the pole (or its intensity) $Z_{\boldsymbol{p}}$ shows what is the relative weight of a single electron in the total wavefunction of the excitation. For Fermi liquids there is a jump of the distribution function of (real) electrons at the Fermi momentum $|\boldsymbol{p}| = p_{\mathrm{F}}$, coinciding with $p_{\mathrm{F}}$ of noninteracting electrons; $Z_{p_{\mathrm{F}}} \equiv Z$ is the magnitude of this jump. The spectral function $A(\boldsymbol{p}, \omega)$ describes the probability of observing the system in a state with energy $\omega$ after an electron with momentum $\boldsymbol{p}$ was added ($\omega > 0$) to or removed ($\omega < 0$) from the system; $A(\boldsymbol{p}, \omega)$ is directly related to the spectrum measured by photoemission and inverse photoemission.

## 8.6 Phonon Green functions

The Green function method and diagram techniques can also be formulated for a system of interacting electrons and phonons. The electron–phonon interaction originates from the change of ionic charge density when ions shift from their equilibrium positions. Generally speaking, the interaction is proportional to

$$-e\rho_{\mathrm{el}}(\boldsymbol{r})V(\boldsymbol{r}, \boldsymbol{r}')\,\mathrm{div}\left(Ze\frac{N}{V}\boldsymbol{u}(\boldsymbol{r}')\right), \tag{8.57}$$

where $\rho_{\mathrm{el}}$ is the electron density at point $\boldsymbol{r}$, $N/V$ is the density of ions, $Ze$ is their charge and $\boldsymbol{u}(\boldsymbol{r})$ is the shift of ions from their equilibrium position. The kernel $V(\boldsymbol{r}, \boldsymbol{r}')$ describes the Coulomb interaction between an electron charge at point $\boldsymbol{r}$ and an 'extra' ionic charge at point $\boldsymbol{r}'$. In metals, due to strong Debye screening, this interaction is very short-range, and we can replace this interaction by one at the same point, $\boldsymbol{r} = \boldsymbol{r}'$. By using second quantization and going into the momentum representation, one can finally reduce this interaction to the standard form (often called the Fröhlich interaction):

$$\mathcal{H}_{\mathrm{e-ph}} = \sum_{\boldsymbol{p},\boldsymbol{q},\sigma} g_{\boldsymbol{q}}\, c^{\dagger}_{\boldsymbol{p}+\boldsymbol{q},\sigma}\, c_{\boldsymbol{p},\sigma}\,(b_{\boldsymbol{q}} + b^{\dagger}_{-\boldsymbol{q}}), \tag{8.58}$$

so that the total electron–phonon Hamiltonian (Fröhlich Hamiltonian) has the form

$$\mathcal{H} = \sum_{\boldsymbol{p},\sigma} \varepsilon(\boldsymbol{p})\, c^{\dagger}_{\boldsymbol{p},\sigma}\, c_{\boldsymbol{p},\sigma} + \sum_{\boldsymbol{q}} \omega_0(\boldsymbol{q})\, b^{\dagger}_{\boldsymbol{q}} b_{\boldsymbol{q}} + \sum_{\boldsymbol{p},\boldsymbol{q},\sigma} g_{\boldsymbol{q}}\, c^{\dagger}_{\boldsymbol{p}+\boldsymbol{q},\sigma}\, c_{\boldsymbol{p},\sigma}\,(b_{\boldsymbol{q}} + b^{\dagger}_{-\boldsymbol{q}}). \tag{8.59}$$

Here, for simplicity, we have left only the interaction of electrons with one phonon mode, e.g. with longitudinal phonons. The interaction with other phonon modes – transverse acoustical phonons, optical phonons – can be also written in the same way as (8.58); all the specifics will be contained in the particular form of the coupling constant, or the corresponding electron–phonon matrix element $g$ (which can in principle depend on both momenta $\boldsymbol{p}$ and $\boldsymbol{q}$).

For the interaction with longitudinal acoustical phonons one can estimate the coupling constant $g$ as

$$g \sim \frac{ea^2 Ze\frac{N}{V}}{s\sqrt{\rho}} , \tag{8.60}$$

where $a$ is the lattice constant, $s$ is the sound velocity and $\rho$ is the density of the metal. One often introduces the dimensionless electron–phonon coupling constant

$$\lambda = \frac{mp_{\mathrm{F}} g^2}{2\pi^2} , \tag{8.61}$$

where $p_{\mathrm{F}}$ is the Fermi momentum, and, according to equation (7.13), the quantity $mp_{\mathrm{F}}/\pi^2 = \rho(\varepsilon_{\mathrm{F}})$ is the electron density of states at the Fermi level (do not confuse this with the density of the metal $\rho$ in (8.60)). The dimensionless electron–phonon coupling constant $\lambda$ thus defined is, in typical cases, $\sim$1. It enters into many important expressions describing different properties of the metal, such as resistivity due to electron–phonon scattering, etc. Probably the most famous is the expression for the critical temperature of conventional superconductors in which superconductivity is due to electron–phonon interactions:

$$T_{\mathrm{c}} = 1.14\, \omega_{\mathrm{D}}\, e^{-1/\lambda} . \tag{8.62}$$

Analogously to the electron Green function one can introduce the phonon Green function

$$D(\boldsymbol{r}_1, t_1; \boldsymbol{r}_2, t_2) = -i \langle 0|\mathrm{T}\{\varphi(\boldsymbol{r}_1, t_1)\, \varphi^\dagger(\boldsymbol{r}_2, t_2)\}|0\rangle , \tag{8.63}$$

where the T-product has the same meaning as before, and $\varphi(\boldsymbol{r}, t)$ are the phonon operators in coordinate space.

In the harmonic approximation, taking for $\varphi$ free operators and making a Fourier transform, we can obtain the Green function of free phonons $D_0(\boldsymbol{q}, \omega)$. Depending on the normalization of phonon operators, there are two different forms of $D_0$ used in the literature:

With the normalization

$$\varphi_q = b_q + b^\dagger_{-q} \tag{8.64}$$

$D_0$ takes the form (see e.g. Schrieffer (1999))

$$D_0(\boldsymbol{q}, \omega) = \left[ \frac{1}{\omega - \omega_0(\boldsymbol{q}) + i\delta} - \frac{1}{\omega + \omega_0(\boldsymbol{q}) - i\delta} \right] = \frac{2\omega_0(\boldsymbol{q})}{\omega^2 - \omega_0^2(\boldsymbol{q}) + i\delta} , \tag{8.65}$$

where $\omega_0(\boldsymbol{q})$ is the bare phonon spectrum.

Another often used normalization is

$$\tilde{\varphi}(\boldsymbol{q}) = \sqrt{\frac{\omega_0(\boldsymbol{q})}{2}}\,(b_{\boldsymbol{q}} + b^{\dagger}_{-\boldsymbol{q}})\,; \tag{8.66}$$

then using the definition (8.63) but with $\varphi \to \tilde{\varphi}$, we obtain the phonon Green function in the form

$$\tilde{D}_0(\boldsymbol{q}, \omega) = \frac{\omega_0^2(\boldsymbol{q})}{\omega^2 - \omega_0^2(\boldsymbol{q}) + i\delta}\,; \tag{8.67}$$

see, e.g. Abrikosov *et al.* (1975).

Accordingly, there will be certain differences in the matrix element of the electron–phonon interaction in the diagram techniques which compensate for this difference and make the physical results independent of the scheme used, as it should be; but one should be aware of these different conventions.

The fact that it is $\omega^2$ which enters into the expression for the phonon Green function, in contrast to $\omega$ in the electron Green function (8.41), (8.54), is actually connected with the fact that phonons are bosons, and in particular that the number of phonons is not conserved, so that phonons can be created and annihilated independently. Consequently, for example, for the electron–phonon coupling the results would be the same if, say, a phonon with momentum $\boldsymbol{q}$ is emitted or if a phonon with momentum $-\boldsymbol{q}$ is absorbed; these processes always enter together, see equation (8.59). Therefore it is always the combination $b_{\boldsymbol{q}} + b^{\dagger}_{-\boldsymbol{q}}$ that enters the expressions (8.64), (8.66), (8.59), and in the Green function (8.65), (8.67) both terms, with $+\omega_0$ and with $-\omega_0$ in the denominator ($\omega(\boldsymbol{q}) = \omega(-\boldsymbol{q})$!), enter on an equal footing.

## 8.7 Diagram techniques

The Green function for free electrons is known, see (8.41). For interacting systems one has to calculate it, usually in a certain approximation.

There exist several methods to do this. One is the method of the equations of motion, similar to the one used in Chapter 6, equations (6.87)–(6.92): one writes down the equations of motion for the Heisenberg operators $\hat{\Psi}(\boldsymbol{r}, t)$ or $\hat{\Psi}(\boldsymbol{p}, \omega)$ entering the definition (8.31), (8.32) using the rule (8.18). Usually one obtains, after commutation with the Hamiltonian, the Green functions of higher order (two-particle Green functions, etc.). To solve this, in principle infinite, set of equations, one has to truncate these equations, making certain *decouplings* dictated by some physical arguments (such as the mean field decoupling used in going from (6.88) to (6.89)). Usually this is an uncontrolled approximation (there is no small parameter here), although it may be physically quite sound. This method is described e.g.

by Zubarev (1960). (Yet another example of such an approach will be given later in the treatment of the Hubbard model, see Section 12.4.)

Another, very widely used method is perturbation theory, treating the interaction as weak. Here we can use the *interaction representation*, see (8.20), (8.21), and use diagram techniques which permit us to represent separate terms of the perturbation expansion by *Feynman diagrams* and which greatly simplifies the calculations.

The idea of the method is the following. The Green functions (8.31), (8.32) are defined using the operators $\hat{\Psi}$, $\hat{\Psi}^{\dagger}$ in the Heisenberg representation, i.e. according to (8.15), $\hat{\Psi}_{\mathrm{H}}(t) = e^{i\mathcal{H}t}\hat{\Psi}(0)e^{-i\mathcal{H}t}$. For free, noninteracting electrons $\mathcal{H} = \mathcal{H}_0$, and we know both $\hat{\Psi}_{\mathrm{H}}(t)$ and the corresponding Green function $G_0$. Now, suppose we have an interacting system, $\mathcal{H} = \mathcal{H}_0 + \mathcal{H}'$, and we want to treat the interaction $\mathcal{H}'$ as a perturbation. Thus we can keep $\mathcal{H}_0$ in the exponent in $\hat{\Psi}_{\mathrm{H}}$ and $G$, and make an expansion in $\mathcal{H}'$, e.g. writing in the lowest order

$$\begin{aligned}\hat{\Psi}_{\mathrm{H}}(t) &= e^{i\mathcal{H}_0 t}e^{i\mathcal{H}'t}\hat{\Psi}(0)e^{-i\mathcal{H}'t}e^{-i\mathcal{H}_0 t}\\ &\Longrightarrow e^{i\mathcal{H}_0 t}(1+i\mathcal{H}'t)\big[e^{-i\mathcal{H}_0 t}e^{i\mathcal{H}_0 t}\big]\hat{\Psi}(0)[e^{-i\mathcal{H}_0 t}e^{i\mathcal{H}_0 t}](1-i\mathcal{H}'t)e^{-i\mathcal{H}_0 t}\\ &\qquad = (1+i\mathcal{H}'_I t)\hat{\Psi}_I(t)(1-i\mathcal{H}'_I t)\,, \end{aligned} \tag{8.68}$$

etc. where we have inserted the unity operator $[e^{-i\mathcal{H}_0 t}e^{i\mathcal{H}_0 t}]$ and used (8.20). (Note that we have to take care of the order of operators $\mathcal{H}_0$, $\mathcal{H}'$, because in general they do not commute.) Keeping $\mathcal{H}_0$ in the exponent means that we can deal with the operators $e^{i\mathcal{H}_0 t}\hat{\Psi}(0)e^{-i\mathcal{H}_0 t}$ and with the corresponding known noninteracting Green functions ($G_0$) and can build our perturbation theory using these as the basis. Different terms of the perturbation theory expansion can then be conveniently depicted as different diagrams, $n$-th order terms containing $n$ interaction lines, or $n$ interaction vertices.

Note also that what we did here, e.g. the transition (8.68) from $\hat{\Psi}_{\mathrm{H}} = e^{i\mathcal{H}t}\hat{\Psi}e^{-i\mathcal{H}t}$ to $e^{i\mathcal{H}_0 t}\hat{\Psi}e^{-i\mathcal{H}_0 t}$ is nothing else but the transition to the interaction representation described above. Indeed, by putting products $e^{-i\mathcal{H}_0 t}e^{i\mathcal{H}_0 t} = 1$ in (8.68) we see that in each term of the perturbation theory we have both $\hat{\Psi}$-operators and the perturbation $\mathcal{H}'$ in the form $e^{i\mathcal{H}_0 t}\hat{\Psi}(0)e^{-i\mathcal{H}_0 t}$ and $e^{i\mathcal{H}_0 t}\mathcal{H}'e^{-i\mathcal{H}_0 t}$, but these are exactly the operators in the interaction representation $\hat{\Psi}_{\mathrm{I}}(t)$, $\mathcal{H}'_{\mathrm{I}}(t)$, cf. (8.20), (8.24).

The possibility to formulate conveniently perturbation theory in the interaction representation was actually the main reason for introducing this representation. The rigorous derivation of the corresponding rules is described in many books, e.g. Abrikosov *et al.* (1975), Fetter and Walecka (2003), Mahan (2000), Mattuck (1992) and Schrieffer (1999); see also (8.23)–(8.30). For us here it is sufficient just to formulate the corresponding rules, so as to be able to use them afterwards. These rules are the following:

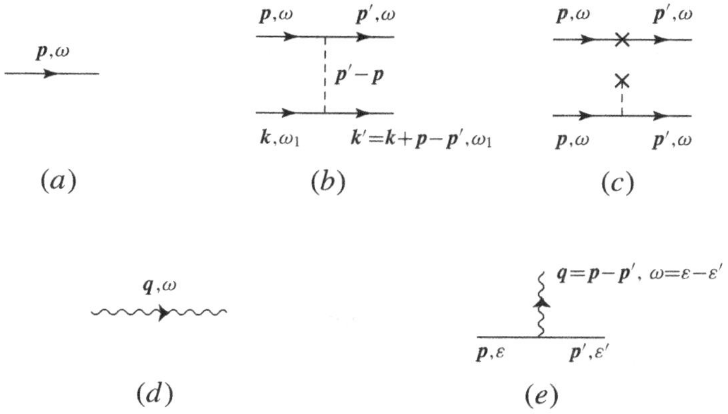


Fig. 8.3

Each term of the perturbation theory expansion for a Green function is represented by a certain diagram. The contribution of each diagram, for an interaction between electrons or an interaction with an external field, is calculated in the following way:

1. Write down all topologically inequivalent diagrams with one external electron line (one going in and one going out). Only connected diagrams are to be included. Electrons are denoted by solid lines (Fig. 8.3(*a*)).
2. Associate with each electron line with momentum $\boldsymbol{p}$ and energy $\omega$ the bare Green function $iG_0(\boldsymbol{p}, \omega)$ (8.41) or (8.44).
3. The Coulomb interaction is denoted by a dashed line; it connects two electron lines, Fig. 8.3(*b*). In each vertex the momentum and energy are conserved. Associate with each interaction line the factor $V(\boldsymbol{p} - \boldsymbol{p}')$; for the Coulomb interaction (8.3) the Fourier transform is

$$V(\boldsymbol{p} - \boldsymbol{p}') = \frac{4\pi e^2}{(\boldsymbol{p} - \boldsymbol{p}')^2} \, . \tag{8.69}$$

4. The interaction with the external potential $\tilde{v}$ (8.2), e.g. that of an impurity, is denoted by a cross $\times$, or a dotted line with a cross (Fig. 8.3(*c*)). Associate with it the factor $\tilde{v}(\boldsymbol{p} - \boldsymbol{p}')$ (the Fourier transform of the potential $\tilde{v}(x)$). Note that at this vertex the electron momentum $\boldsymbol{p}$ is not conserved (an impurity can take extra momentum); for elastic scattering the energy $\omega$ is of course conserved.
5. Similarly, one can also include phonons (denoted by wavy lines, Fig. 8.3(*d*)): one associates with each phonon line with $(\boldsymbol{q}, \omega)$ the phonon Green function $D_0(\boldsymbol{q}, \omega)$ (8.65) or (8.67). In the electron–phonon vertex, Fig. 8.3(*e*), one puts the corresponding matrix element $g_q$.

6. Multiply by the extra factor $(-i)^n$, where $n$ is the number of internal interaction lines, and by $(-1)^l$, where $l$ is the number of closed electron loops in the diagram.
7. Multiply all these factors together and integrate over all intermediate momenta and energies,

$$\int \frac{d^3\boldsymbol{p}_1 d\omega_1}{(2\pi)^4} \frac{d^3\boldsymbol{p}_2 d\omega_2}{(2\pi)^4} \cdots ,$$

and sum over spin directions at the internal lines.

These rules and examples of the simplest diagrams are presented in many books (e.g. Abrikosov (1975), Mahan (2000), Schrieffer (1999), and in many others). Thus, e.g. the lowest-order diagrams for the electron Green function for the electron–electron interaction have the form

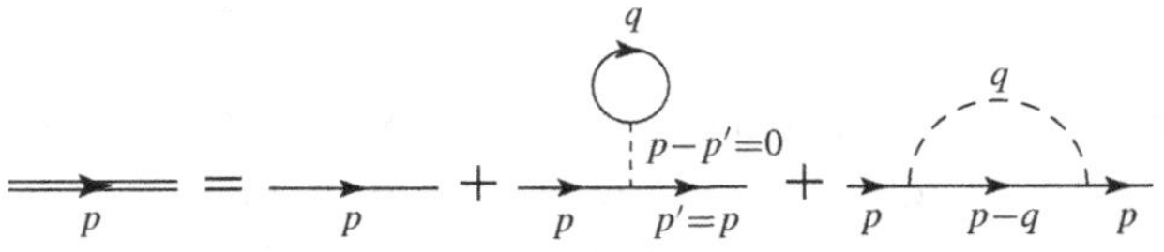

(8.70)

Note that this is actually a *formula*, an expression for the Green function (double line) (the double line, or thick solid line denotes the *full* Green function, in contrast to the thin solid line which represents the bare Green function $G_0$). Thus equation (8.70) may be rewritten as

$$G(p) = G_0(p) + (-2i)G_0(p) \int V(0)\, G_0(q) \frac{d^4q}{(2\pi)^4}\, G_0(p)$$

$$+\, iG_0(p) \int G_0(p-q)\, V(q) \frac{d^4q}{(2\pi)^4}\, G_0(p)\,. \qquad (8.71)$$

In the second term one minus sign comes from the existence of one loop; the factor $(-i)$ comes from one interaction line, and the factor of 2 from summation over spins. We have introduced here the shorthand notation $p = (\boldsymbol{p}, \omega)$, using the four-vector $p$.

The second term (diagram ) contains formally $V(0)$ which for the Coulomb interaction (8.69) is infinite. Actually this term describes the Coulomb interaction of a given electron with the *average* density of all other electrons – this is the Hartree term of the Hartree–Fock approximation. In real systems the condition of electroneutrality should be obeyed, i.e. there should exist an equal positive charge density (ions, or structureless positive background – jellium). It can be incorporated in the external potential $\tilde{U}$ in (8.1), and it will compensate the contribution of this diagram in a homogeneous system. Therefore in the following we will always omit this and similar diagrams. One should be aware, however, that they may be important in inhomogeneous systems (e.g. in an atom, or at a surface, or close to a contact of different materials).

As mentioned above, the second term in (8.70) is the Hartree, or density–density term of the mean field approximation. The last term, , represents an *exchange*, or Fock contribution in the Hartree–Fock approximation.

### 8.7.1 Dyson equations, self-energy and polarization operators

The diagrams presented in equation (8.70) are only the lowest-order contributions to the full Green function. One should in principle consider also higher order contributions, the choice of which is dictated by the concrete situation. Often it is necessary to perform even the summation of an infinite number of diagrams of a certain class.

There exists a convenient way to carry out such a programme. It is called the method of Dyson equations; it uses objects known as the *self-energy* and *polarization* operators. Suppose we start from the lowest-order diagrams

(as explained above we can omit the Hartree diagram). At the next step we can repeat this contribution yet another time, and so on, and we will get a series

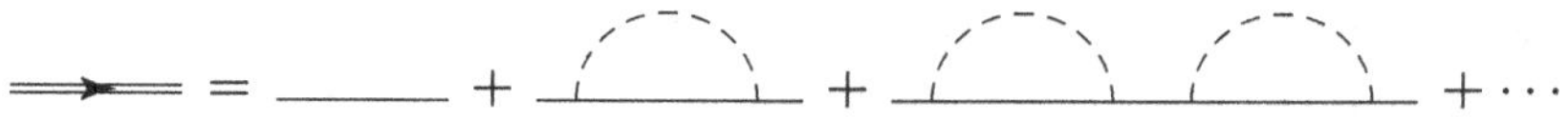

(8.72)

or

$$G = G_0 + G_0\Sigma_0 G_0 + G_0\Sigma_0 G_0\Sigma_0 G_0 + \cdots \ , \tag{8.73}$$

where we have denoted the contribution  by $\Sigma_0(\boldsymbol{p}, \omega)$, so that

$$\Sigma_0(\boldsymbol{p}, \omega) = \int V(\boldsymbol{q}, \omega')\, G_0(\boldsymbol{p} - \boldsymbol{q}, \omega - \omega')\, \frac{d^3\boldsymbol{q}\, d\omega'}{(2\pi)^4} \ . \tag{8.74}$$

One immediately sees that the sum (8.72) is a geometric series which can be easily summed to give

$$G(\boldsymbol{p}, \omega) = \frac{G_0}{1 - \Sigma_0 G_0} = \frac{1}{G_0^{-1} - \Sigma_0} = \frac{1}{\omega - \varepsilon_{\boldsymbol{p}} - \Sigma_0(\boldsymbol{p}, \omega) + i\delta_{\boldsymbol{p}}} \ . \tag{8.75}$$

Another way to perform this summation is to notice that equation (8.72) can be rewritten as

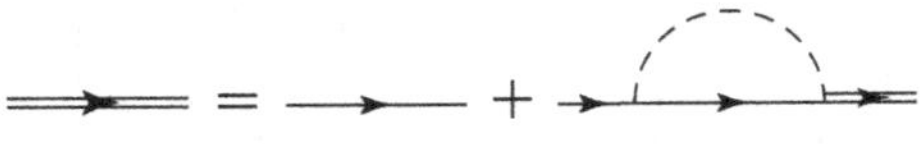

$$G = G_0 + G_0\Sigma_0 G \; ; \tag{8.76}$$

The solution of this equation again gives the result (8.75). Equation (8.76) already has the form of a Dyson equation. However, diagrams of the type (8.72) do not exhaust all the possibilities. We can have, e.g. terms of the type , which can also be repeated, 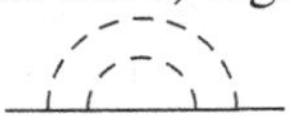 + 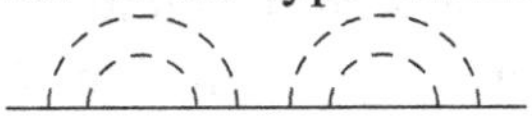 + ⋯; or , which would give the sum  +  + ⋯, or  + 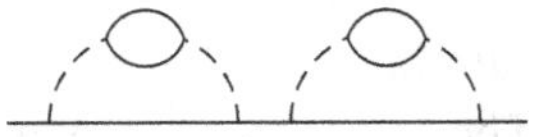 + ⋯, etc. Each of these series can be summed up, giving a result similar to (8.75), but with different Σ's.

One can formally incorporate all such terms by introducing the object $\Sigma(\boldsymbol{p}, \omega)$ which is called the *self-energy*, or *mass operator*, and which is the sum of all diagrams of the type

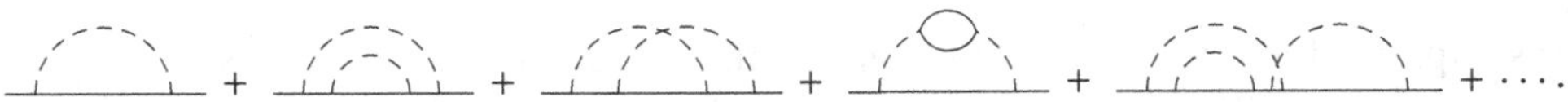

The term $\Sigma_0$ (8.74) is the lowest-order contribution to the total self-energy.

One can express $G$ through this new function Σ as

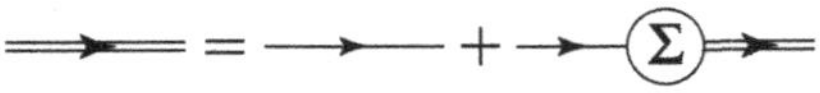

$$G(\boldsymbol{p}, \omega) = G_0(\boldsymbol{p}, \omega) + G_0(\boldsymbol{p}, \omega)\,\Sigma(\boldsymbol{p}, \omega)\,G(\boldsymbol{p}, \omega) \; , \tag{8.77}$$

which can be formally solved, giving

$$G^{-1} = G_0^{-1} - \Sigma \; , \tag{8.78}$$

or

$$\boxed{G(\boldsymbol{p}, \omega) = \frac{1}{\omega - \varepsilon_{\boldsymbol{p}} - \Sigma(\boldsymbol{p}, \omega) + i\delta_{\boldsymbol{p}}}} \; . \tag{8.79}$$

Here the self-energy $\Sigma(\boldsymbol{p}, \omega)$ contains all the diagrams which cannot be cut across one electron line. Thus, e.g. the diagrams  or  should not be included in Σ – they are already taken into account in the summation leading to (8.75) or to its generalization (8.79). By comparing (8.75), (8.79) with (8.41), (8.44) one can immediately understand the origin of the terms 'self-energy' and 'mass operator': the contribution $\Sigma$ is added to the bare energy of the electron, and, if the pole structure of the Green function (8.79) is preserved, it will modify the effective mass of the quasiparticle (see below, (8.85)–(8.91)). The equation (8.77) or (8.78), (8.79) is called the Dyson equation. The Dyson equation (8.77) is not yet a closed equation for the Green function: the self-energy $\Sigma$ should be calculated separately. Despite this, the use of the Dyson equation proves to be very useful: the virtue of this approach is that it permits one actually to sum up an infinite number of terms in perturbation theory even by including only the lowest-order terms in the self-energy $\Sigma$.

This treatment can also be generalized for the case of electron–phonon interactions: one should only substitute everywhere instead of the electron interaction line the phonon line  with which we associate the phonon Green function (8.65) or (8.67), and at each electron–phonon vertex  we put the electron–phonon coupling constant $g_q$.

Analogously to the electron Green function and self-energy, one can introduce similar objects also for the interaction lines and for phonons

 (8.80)

or

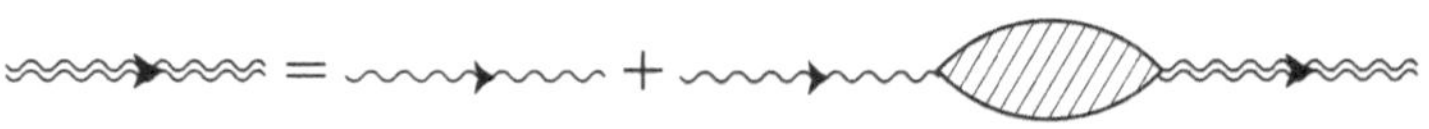 (8.81)

Here  is the screened Coulomb interaction (which now becomes also frequency dependent, or retarded), and  is the full (dressed) phonon Green function. The 'bubble' 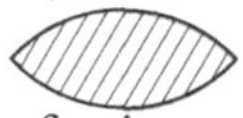 $\Pi(\boldsymbol{q}, \omega)$ is the so-called *polarization operator*. The simplest diagrams for it are

 (8.82)

etc. (and the same with phonons). Again, it is an *irreducible* operator (in the sense that it should not contain terms like 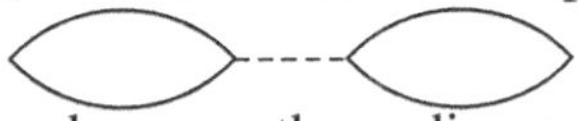 which can be cut through one Coulomb (or phonon) line, because these diagrams are already accounted for when we put the full Green functions  or  in the right-hand side of (8.80), (8.81)).

The diagrams shown in (8.80), (8.81) are again *equations* which can be formally solved similarly to equation (8.79). We do not present the corresponding results here, but they will be discussed in detail in Chapter 9.

### 8.7.2 Effective mass of the electron excitation

From the general treatment given above we can obtain some important relations even without detailed calculation of particular diagrams. Let us start from equation (8.79) expressing the one-electron Green function through the self-energy $\Sigma$. The bare Green function $G_0(\boldsymbol{p}, \omega)$ (8.41) has a pole at the bare spectrum $\varepsilon_p = \boldsymbol{p}^2/2m - \mu$. Suppose that there is a pole (renormalized electron spectrum) $\tilde{\varepsilon}_p$ also in the interacting system, so that

$$G(\boldsymbol{p}, \omega) \sim \frac{1}{\omega - \tilde{\varepsilon}_p} \,. \tag{8.83}$$

From (8.79) the spectrum $\tilde{\varepsilon}_p$ should be a solution of the equation

$$\omega - \varepsilon_p - \operatorname{Re}\Sigma(\boldsymbol{p}, \omega) = 0 \,, \quad \text{or} \quad \tilde{\varepsilon}_p - \varepsilon_p - \operatorname{Re}\Sigma(\boldsymbol{p}, \tilde{\varepsilon}_p) = 0 \,. \tag{8.84}$$

We ignore for a while the imaginary part of $\Sigma$, which in principle will determine the finite lifetime of the excitation (assume that close to the pole $\operatorname{Im}\Sigma < \operatorname{Re}\Sigma$, or that the lifetime is long enough).

Let us expand the Green function close to the pole $\tilde{\varepsilon}_p$:

$$\begin{aligned} G(\boldsymbol{p}, \omega) &= \frac{1}{\omega - \varepsilon_p - \Sigma(\boldsymbol{p}, \omega)} = \frac{1}{\omega - \varepsilon_p - \left[\Sigma(\boldsymbol{p}, \tilde{\varepsilon}_p) + \frac{\partial \Sigma}{\partial \omega}\Big|_{\omega=\tilde{\varepsilon}_p}(\omega - \tilde{\varepsilon}_p)\right]} \\ &= \frac{1}{\omega - \varepsilon_p - \Sigma(\boldsymbol{p}, \tilde{\varepsilon}_p) - \frac{\partial \Sigma}{\partial \omega}(\omega - \tilde{\varepsilon}_p)} \,. \end{aligned} \tag{8.85}$$

As, according to (8.84), $\varepsilon_p + \Sigma(\boldsymbol{p}, \tilde{\varepsilon}_p) = \tilde{\varepsilon}_p$, we have (cf. (8.54))

$$G(\boldsymbol{p}, \omega) = \frac{1}{\omega - \tilde{\varepsilon}_p - \frac{\partial \Sigma}{\partial \omega}(\omega - \tilde{\varepsilon}_p)} = \frac{\frac{1}{1 - \frac{\partial \Sigma}{\partial \omega}\big|_{\omega=\tilde{\varepsilon}_p}}}{\omega - \tilde{\varepsilon}_p} \equiv \frac{Z_p}{\omega - \tilde{\varepsilon}_p} \,. \tag{8.86}$$

Thus we see that because of the interaction the spectrum $\varepsilon$ is renormalized, $\varepsilon_p \longrightarrow \tilde{\varepsilon}_p$, and there appears a factor $Z_p$: the strength of the pole, the residue, is no longer 1 but is

$$Z_p = \frac{1}{1 - \frac{\partial \Sigma(\boldsymbol{p}, \omega)}{\partial \omega}\Big|_{\omega=\tilde{\varepsilon}_p}} \,. \tag{8.87}$$

We have thus managed to express the $Z$-factor, introduced phenomenologically in (8.54), through the self-energy $\Sigma(\boldsymbol{p}, \omega)$, which gives in principle a way to calculate it.

One can also obtain a useful expression for the change of the effective mass of the electron due to interaction. From (8.84), assuming that

$$\tilde{\varepsilon}_p = \frac{p^2}{2m^*}\,, \tag{8.88}$$

with the bare spectrum being $\varepsilon_p = p^2/2m$ (actually this is the definition of the effective mass $m^*$), we obtain:

$$\begin{aligned} \frac{1}{2m^*} = \frac{\partial \tilde{\varepsilon}_p}{\partial (p^2)} &= \frac{\partial \varepsilon_p}{\partial (p^2)} + \left( \frac{\partial \Sigma}{\partial (p^2)} + \frac{\partial \Sigma}{\partial \tilde{\varepsilon}_p} \frac{\partial \tilde{\varepsilon}_p}{\partial (p^2)} \right) \\ &= \frac{1}{2m} + \frac{\partial \Sigma}{2m\,\partial\left(\frac{p^2}{2m}\right)} + \left.\frac{\partial \Sigma}{\partial \omega}\right|_{\omega=\tilde{\varepsilon}_p} \frac{\partial \tilde{\varepsilon}_p}{\partial (p^2)} \end{aligned} \tag{8.89}$$

or, collecting different terms, we obtain

$$\frac{1}{m^*}\left(1 - \left.\frac{\partial \Sigma}{\partial \omega}\right|_{\tilde{\varepsilon}_p}\right) = \frac{1}{m}\left(1 + \frac{\partial \Sigma}{\partial \varepsilon_p}\right)\,, \tag{8.90}$$

$$\boxed{\frac{m^*}{m} = \frac{1 - \left.\frac{\partial \Sigma}{\partial \omega}\right|_{\tilde{\varepsilon}_p}}{1 + \frac{\partial \Sigma}{\partial \varepsilon_p}} = \frac{1}{Z_p}\frac{1}{1 + \frac{\partial \Sigma(p,\omega)}{\partial \varepsilon_p}}}\,. \tag{8.91}$$

This is a very important formula which connects the effective mass renormalization $m^*/m$ with the pole strength $Z_p$. One can show that if the interaction leading to mass renormalization is retarded (for instance, is carried out by low-energy excitations, e.g. is an electron–phonon interaction with $\omega_{\mathrm{ph}} \ll \varepsilon_{\mathrm{F}}$), then one can neglect the momentum dependence of $\Sigma(p, \omega)$,[1] and we have $m^*/m = 1/Z_p$, which is necessarily $\geq 1$ (as $Z_p \leq 1$). In particular, for the electron–phonon interaction we get (see below)

$$m^* = m(1 + \lambda)\,, \tag{8.92}$$

where $\lambda$ is the dimensionless electron–phonon coupling constant (8.61), entering also, e.g. into the expression (8.62) for the superconducting $T_c$ in ordinary superconductors, in which superconductivity is due to electron–phonon interactions.

[1] This is connected with the Migdal theorem which states that in typical metals, due to the existence of a small parameter $\omega_{\mathrm{D}}/\varepsilon_{\mathrm{F}} \ll 1$, one can keep in the electron–phonon self-energy only the simplest diagrams of the type 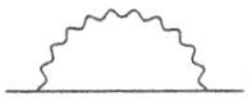 and ignore the so-called vertex corrections, e.g. 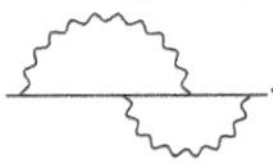. Similarly, we can keep only the simplest bubble diagrams (but in general with the full electron Green functions) in the polarization operator, i.e. we can ignore the diagrams of the type  which are small if $\omega_{\mathrm{D}}/\varepsilon_{\mathrm{F}} \ll 1$.

Even more interesting is the situation in the so-called *heavy fermion* systems, see below, Chapter 13. In this case experimentally $m^*/m$ (and the coefficient $\gamma$ in the linear specific heat $c = \gamma T$ which is proportional to $m^*$, cf. (7.25)), is extremely large, $\sim 10^2$–$10^3$. One can also describe the renormalization of the effective mass here as occurring due to interaction with very soft spin fluctuations, and the value $m^*/m$ is also directly related to the pole strength $Z \ll 1$.

If, on the other hand, both $\omega$ and the $\boldsymbol{p}$-dependence of $\Sigma(\boldsymbol{p}, \omega)$ are important (e.g. for the Hubbard on-site Coulomb interaction), one cannot draw any general conclusion about the mass renormalization $m^*/m$.

# 9

# Electrons with Coulomb interaction

Using the techniques described in the previous chapter, we can in a unified way discuss properties of the electron gas with Coulomb interaction and consider such effects as optical response, screening, plasmons, etc. In many textbooks these properties are obtained using a variety of methods. The virtue of the Green function method is its universality and, I would say, not much simpler, but standardized form. This method permits one to obtain all the properties mentioned above in the form of one general expression, and it also gives the possibility to generalize the results quite easily to the cases of low-dimensional (1d, 2d) systems, or to take into account the details of the band structure of the material, etc. But more important is the fact that it leads naturally to a number of special interesting consequences which would be rather difficult to obtain with the usual classical methods. In this and in the next two chapters I will demonstrate how to reproduce, using this method, the familiar results such as Debye screening or the plasmon energy, but I will mostly concentrate on less frequently discussed effects which are quite naturally obtained using this technique.

## 9.1 Dielectric function, screening: random phase approximation

We start by studying the form of the effective electron–electron interaction in metals. The ordinary Coulomb interaction $V(\boldsymbol{q}) = 4\pi e^2/\boldsymbol{q}^2$ is modified by the reaction of the electronic system. The first, well-known effect is just the screening of the electric charge. But there are other interesting effects as well.

One can describe the modification of the Coulomb interaction using the corresponding Dyson equation (8.80)

(9.1)

or

$$v(\boldsymbol{q}, \omega) = V(\boldsymbol{q}) + V(\boldsymbol{q})\,\Pi(\boldsymbol{q}, \omega)\,v(\boldsymbol{q}, \omega)\,. \tag{9.2}$$

Here $\Pi(\boldsymbol{q}, \omega)$ is the irreducible polarization operator (8.82), and $\mathcal{V}(\boldsymbol{q}, \omega)$ is the renormalized interaction which now becomes frequency dependent (this corresponds to *retardation* effects: the reaction of electrons has a certain characteristic time-scale, which leads to the retardation of the effective interaction, in contrast to the initial Coulomb interaction $V(\boldsymbol{q})$ which is, in nonrelativistic theory, instantaneous or frequency independent). Note that sometimes the polarization operator is defined with the opposite sign, e.g. our definition of $\Pi(\boldsymbol{q}, \omega)$ differs by a sign from that in the book by Schrieffer (1999): $\Pi(\boldsymbol{q}, \omega) = -P_{\text{Schrieffer}}(\boldsymbol{q}, \omega)$. Solving equation (9.1) formally we obtain:

$$\mathcal{V}(\boldsymbol{q}, \omega) = \frac{V(\boldsymbol{q})}{1 - V(\boldsymbol{q})\,\Pi(\boldsymbol{q}, \omega)} \equiv \frac{V(\boldsymbol{q})}{\epsilon(\boldsymbol{q}, \omega)} \,. \tag{9.3}$$

Here we have introduced the dielectric function

$$\epsilon(\boldsymbol{q}, \omega) = 1 - V(\boldsymbol{q})\,\Pi(\boldsymbol{q}, \omega) \,. \tag{9.4}$$

Taking the lowest approximation for the polarization operator

$\Pi_0(\boldsymbol{q}, \omega) =$ 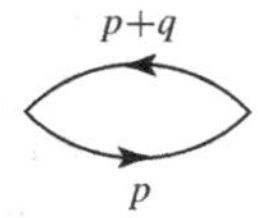 $= -2i \int G_0(p+q)\, G_0(p) \frac{d^4p}{(2\pi)^4} \,,$ (9.5)

we obtain what is known as the random phase approximation (RPA) for the dielectric function and for the effective interaction. The Dyson equation (9.1) in this approximation corresponds to the summation of an infinite set of diagrams of the type

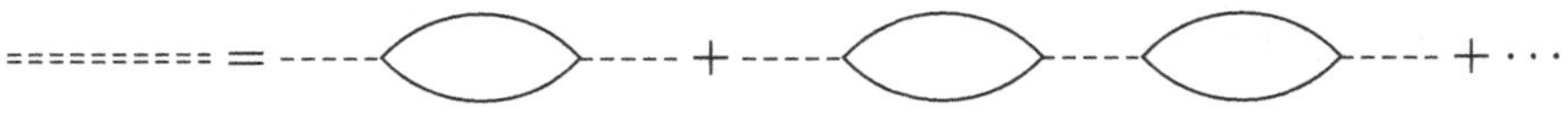

(9.6)

i.e. we dress the Coulomb line -------- by electron–hole 'bubbles' .
One can show that this is a good approximation for high-density electron systems, $r_s \ll 1$ (where $r_s$ is the dimensionless parameter characterizing the electron density, see Chapter 7). The justification of this approximation is connected with the fact that the Coulomb interaction $V(\boldsymbol{q})$ is large for $q \to 0$, so that in the perturbation theory (9.6) higher-order terms contain extra factors $e^2$ but also $(1/\boldsymbol{q}^2)^2$, and we have to sum all terms of this kind which gives (9.3) with $\Pi = \Pi_0$ given by (9.5), i.e. the RPA result.

The RPA form for the dielectric response function, which corresponds to keeping only the sum of bubble diagrams, may be presented in two equivalent forms:

One can say that the total interaction is

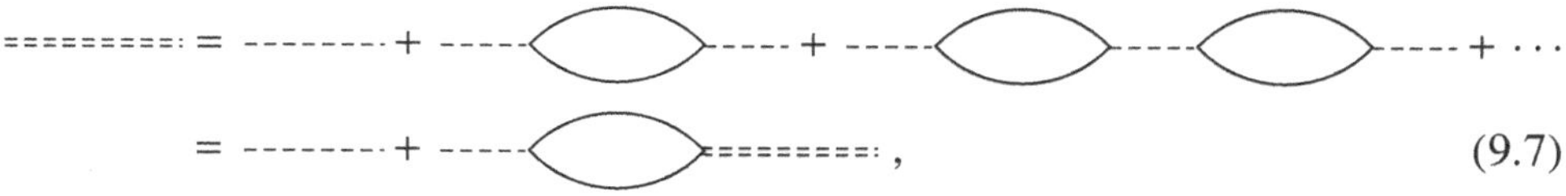

(9.7)

i.e.

$$v = V + V\Pi_0 v \tag{9.8}$$

from which we get the result (9.3) with $\Pi \to \Pi_0$. Or we can write this also as

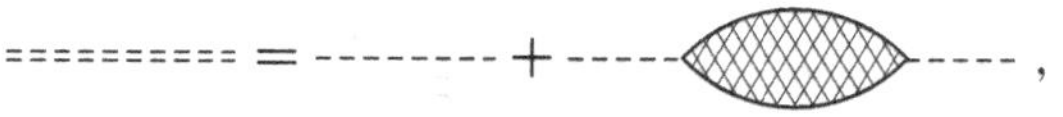

(9.9)

i.e.

$$v = V + V\tilde{\Pi}V\ , \tag{9.10}$$

where

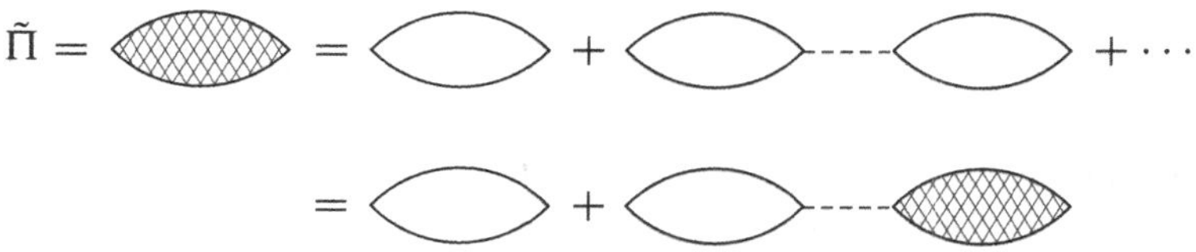

(9.11)

or

$$\tilde{\Pi} = \Pi_0 + \Pi_0 V\tilde{\Pi}\ . \tag{9.12}$$

(Do not confuse $\tilde{\Pi} =$ 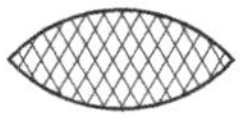 (9.11) with the irreducible polarization operator $\Pi =$  defined by equation (8.82)! As is seen from (9.11), in $\tilde{\Pi}$ we include diagrams which can be cut across one electron line, which are excluded in the irreducible polarization operator $\Pi$ (8.82). On the other hand, in $\tilde{\Pi}$ we included only simple electron–hole bubbles, ignoring vertex corrections included in $\Pi$ (8.82).)

From (9.11) we find

$$\tilde{\Pi} = \frac{\Pi_0}{1 - \Pi_0 V}\ , \tag{9.13}$$

and then from (9.9) we obtain the same expression (9.3) for $v$:

$$v = V(1 + \tilde{\Pi}V) = V\left(1 + \frac{\Pi_0 V}{1 - \Pi_0 V}\right) = \frac{V}{1 - \Pi_0 V} = \frac{V}{\epsilon}\ . \tag{9.14}$$

The expression (9.13) describes the polarizability of the system (still in the RPA).

One can obtain a similar expression also for the magnetic susceptibility. We have to consider the response of our system to a magnetic field, causing, e.g. a reversal of spin directions. Then the response function is given by electron–hole diagrams with opposite spins

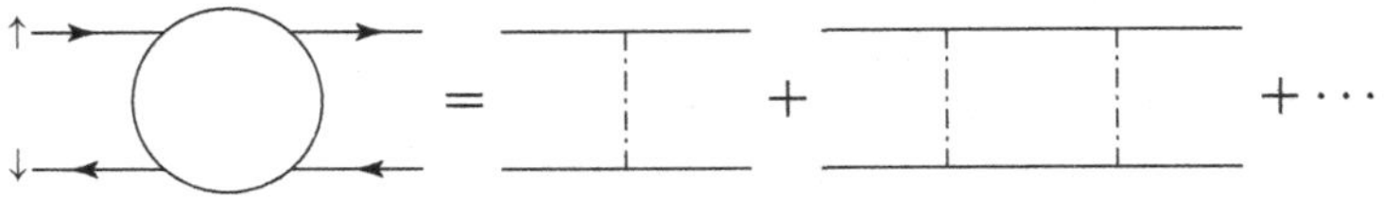
(9.15)

where -·-·-·-·- denotes the effective electron–electron interaction (not necessarily long-range Coulomb interaction). Thus, for a local interaction $Un_{i\uparrow}n_{i\downarrow}$ (the Hubbard interaction) one can easily sum these diagrams and obtain

$$\chi(\boldsymbol{q},\omega) = \frac{\chi_0(\boldsymbol{q},\omega)}{1 + U\Pi_0(\boldsymbol{q},\omega)}\,, \tag{9.16}$$

where $\chi_0(\boldsymbol{q},\omega)$ is proportional to the same expression (9.5) (the block repeated in (9.15) is in fact the same product of electron and hole Green functions as in (9.5)),

$$\chi_0(\boldsymbol{q},\omega) = \tfrac{1}{2}g^2\mu_{\mathrm{B}}^2\Pi_0(\boldsymbol{q},\omega)\,. \tag{9.17}$$

Note the change of sign in the denominator of (9.16) as compared to (9.13): it comes from the fact that for density–density correlations described by $\Pi_0$ or $\epsilon$ the closed loops in the diagrams give an extra factor $-1$, according to the general rules of Chapter 8. Here, for the magnetic response, we have rather the so-called ladder diagrams (9.15). Closed loops here are forbidden because of spin conservation: electron and hole lines in (9.15) correspond to opposite spins, and these cannot simply recombine. Nevertheless for a local (or $q$-independent) interaction the basic expressions are similar. The expressions (9.16), (9.17) will be very useful later on in the discussion of magnetic instabilities of electron systems.

The calculation of the polarization operator $\Pi_0$ (9.5) is a straightforward but tedious task. The first step is relatively easy. We write $\Pi_0$ (9.5) as

$$\Pi_0(\boldsymbol{q},\omega) = -\frac{2i}{(2\pi)^4}\int\frac{d^3\boldsymbol{p}\,d\omega'}{\left[\omega+\omega'-\varepsilon(\boldsymbol{p}+\boldsymbol{q})+i\delta\,\mathrm{sign}\big(\varepsilon(\boldsymbol{p}+\boldsymbol{q})-\mu\big)\right]\left[\omega'-\varepsilon(\boldsymbol{p})+i\delta\,\mathrm{sign}\big(\varepsilon(\boldsymbol{p})-\mu\big)\right]} \tag{9.18}$$

and integrate over $\omega'$ using contour integration, closing the contour in the upper half-plane. The result has the form

$$\Pi_0(\boldsymbol{q},\omega) = -\frac{2}{(2\pi)^3}\int d^3\boldsymbol{p}\,\frac{n(\boldsymbol{p}) - n(\boldsymbol{p}+\boldsymbol{q})}{\omega + \varepsilon(\boldsymbol{p}+\boldsymbol{q}) - \varepsilon(\boldsymbol{p}) + i\delta}\,, \tag{9.19}$$

where $n(\boldsymbol{p})$ and $n(\boldsymbol{p}+\boldsymbol{q})$ are the usual Fermi functions, $n(\boldsymbol{p}) = 1$ for $|\boldsymbol{p}| < p_{\mathrm{F}}$ and $= 0$ for $|\boldsymbol{p}| > p_{\mathrm{F}}$. Thus the integration region is the hatched area in Fig. 9.1,

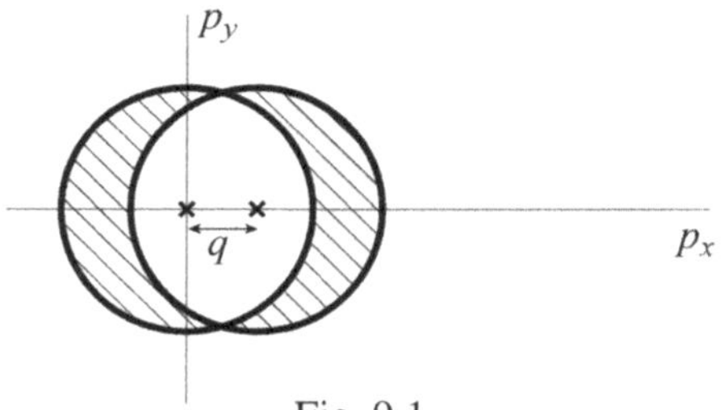


Fig. 9.1

i.e. it lies inside the sphere $|\boldsymbol{p}| = p_{\rm F}$, but outside $|\boldsymbol{p}+\boldsymbol{q}| = p_{\rm F}$, or vice versa. The origin of the factor $n(\boldsymbol{p}) - n(\boldsymbol{p}+\boldsymbol{q})$ in the numerator of equation (9.19) mathematically follows from the position of the poles in the Green functions entering (9.18) (here the imaginary parts in the denominators of (9.18) play a role!). Physically this is easy to understand: the polarization operator (9.5) is given by the electron–hole bubble, and if one particle, e.g. with momentum $\boldsymbol{p}$, is a hole, i.e. $|\boldsymbol{p}| < p_{\rm F}$, then the second one, with momentum $\boldsymbol{p}+\boldsymbol{q}$, should be an electron, i.e. $|\boldsymbol{p}+\boldsymbol{q}| > p_{\rm F}$. The factor $n(\boldsymbol{p}) - n(\boldsymbol{p}+\boldsymbol{q})$ reflects just this fact.

Further integration is straightforward but rather elaborate. The result is known as the Lindhardt function. I will present here only the important limiting cases; for the full expression see, e.g. Schrieffer (1999).

The static dielectric function $\epsilon(\boldsymbol{q}, 0)$ is given by the expression

$$\epsilon(\boldsymbol{q},0) = 1 - \Pi_0 V = 1 + \frac{4me^2 p_{\rm F}}{\pi q^2}\, u\left(\frac{q}{2p_{\rm F}}\right) = 1 + \left(\frac{4}{9\pi^4}\right)^{1/3} \frac{r_s}{x^2}\, u(x)$$
$$= 1 + 0.66\, r_s \left(\frac{p_{\rm F}}{q}\right)^2 u\left(\frac{q}{2p_{\rm F}}\right), \qquad (9.20)$$

where $x = q/2p_{\rm F}$ and

$$u(x) = \frac{1}{2}\left[1 + \frac{(1-x^2)}{2x} \ln\left|\frac{1+x}{1-x}\right|\right]. \qquad (9.21)$$

From this expression we can easily obtain the Debye screening length $r_{\rm D} = \kappa_{\rm D}^{-1}$:

$$\epsilon(\boldsymbol{q},0)\Big|_{\boldsymbol{q}\to 0} = 1 + \frac{\kappa_{\rm D}^2}{q^2}, \qquad \kappa_{\rm D}^2 = 4\pi e^2 \rho(\varepsilon_{\rm F}) = \frac{6\pi n e^2}{\varepsilon_{\rm F}}, \qquad (9.22)$$

(see (7.13)), so that the screened Coulomb interaction (9.3), (9.14) would have the usual form

$$v(r) = \frac{e^2}{r} e^{-\kappa_{\rm D} r}. \qquad (9.23)$$

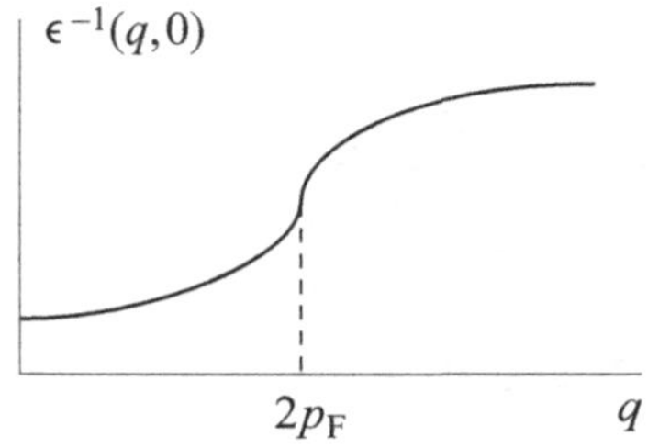


Fig. 9.2

The screening length $\kappa_D^{-1}$ coincides with that obtained in the usual semiclassical approximation (Thomas–Fermi screening). (Note also the convenient result:

$$\Pi_0(\boldsymbol{q},0)\Big|_{\boldsymbol{q}\to 0} = -\rho(\varepsilon_F)\,, \tag{9.24}$$

where $\rho(\varepsilon_F)$ is the density of states at the Fermi level (7.13).)

However, this is not really the full story. If we look at the detailed behaviour of $\epsilon(\boldsymbol{q},0)$ for arbitrary $\boldsymbol{q}$, we see from equations (9.20), (9.21) that there is a *singularity* in $\epsilon(\boldsymbol{q},0)$ at $q \to 2p_F$: $\partial\epsilon^{-1}/\partial q \to \infty$ at this point, see Fig. 9.2. Mathematically this singularity is connected with the fact that starting from $q = 2p_F$ the region of integration in (9.19) no longer changes, see Fig. 9.1. Correspondingly, when we study how the screened potential $\mathcal{V}(\boldsymbol{q},0)$ (9.3) behaves in real space, $\mathcal{V}(\boldsymbol{r}-\boldsymbol{r}')$, we have to make a Fourier transform of $\mathcal{V}(\boldsymbol{q}) = 4\pi e^2/q^2\epsilon(\boldsymbol{q},0)$ with $\epsilon(\boldsymbol{q},0)$ given by (9.20), (9.21), and the presence of the singularity of $\epsilon(\boldsymbol{q},0)$ at $q = 2p_F$ modifies the behaviour of the screened Coulomb potential.

Usually the asymptotic behaviour of Fourier transforms (the behaviour of integrals of the type $\tilde{f}(x) = \int_a^b e^{iqx} f(q)\,dq$ at large $x$) is determined by the behaviour of $f(q)$ at the limits of integration $(a,b)$ (here at $q \to 0$), but also by the *special points* (singularities in $f(q)$ and its derivatives) *inside* the interval $(a,b)$, if they exist. Thus, e.g. if $f(q)$ itself $\to \infty$ at $q = q_0$, for example, if $f(q)$ contains a $\delta$-function, $c\delta(q-q_0)$, then $\tilde{f}(x)$ would evidently have a contribution $ce^{iq_0x}$, or it will be an oscillating function not decaying at large $x$ at all. A similar situation exists in our case: because of the singularity of $\partial\epsilon^{-1}/\partial q$ at $q = 2p_F$ the screened Coulomb interaction $\mathcal{V}(r)$ is not simply

$$\mathcal{V}(r) = \frac{e^2}{r}\,e^{-\kappa_D r} \tag{9.25}$$

as one would expect classically and as is usually taken, but there exists also a *long-range oscillating term*

$$\mathcal{V}(r)\Big|_{r\to\infty} \sim \frac{\cos(2p_F r + \varphi)}{r^3}\,, \tag{9.26}$$

where $\varphi$ is a certain phase.

Correspondingly, the screening charge around a charged impurity in a metal would also follow the law (9.26) at large distances. This is known as *Friedel oscillations*, and they occur due to the sharp cut-off of the Fermi distribution function $n(\boldsymbol{p})$ at $p = p_{\mathrm{F}}$. These oscillations are a real effect; they were observed experimentally.

As at finite temperatures the sharp Fermi surface is somewhat washed out, these oscillations will start to decay with increasing temperature. Impurities in the metal also lead to a similar effect: in the presence of impurities the Friedel oscillations start to decay as $e^{-r/l}$, where $l$ is the mean free path.

This phenomenon is even more important not for charge but for magnetic interaction in metals. It turns out that in the first approximation (RPA) practically the same formulae describe the screening of a localized magnetic moment inserted into the metal (i.e. the magnetic polarization of conduction electrons by a magnetic impurity). The magnetic susceptibility $\chi(\boldsymbol{q}, \omega)$ is also given in this approximation by the summation of the same 'bubble'-like (in fact ladder) series (9.15) (see, however, the discussion of the Kondo effect below, Chapter 13). The result has the form (in the static case $\omega = 0$):

$$\chi(\boldsymbol{q}, \omega = 0) = \frac{3g^2\mu_{\mathrm{B}}^2 n}{8\varepsilon_{\mathrm{F}}}\, u\left(\frac{q}{2p_{\mathrm{F}}}\right) \tag{9.27}$$

where $g$ is the $g$-factor of the electron (usually $g = 2$), $\mu_{\mathrm{B}} = e\hbar/2mc$ is the Bohr magneton, and $u(x)$ is the same function (9.21) as in the expression for the dielectric function. Correspondingly, the spin density $\boldsymbol{s}(\boldsymbol{r})$ at distance $\boldsymbol{r}$ from the impurity spin $\boldsymbol{S}_a$,

$$\boldsymbol{s}(\boldsymbol{r}) = \frac{J}{g^2\mu_{\mathrm{B}}^2 V}\sum_{\boldsymbol{q}} \chi(\boldsymbol{q}) e^{i\boldsymbol{q}\cdot\boldsymbol{r}} \boldsymbol{S}_a\,, \tag{9.28}$$

will behave as (9.26) (here $J$ is the contact exchange interaction between an impurity and a conduction electron, $-J\boldsymbol{S}_a \cdot \boldsymbol{s}$). If we now put another spin $\boldsymbol{S}_b$ (another magnetic impurity) at some point, it will have a similar local exchange interaction with the conduction electrons polarized by the first impurity $\boldsymbol{S}_a$. As a result there will be an effective exchange interaction between $\boldsymbol{S}_a$ and $\boldsymbol{S}_b$ mediated by the (oscillating) spin polarization of conduction electrons. This interaction is called the Ruderman–Kittel–Kasuya–Yosida or simply RKKY interaction, and it has the form

$$\mathcal{H}_{\mathrm{RKKY}} \sim \frac{J^2}{g^2\mu_{\mathrm{B}}^2 V}\sum_{\boldsymbol{q}} \chi(\boldsymbol{q})\, e^{i\boldsymbol{q}\cdot\boldsymbol{r}}\, \boldsymbol{S}_a \cdot \boldsymbol{S}_b \sim \frac{J^2}{\varepsilon_{\mathrm{F}}}\frac{\cos(2p_{\mathrm{F}}r + \varphi)}{r^3}\,. \tag{9.29}$$

This oscillating long-range interaction has a lot of consequences. It gives, e.g. a broadening of NMR lines. A special form of interaction (9.29) with alternating

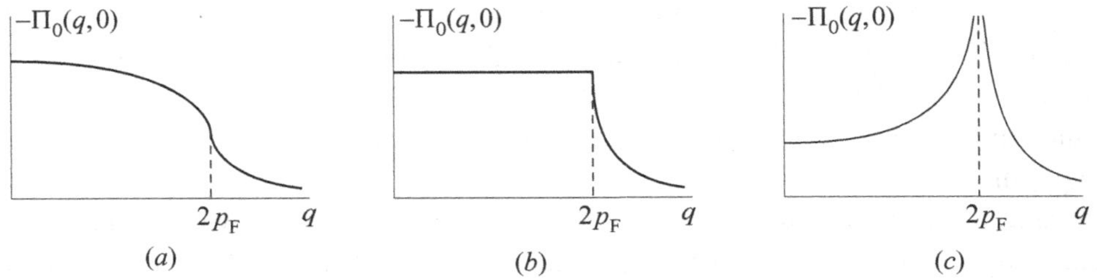


Fig. 9.3

positive (antiferromagnetic) and negative (ferromagnetic) parts gives rise to often complicated magnetic structures in many metallic magnets, in particular to spiral or helicoidal structures in rare earth metals. It is also in most cases responsible for the formation of the so-called *spin-glass phase*: when we put random magnetic impurities in a metal, due to the oscillating form of the RKKY interaction (9.29) the effective exchange interaction, or the effective (exchange) field produced at the position of a given impurity by all the others, has random orientation. This leads to a specific state: at low temperatures all spins are *frozen*, but in random orientations, so that there is neither total magnetization (no ferromagnetic moment), nor any other spatial order. However, the susceptibility in this case usually has an anomaly at the *spin freezing* temperature, which is different for zero-field and field-cooled measurements. This anomaly also depends on the measuring frequency. These are the 'fingerprints' which allow one to identify the spin glass state.

## 9.2 Nesting and giant Kohn anomalies

The same feature – a singularity in the response functions – becomes even more pronounced in low-dimensional systems or for special shapes of Fermi surfaces. Both $\epsilon(\boldsymbol{q},\omega)$ and $\chi(\boldsymbol{q},\omega)$ are called *response functions*, because they describe the response of our system to external perturbations – external electric or magnetic fields, charged or magnetic impurities, etc. (By the way, one can easily understand why we have summation over $\boldsymbol{q}$ in the RKKY interaction: the response to an external field, e.g. to the electric field $\boldsymbol{E_q}$ in optics, is described by the corresponding Fourier component $\epsilon(\boldsymbol{q},0)$; however when we are dealing with a perturbation *localized* in space, as in the case of impurities, we have to integrate over all harmonics.)

Whereas in the usual 3d systems the polarization operator $\Pi_0(\boldsymbol{q},0)$ has a logarithmic singularity in the derivative $\partial\Pi_0/\partial q|_{q=2p_F}$, see equations (9.20), (9.21) and Fig. 9.3(*a*), the singularities in 2d and in 1d cases have the form shown in Figs. 9.3(*b*), (*c*) (note that with our convention (9.3), $\Pi_0(\boldsymbol{q},0)$ is negative). Thus, for example, in the 1d case the *polarization operator itself* has a logarithmic divergence,

$$\Pi_0(q, 0) \sim \ln|q - 2p_\mathrm{F}| \,. \tag{9.30}$$

This divergence has important consequences, in particular it leads to instabilities of one-dimensional metals with respect to the formation of superstructures with wavevector $Q = 2p_\mathrm{F}$ or period $l = 2\pi a/Q$. There will be a special chapter devoted to the detailed discussion of this and other instabilities (Chapter 11); here, anticipating this material, we will only point out that, e.g. the renormalization of the phonon spectrum is determined in the first approximation by the same diagrams entering the expression for $\epsilon(\boldsymbol{q}, \omega)$,

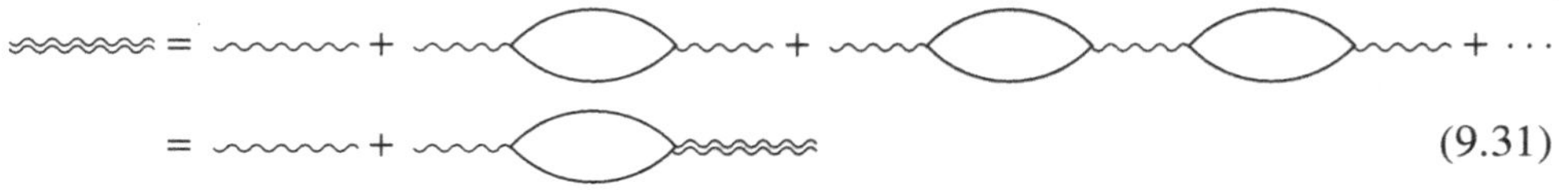

(9.31)

(cf. (9.7)), i.e.

$$D(\boldsymbol{q}, \omega) = D_0 + D_0\, g^2 \Pi_0\, D \,, \tag{9.32}$$

where $g$ is the electron–phonon coupling constant (vertex ). This gives

$$D(\boldsymbol{q}, \omega) = \frac{1}{D_0^{-1}(\boldsymbol{q}, \omega) - g^2 \Pi_0(\boldsymbol{q}, \omega)} = \frac{\omega_0^2(\boldsymbol{q})}{\omega^2 - \omega_0^2(\boldsymbol{q}) - g^2 \omega_0^2(\boldsymbol{q})\, \Pi_0(\boldsymbol{q}, \omega)} \tag{9.33}$$

(with the normalization (8.67)). Thus, the new renormalized phonon spectrum (pole of the Green function (9.33)) is given by the equation

$$\omega^2(\boldsymbol{q}) = \omega_0^2(\boldsymbol{q})\Big[1 + g^2 \Pi_0(\boldsymbol{q}, \omega)\Big] \,, \tag{9.34}$$

so that if $\Pi_0(\boldsymbol{q}, \omega) \to -\infty$ (as it does at $q = 2p_\mathrm{F}$ in the 1d case), the phonon frequency becomes imaginary ($\omega^2 < 0$) which means absolute instability of the system (the phonon mode with $q = 2p_\mathrm{F}$ would 'accelerate', leading to macroscopic occupation of the corresponding mode, i.e. to a real deformation).[1] Such instability in one-dimensional systems is called *Peierls instability* (see also Section 11.1 below). The corresponding Peierls distortion is often observed in real one-dimensional materials, e.g. in some long organic molecules or polymers,

[1] By the way, we see here why in discussing possible instabilities of an anharmonic lattice and melting in Section 4.4.2, I said without proof that the actual equation for phonon softening contains not the phonon frequency $\omega$, but $\omega^2$: the phonon Green function $D(\boldsymbol{q}, \omega)$ and the renormalized spectrum always contain $\omega^2$. Equation (9.34) is written for the phonon renormalization due to electron–phonon coupling, but a similar expression can also be obtained for the case of the anharmonic phonon–phonon interaction of Chapter 4.

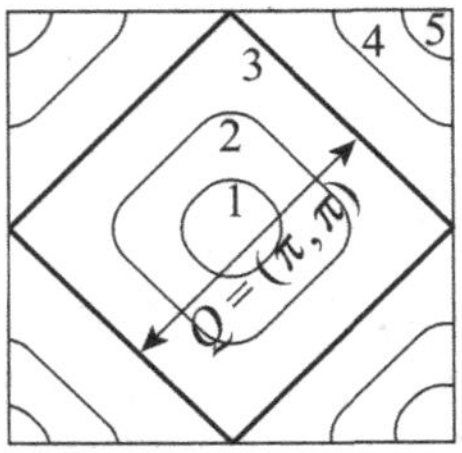


Fig. 9.4

the best-known example being polyacetylene $C_nH_n$. Note also that the deformation with the wavevector $Q = 2p_F$ gives new periodicity with a new Brillouin zone and, according to the usual picture of nearly free electrons in a periodic potential, it produces a gap exactly at the position of our initial Fermi surface $(-p_F, +p_F)$, so that the quasi-1d material after such transition acquires a gap and becomes an insulator.

The anomaly of the phonon spectrum (9.34) in the 1d case is really extreme; but even if the anomaly is not so strong (in 2d or 3d cases, for instance), still the singularity of $\Pi_0$ at $q = 2p_F$ is reflected in the phonon spectrum $\omega(q)$ at the same wavevector. This anomaly in the phonon spectrum is called a *Kohn anomaly.* If such an anomaly is strongly enhanced, as happens in low-dimensional materials, one speaks of a *giant Kohn anomaly.* As is clear from the preceding discussion, this Kohn anomaly is closely related to Friedel oscillations and to the oscillating long-range character of forces in metals, see (9.26), (9.29).

Up to now when considering 1d, 2d or 3d systems we always had in mind the situation with isotropic electron spectrum, e.g. $\varepsilon(\boldsymbol{p}) = p^2/2m$. However, in real crystalline materials the spectrum may be anisotropic, and the corresponding Fermi surface need not be a sphere (or a circle in the 2d case); it can and often does have a very complicated shape. In particular there may be cases when the whole or parts of the Fermi surface are flat. Such is, for example, the Fermi surface of the two-dimensional square lattice in the tight-binding approximation for a half-filled band (one electron per site). In this case the spectrum is (we put here the lattice constant $a = 1$)

$$\varepsilon(\boldsymbol{p}) = -2t(\cos p_x + \cos p_y) , \tag{9.35}$$

and the Fermi surfaces for different band fillings have the form shown in Fig. 9.4; the numbers 1, ..., 5 correspond to increasing electron concentration, such that the case of the half-filled band (one electron per site, $n = 1$) is represented by the *square* Fermi surface 3. One can check that in this case the polarization operator $\Pi_0(\boldsymbol{q}, 0)$ will behave for $\boldsymbol{q} \to (\pi, \pi)$ exactly as in the one-dimensional case:

$$\Pi_0(\boldsymbol{q}, 0) \sim \ln|\boldsymbol{q} - \boldsymbol{Q}| , \qquad \boldsymbol{Q} = (\pi, \pi) . \tag{9.36}$$

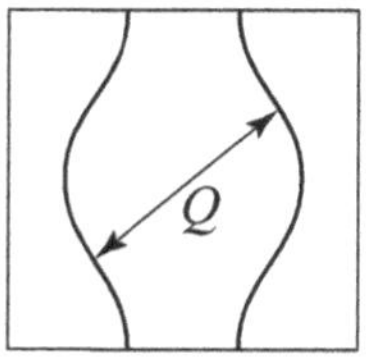


Fig. 9.5

(This is connected with the fact that for this Fermi surface the region of integration in (9.19) (cf. Fig. 9.1) will change very rapidly as $\boldsymbol{q} \to (\pi, \pi)$, as in the 1d case.) Consequently, many properties of such systems will resemble those of one-dimensional systems, in particular they will be very susceptible to different instabilities, there may exist giant Kohn anomalies in them, etc.

More generally one needs for this not necessarily flat Fermi surfaces (or flat parts of Fermi surfaces) but what is known as *nesting*. Nested Fermi surfaces are such that different parts of them will coincide when shifted by a certain wavevector. For example, the Fermi surface for the spectrum (9.35) for $n = 1$ is a nested Fermi surface, with the nesting vector $\boldsymbol{Q} = (\pi, \pi)$. But in general we can have nesting even for curved Fermi surfaces, for example, such as the one shown in Fig. 9.5. Mathematically what we need is that the energy spectrum satisfies the condition

$$\boxed{\varepsilon(\boldsymbol{p} + \boldsymbol{Q}) = -\varepsilon(\boldsymbol{p})} \tag{9.37}$$

where the energies are counted from the chemical potential. Thus the spectrum (9.35) definitely obeys this condition at half-filling ($\mu = 0$), with $\boldsymbol{Q} = (\pi, \pi)$. Similarly, the spectrum of a three-dimensional cubic lattice in the tight-binding approximation,

$$\varepsilon(\boldsymbol{p}) = -2t(\cos p_x + \cos p_y + \cos p_z)\,, \tag{9.38}$$

has the same property, with $\boldsymbol{Q} = (\pi, \pi, \pi)$, and its Fermi surface for $n = 1$ is also nested, although it does not actually contain flat parts. In this sense a one-dimensional metal always has a nested Fermi surface: the Fermi surface consists simply of two 'Fermi points' which of course will coincide when shifted by the 'nesting wavevector' $2p_{\mathrm{F}}$. The property of nesting and respective divergence of the polarization operator will play a crucial role in discussions of different instabilities in metals, see Chapter 11.

## 9.3 Frequency-dependent dielectric function; dynamic effects

Let us now turn to the opposite limit: consider the polarization operator and dielectric function at small $q$ as functions of $\omega$. This limit is especially important in optics, as the wavevector of light (photons) is small (it can usually be taken as zero), and we are mostly interested in the optical spectrum, i.e. in the frequency dependence.

For large $\omega$ ($|\omega| \gg (q^2 + 2qp_{\mathrm{F}})/2m$) one can get from the general expression for $\epsilon$, obtained from (9.19), that

$$\mathrm{Re}\,\epsilon(\boldsymbol{q}, \omega) = 1 - \frac{\omega_{\mathrm{pl}}^2}{\omega^2}\,, \tag{9.39}$$

where $\omega_{\mathrm{pl}}$ is the usual plasma frequency

$$\omega_{\mathrm{pl}}^2 = \frac{4\pi n e^2}{m}\,. \tag{9.40}$$

(In the case of finite $\omega$ we have to be careful: the polarization operator $\Pi_0(\boldsymbol{q}, \omega)$ and $\epsilon(\boldsymbol{q}, \omega)$ have imaginary parts.)

Actually the plasma frequency is an eigenfrequency of the density oscillations of electrons on a positive background. This is how it is usually obtained in most textbooks; it is essentially a classical notion. However, one can also give it another interpretation. We see from (9.39) that the plasma frequency is a pole of the corresponding response function or of the effective interaction $\mathcal{V}(\boldsymbol{q}, \omega)$ (9.3).

Let us consider the process of scattering of an electron and a hole. During this process they can recombine, emitting a photon, which in turn can be absorbed, creating again an electron and a hole. One can represent this process by the diagram

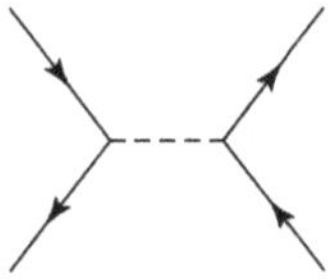

Seemingly another but actually equivalent process is the usual scattering, when an electron emits a photon, and a hole absorbs it:

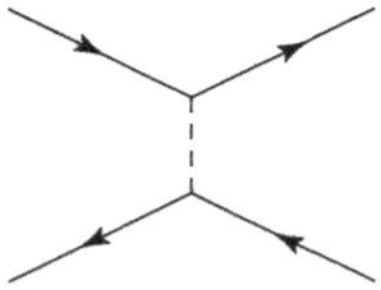

Topologically these diagrams are equivalent. They represent the first (in the interaction) term in the perturbation expansion of the so-called two-particle Green

function

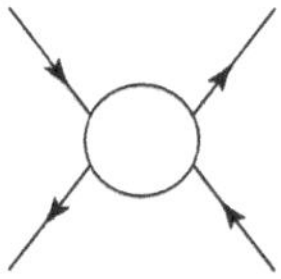

The next most important terms are our old ‘bubbles’:

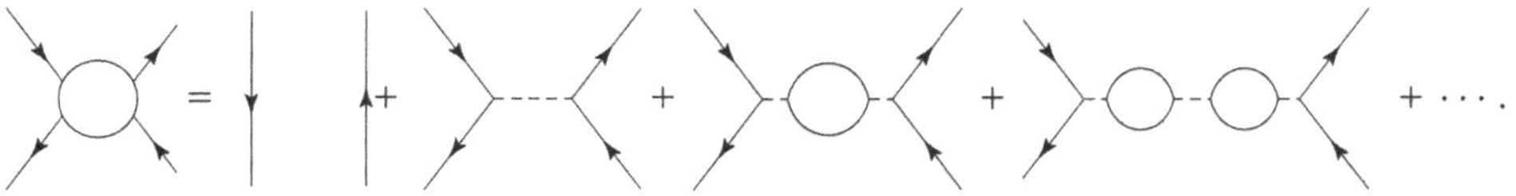


(9.41)

Or, using the summation (9.6), we can write this contribution (RPA contribution) as

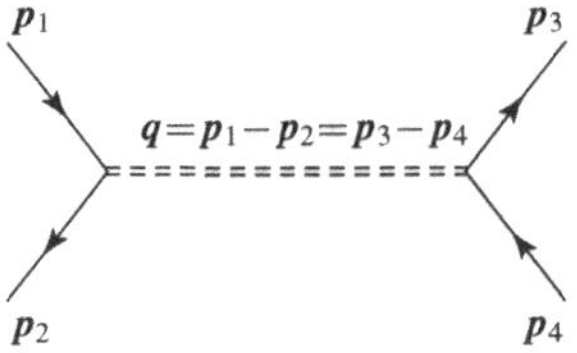


i.e. we put inside the full renormalized interaction $\mathcal{V}(\boldsymbol{q}, \omega) = \mathcal{V}(\boldsymbol{p}_1 - \boldsymbol{p}_2, \omega_1 - \omega_2)$. Consequently the poles of $\mathcal{V}(\boldsymbol{q}, \omega)$ are simultaneously poles of the two-particle Green function in the corresponding total momentum of the incoming electron ($\boldsymbol{p}_1$) and hole ($-\boldsymbol{p}_2$). They describe actually something like bound (or antibound) states of an electron and a hole.

The diagrams (9.41) are of course not all possible contributions to the two-particle Green functions; there may be, for example, diagrams of the type

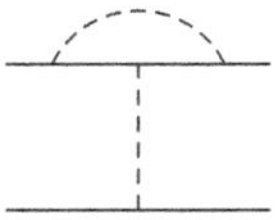

or

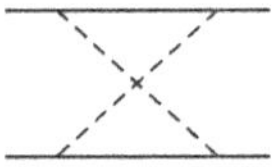

etc. Thus the full solution of the two-particle problem in the presence of an interaction is a very complicated problem which can be solved only in a few simple cases. However, already the ‘simple’ RPA approximation gives a lot of information and provides if not the full description, then at least a very useful starting point (and often it is really sufficient for many purposes).

The plasmon pole does not exhaust all eigenstates of a two-particle (two excitation) system, or all zeros of $\epsilon(\boldsymbol{q}, \omega)$. Actually from the expression (9.19) we see that for a fixed $\boldsymbol{q}$ there are also eigenstates (‘poles’, or zeros of the denominator, which

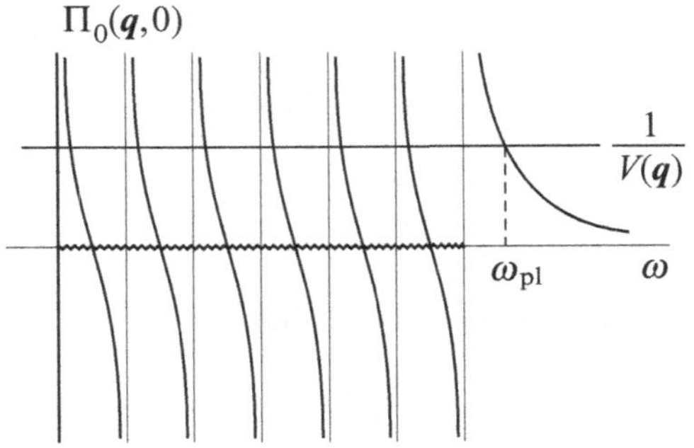


Fig. 9.6

in the limit of an infinite system come close together and form mathematically a *cut* in the complex $\omega$-plane) at all $\varepsilon(\boldsymbol{p}) - \varepsilon(\boldsymbol{p}+\boldsymbol{q})$ for different $\boldsymbol{p}$. In terms of the two-particle system these states correspond to scattering states of an electron and a hole, having a continuous spectrum, whereas a plasmon pole is a bound (or rather antibound) state. The situation here and the mathematical treatment is rather similar to that of impurity states in Section 6.5. One can write the expression (9.19) for $\Pi_0(\boldsymbol{q}, \omega)$ in the discrete case as

$$\Pi_0(\boldsymbol{q}, \omega) = \sum_{\substack{|\boldsymbol{p}|<p_F \\ |\boldsymbol{p}+\boldsymbol{q}|>p_F}} \left( \frac{1}{\omega - \varepsilon(\boldsymbol{p}+\boldsymbol{q}) + \varepsilon(\boldsymbol{p}) + i\delta} - \frac{1}{\omega + \varepsilon(\boldsymbol{p}+\boldsymbol{q}) - \varepsilon(\boldsymbol{p}) + i\delta} \right), \tag{9.42}$$

and then the poles of $\epsilon(\boldsymbol{q}, \omega)$ (9.4) are the solutions of the equation

$$1 = V(\boldsymbol{q})\,\Pi_0(\boldsymbol{q}, \omega) \tag{9.43}$$

which can be analysed graphically, Fig. 9.6, as we did for equation (6.132), cf. Fig. 6.47: all states with $0 < \omega < \omega_{\mathrm{max}} = p_{\mathrm{F}}q/m + q^2/2m$ describe scattering states and form for an infinite system a continuum spectrum (wavy line), and the special solution lying above the continuum is the plasmon. One can show that the plasmon pole exists (i.e. does not decay) only for a certain limited range of $q$. The total spectrum has the form shown in Fig. 9.7: plasmons exist as real excitations only for $q < q^*$, and for $q > q^*$ plasmons merge with the continuum (the hatched area in Fig. 9.7) and decay into independent particle–hole excitations.

It is also interesting and instructive (and important practically) to consider what would become of plasmons in low-dimensional (e.g. quasi-two-dimensional) materials, such as layered systems or thin films. It turns out that the general treatment remains qualitatively the same, but the plasmon spectrum starts from zero, Fig. 9.8; in a pure 2d system it is $\omega_{\mathrm{pl}}(q) \sim \sqrt{q}$, and in quasi-2d systems, depending on

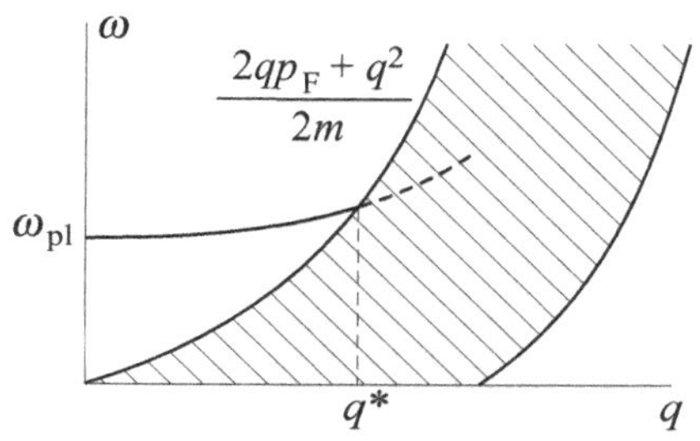


Fig. 9.7

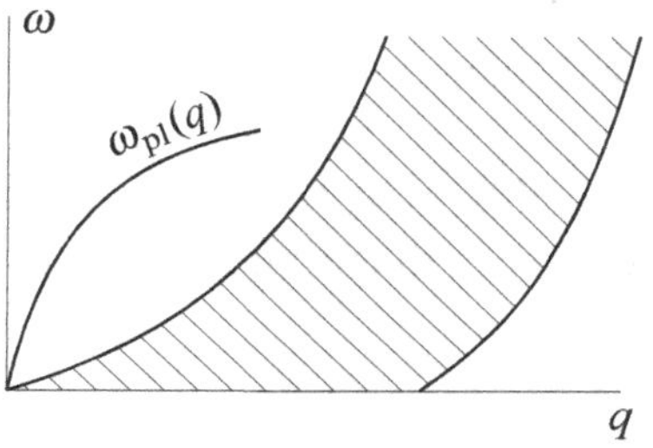


Fig. 9.8

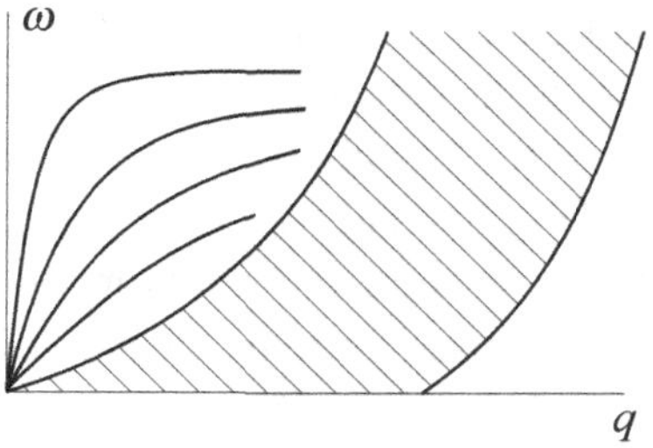


Fig. 9.9

the perpendicular component of the total wavevector $\boldsymbol{q} = (q_\perp, q_\parallel)$, it occupies the whole region, Fig. 9.9 (the curves above the hatched electron–hole continuum represent plasmon dispersion for different values of $q_\perp$). And the most straightforward and standard way to study this problem is to use the Green functions and diagram technique described above.

One can extend this approach and use the results obtained in the RPA to discuss the modifications of the electron spectrum brought about by the interaction. This spectrum is given by the poles of the one-electron Green function (8.79) and can be expressed through the self-energy or mass operator $\Sigma$, see (8.85)–(8.91). We can calculate $\Sigma$ in the RPA using the diagrams

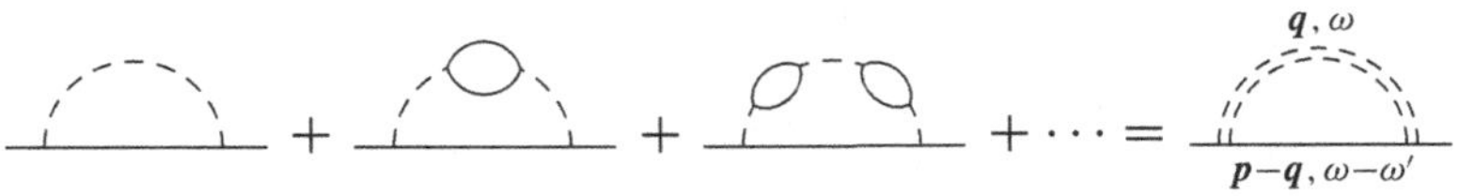


(9.44)

or

$$\Sigma(\boldsymbol{p}, \omega) = i \int v(\boldsymbol{q}, \omega')\, G_0(\boldsymbol{p} - \boldsymbol{q}, \omega - \omega')\, \frac{d^3\boldsymbol{q}\, d\omega'}{(2\pi)^4} \,. \tag{9.45}$$

The calculation finally gives the value of the effective mass

$$\frac{1}{m^*} = \frac{1}{m}\left[1 - 0.083\, r_s(\ln r_s + 0.203)\right] \tag{9.46}$$

which shows that the Coulomb interaction leads to a certain increase of the effective mass.

One can also calculate in this limit the total energy of the electron system (M. Gell-Mann and K. Brueckner); the result is

$$\frac{E_0}{N} = \left\{\frac{2.21}{r_s^2} - \frac{0.916}{r_s} + 0.062\ln(r_s) - 0.096\right\}\mathrm{Ry} \tag{9.47}$$

where $1\,\mathrm{Ry} = 13.6\,\mathrm{eV}$. The first term in equation (9.47) represents the kinetic energy of electrons (it is equal to $\frac{3}{5}\varepsilon_{\mathrm{F}}$, with $\varepsilon_{\mathrm{F}}$ given by (7.9)), and the second is the exchange part of the average Coulomb energy (the direct part of the electron–electron interaction is cancelled by interaction with the positive background). The first two terms together represent the Hartree–Fock energy of the electron gas, and the remaining terms give the so-called *correlation energy* (by definition the correlation energy is 'everything beyond Hartree–Fock').

These results, and the RPA in general, are valid for dense systems, $r_s \ll 1$. In the opposite limit of low-density electron systems the situation is changed drastically, and perturbation theory in the interaction is no longer valid. In this case we enter the field of strong electron correlations, which will be extensively discussed in the last section of Chapter 11 and in Chapter 12.

# 10

# Fermi-liquid theory and its possible generalizations

As we have seen in the previous chapters, interactions, if they are not too strong, preserve many features of the electronic system which are present in the noninteracting case (Fermi gas, Chapter 7). In general, however, the interactions are not at all small: for instance in typical metals, $r_s \sim 2–3$, and not $r_s \ll 1$ as was implicitly assumed in Chapter 9 and which was actually the condition for the applicability of perturbation theory used there. Nevertheless we know that the description of normal metals using the concepts developed for the free Fermi gas or Fermi systems with weak interactions (such as Drude theory, for example) is very successful.

An explanation of the success of the conventional theory of metals, and the generalization of the corresponding description to a more general situation, was given by Landau in his theory of Fermi liquids. This theory is very important conceptually, although in the usual metals there are only few special effects which indeed require this treatment. However, there exist also systems ($^3$He, or rare earth systems with mixed valence and heavy fermions) for which this approach is really vital. Also the emerging new field of non-Fermi-liquid metallic systems requires first an understanding of what *is* the normal Fermi liquid.

## 10.1 The foundations of the Fermi-liquid theory

The main assumptions of the Landau Fermi liquid theory are completely in line with our general approach. Landau postulated that, given the ground state of a metallic system, weakly excited states can be described in terms of elementary excitations, or *quasiparticles*, which are very similar to the usual electrons and holes in a Fermi gas, despite strong interactions. In particular, these quasiparticles are fermions with spin $\frac{1}{2}$ and with charge $-e$, which can be characterized by momentum $\boldsymbol{p}$ and energy $\varepsilon(\boldsymbol{p})$. A crucial assumption is that there exists a well-defined Fermi surface, with the radius $p_F$ satisfying the same relation

$$n = \frac{N}{V} = \frac{p_F^3}{3\pi^2\hbar^3} \tag{10.1}$$

as for the free Fermi gas, cf. (7.2). Thus, the value of the Fermi momentum $p_F$ and the volume of the Fermi surface are determined by the total particle density $N/V$ and do not depend on the interaction. This statement can actually be proven in every order of perturbation theory, using the Green function technique, and is known as the *Luttinger theorem*. Despite its apparent simplicity, this is a very strong statement, which leads to a number of experimental consequences and which can be really checked experimentally, e.g. by angular-resolved photoemission. One should say right away that it is not the only possible state of an electronic system. Thus, we know for sure that there exists an alternative state – the superconducting state – in which the Luttinger theorem formally does not hold and there is no Fermi surface left (there is an energy gap at the position of the former Fermi surface). There are also nowadays a lot of discussions about possible breakdown of the Fermi liquid description and of non-Fermi-liquid states in cases of strong electron correlations which may (but also may not) be accompanied by a breakdown of the Luttinger theorem. And finally the whole big field of *strongly correlated electrons*, discussed in Chapter 12 below, is a bona fide example of the inapplicability of Fermi liquid theory. These questions will be discussed later on; now however, having made these remarks, we continue to discuss 'positive' aspects and the formulation of the Landau theory of Fermi liquids, in cases when this description is indeed valid.

What are the physical reasons which permit one in principle to retain many essential features of free fermions in a strongly interacting system? The main factor is the very fact that electrons are fermions, i.e. the Pauli principle. One can show that because of that the Fermi surface remains sharp, and quasiparticles are well defined close to the Fermi surface. If we consider one electron with momentum $p_1 > p_F$ excited above the Fermi surface, one can show that despite interactions with other electrons its lifetime becomes infinite at the very Fermi surface, when $p_1 \to p_F$. Indeed the simplest process which can lead to a finite lifetime is the excitation of another electron from below $\varepsilon_F$ to above it due to the electron–electron interaction, i.e. the creation of an electron–hole pair, Fig. 10.1. (We use here a simplified form of diagram, 'contracting' interaction lines to a point; this is rather convenient in the case of a four-fermion interaction.) In this process we have as usual $\boldsymbol{p}_1 + \boldsymbol{p}_2 = \boldsymbol{p}_3 + \boldsymbol{p}_4$, and $\varepsilon_1 + \varepsilon_2 = \varepsilon_3 + \varepsilon_4$ (here $|\boldsymbol{p}_1|, |\boldsymbol{p}_3|, |\boldsymbol{p}_4| \geq p_F$, $|\boldsymbol{p}_2| \leq p_F$; we also took $\varepsilon_1, \varepsilon_3, \varepsilon_4 \geq \varepsilon_F$, $\varepsilon_2 \leq \varepsilon_F$).

From these conservation laws it is clear that when $|\boldsymbol{p}_1| \to p_F$, all other particles should also have $|\boldsymbol{p}_2|, |\boldsymbol{p}_3|, |\boldsymbol{p}_4| \to p_F$ (this is especially easy to see if we count all

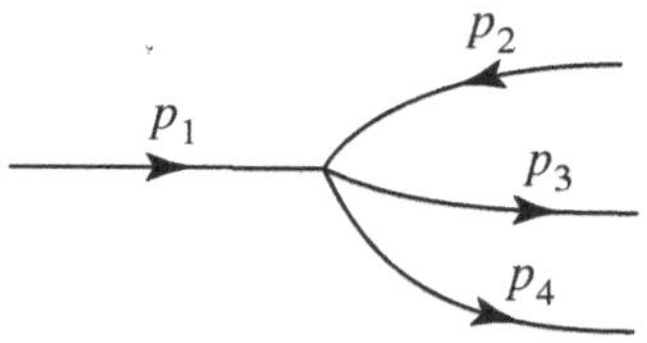


Fig. 10.1

energies from $\varepsilon_F$: then $\varepsilon_1, \varepsilon_3, \varepsilon_4 > 0$, $\varepsilon_2 < 0$, and the condition $\varepsilon_1 + \varepsilon_2 = \varepsilon_3 + \varepsilon_4$ implies that when $\varepsilon_1 \to +0$, all other $\varepsilon_\alpha$ should $\to 0$).

When $p_1$ is slightly larger than $p_F$, all other momenta $|\boldsymbol{p}_\alpha| - p_F$ should be of the same order as $|\boldsymbol{p}_1| - p_F$. Then by the 'golden rule' the probability of the process shown in Fig. 10.1, i.e. the probability of the decay of an electron (which determines its lifetime $\tau$), is proportional to

$$W = \tau^{-1} \sim \int \delta(\varepsilon_1 + \varepsilon_2 - \varepsilon_3 - \varepsilon_4)\, d^3\boldsymbol{p}_2\, d^3\boldsymbol{p}_3 \tag{10.2}$$

($\boldsymbol{p}_1$ is given, and $\boldsymbol{p}_4 = \boldsymbol{p}_1 + \boldsymbol{p}_2 - \boldsymbol{p}_3$, thus there remain two independent integrations). As both $\boldsymbol{p}_2$ and $\boldsymbol{p}_3$ lie close to $p_F$, $(|\boldsymbol{p}_\alpha| - p_F) \sim (p_1 - p_F)$, this integral is $\sim(|\boldsymbol{p}_1| - p_F)^2$, i.e. the inverse lifetime of the electron with momentum $p$ goes to zero as $p \to p_F$ quadratically in $(p - p_F)$:

$$\tau_p^{-1} \sim (p - p_F)^2 \,. \tag{10.3}$$

(Here and below $p = |\boldsymbol{p}|$.) This is essentially the 'phase space' argument: because of the Pauli principle the scattering of weakly excited electrons is confined to the vicinity of the Fermi surface, which gives (10.3). In the language of Green functions the condition (10.3) means that the pole $\omega = \varepsilon(p)$ of the one-electron Green function is well defined when $p \to p_F$: its real part $\mathrm{Re}\, \varepsilon(p) \sim v_F(p - p_F)$ is much larger than its imaginary part $\mathrm{Im}\, \varepsilon(p) \sim \tau^{-1} \sim (p - p_F)^2$. In other words this means that the quasiparticles (quasi-electrons and quasi-holes) are well defined at the Fermi surface and in its vicinity. This does not imply, however, that they are well-defined objects far away from the Fermi surface: in general they are not, and the structure of the spectrum far from the Fermi surface often does not have the character of quasiparticles. However, at low energies and at low enough temperatures it is sufficient to consider only excitations close to $\varepsilon_F$, and for these purposes the Fermi-liquid description usually works fine.

As to the finite temperature, we have seen in Chapter 7 that the Fermi-distribution function $n(p)$ is spread over the region $\sim T$ around $\varepsilon_F$. This gives that, besides the factor $(p - p_F)^2$ in $\tau^{-1}$ (10.3), there may also be contributions of thermally excited quasiparticles. As we have seen in (7.23), (7.24), actually the corrections to the energy and to other properties due to such thermally excited quasiparticles are $\sim T^2 \rho(\varepsilon_F) \simeq T^2/\varepsilon_F$. Consequently, combining this with (10.3) (with the natural scale $\varepsilon_F$ in this term too) we get the estimate

$$\tau^{-1} = a\left(\frac{(\varepsilon - \varepsilon_F)^2}{\varepsilon_F} + \frac{T^2}{\varepsilon_F}\right), \tag{10.4}$$

where the constant $a$ is of order 1.

Already from this simple treatment, which actually used only very general arguments such as the Pauli principle and energy and momentum conservation laws, we can make several important conclusions. First, as the low-temperature excited states have nearly the same nature as ordinary electrons, such quantities as specific heat and entropy will be given by the usual relations (7.25), (7.26), with the same $p_F$, with the only difference being the substitution of the effective mass $m^*$ instead of the bare mass $m$:

$$c = \gamma T\,, \qquad \gamma \sim m^* \tag{10.5}$$

(cf. e.g. (9.46)). In general this mass renormalization can be substantial, and sometimes even extremely large. Thus in heavy-fermion compounds such as $CeCu_6$, $CeAl_3$, $UBe_{13}$, $m^* \sim 10^3 m$, i.e. the effective Fermi energy $\varepsilon_F^* = p_F^2/2m^*$ is $10^3$ times smaller than the typical values for conventional metals like Al; whereas in these simple cases $\varepsilon_F \sim$ 1–10 eV, i.e. $\sim 10^4$–$10^5$ K, in heavy-fermion compounds the effective Fermi energy is $\varepsilon_F^* \sim 10^{-3}\,\text{eV} \sim 10\,\text{K}$.

From the expression for the electron lifetime (10.4) one can also make interesting conclusions about some transport properties of the system, in particular electrical resistivity. From the usual Drude expression for the conductivity

$$\sigma = \frac{ne^2\tau}{m} \tag{10.6}$$

one gets, using $\tau^{-1} \sim T^2/\varepsilon_F$ and also using (10.1), that the resistivity behaves as

$$R = \sigma^{-1} \sim \frac{T^2 m}{\varepsilon_F p_F^3 e^2} \sim \frac{1}{e^2 p_F}\left(\frac{T}{\varepsilon_F}\right)^2. \tag{10.7}$$

One can further transform this expression using the fact that $p_F \sim 1/\tilde{r}_s$ (see (7.6)), so that typically $p_F \sim 1/a_0$, where $a_0$ is the Bohr radius, $a_0 = \hbar^2/me^2$. Then in (10.7) $e^2 p_F \sim e^2/a_0$ ( $\sim$ Rydberg). In general the functional dependence of the resistivity due to electron–electron scattering is

$$R = AT^2\,, \qquad A \sim 1/\varepsilon_F^2\,. \tag{10.8}$$

This is an important result; sometimes it is called the Baber law. In ordinary metals $\varepsilon_F$ is very large, and the contribution of electron–electron scattering is masked by other, much stronger contributions, such as electron–phonon scattering, which typically behaves at low temperatures as $\rho_{e-ph} \sim T^5$, or impurity scattering, giving residual resistivity. The $T^2$-dependence of resistivity becomes noticeable in transition metals where the 3d band is narrower, and it is very clearly seen in heavy-fermion compounds. The scaling between the coefficient $A$ in the resistivity (10.8) with the effective mass or effective Fermi temperature is also confirmed there experimentally: as $\varepsilon_F = p_F^2/2m^*$, $A$ is proportional to $m^{*2}$, or to $\gamma^2$ (10.5).

Up to now we have presented some arguments which helped to justify the applicability for interacting electrons of ordinary relations valid for free fermions. However, one can substantially deepen the description of the system, making the next step and taking into account the fact that the Landau quasiparticles are really not free fermions (with a different mass), but are still strongly interacting objects. It is this next step which permits one to get new results and improve the description considerably. We will describe it only schematically, and will qualitatively discuss some of the consequences; one can find more detailed discussions, e.g. in Abrikosov *et al.* (1975) and Baym and Pethick (1991). Suppose we somehow changed the distribution of quasiparticles, $n(\boldsymbol{p}) = n(\boldsymbol{p}) + \delta n(\boldsymbol{p})$. As quasiparticles interact with each other, the energy of each of them depends on the distribution function of the others. Taking into account only the first, linear term of this dependence, we can write:

$$\varepsilon(\boldsymbol{p}) = \varepsilon_0(\boldsymbol{p}) + \sum_{\boldsymbol{p}'} f(\boldsymbol{p}, \boldsymbol{p}')\,\delta n(\boldsymbol{p}') \,. \tag{10.9}$$

This expression, which looks extremely (and deceptively) simple, actually permits one to treat a lot of effects using only the most general properties, such as Galilean invariance, symmetry, conditions of stability of the system, etc. without specifying the detailed nature of the interaction. The function $f(\boldsymbol{p}, \boldsymbol{p}')$ introduced phenomenologically in (10.9) is called the Landau interaction function; one can show that it is connected with the scattering amplitude of electrons.

In general the quasiparticle energy, the electron distribution function and the interaction function $f$ may depend also on spin indices. For a nonmagnetic system one can decompose $f(\boldsymbol{p}, \sigma; \boldsymbol{p}', \sigma')$ into two terms:

$$f(\boldsymbol{p}, \sigma; \boldsymbol{p}', \sigma') = \varphi(\boldsymbol{p}, \boldsymbol{p}')\mathbf{1} + \psi(\boldsymbol{p}, \boldsymbol{p}')(\boldsymbol{\sigma} \cdot \boldsymbol{\sigma}') \tag{10.10}$$

where $\mathbf{1}$ is a unit matrix in spin space and $\boldsymbol{\sigma}$, $\boldsymbol{\sigma}'$ are the Pauli matrices. As the whole theory is valid close to the Fermi surface, one can also put $|\boldsymbol{p}| = |\boldsymbol{p}'| \simeq p_F$ and keep only the dependence on the angle $\theta$ between $\boldsymbol{p}$ and $\boldsymbol{p}'$. Then one can use a standard expansion in terms of spherical harmonics (Legendre polynomials

$P_n(\cos\theta)$), writing

$$\varphi(\boldsymbol{p},\boldsymbol{p}') = \frac{\pi^2\hbar^3}{m^* p_{\mathrm{F}}} A(\theta) = \frac{\pi^2\hbar^3}{m^* p_{\mathrm{F}}}\left[A_0 + A_1 P_1(\cos\theta) + \cdots\right] \tag{10.11}$$

$$\psi(\boldsymbol{p},\boldsymbol{p}') = \frac{\pi^2\hbar^3}{m^* p_{\mathrm{F}}} B(\theta) = \frac{\pi^2\hbar^3}{m^* p_{\mathrm{F}}}\left[B_0 + B_1 P_1(\cos\theta) + \cdots\right]. \tag{10.12}$$

(Here we have introduced the dimensionless interactions $A_n$, $B_n$, normalizing the interaction with the use of the density of states $\rho(\varepsilon_{\mathrm{F}}) = m^* p_{\mathrm{F}}/\pi^2\hbar^3$ (7.13).)

The coefficients $A_n$, $B_n$ are in general unknown; they are treated as phenomenological parameters, which are usually determined from experiment. The virtue of the approach described above is that just a few parameters are usually needed to explain a lot of experimental data. For electrons in metals, when we more or less know and can describe important interactions, this approach is, maybe, less significant, but it is really necessary and extremely useful for strongly interacting Fermi systems such as, e.g. $^3$He and nuclear matter.

There are several slightly different ways to introduce the Landau parameters. Often one uses a different notation instead of (10.10)–(10.12):

$$f(\boldsymbol{p},\sigma;\boldsymbol{p}',\sigma') = f^s + (\boldsymbol{\sigma}\cdot\boldsymbol{\sigma}')f^a\,, \tag{10.13}$$

$$f^{s,a}(\theta) = \frac{\pi^2\hbar^3}{m^* p_{\mathrm{F}}}\sum_{l=0}^{\infty} F_l^{s,a} P_l(\cos\theta)\,, \tag{10.14}$$

where '$s$, $a$' stands for 'symmetric' and 'antisymmetric'. Sometimes the expansion is also written with different normalization (note the factors $(2l+1)$ in the sum)

$$f^{s,a}(\theta) = \frac{\pi^2\hbar^3}{m^* p_{\mathrm{F}}}\sum_{l=0}^{\infty}(2l+1)\tilde{F}_l^{s,a} P_l(\cos\theta)\,. \tag{10.15}$$

In a one-component Galilean-invariant system such as $^3$He, the effective mass is expressed as

$$\frac{m^*}{m} = 1 + \frac{F_1^s}{3} = 1 + \frac{A_1}{3} = 1 + \tilde{F}_1^s\,. \tag{10.16}$$

(Note that in crystals, if the interaction is predominantly local but retarded, such as the interaction with phonons, the expression for the effective mass may be different and may contain not $F_1^s$ but $F_0^s$, cf. (8.91), (8.92). In terms of Fermi-liquid theory $\partial\Sigma/\partial\omega$ gives the terms $\sim F_0^s$, and $\partial\Sigma/\partial k$ contains $F_1^s$.)

One can also get expressions for some other properties of Fermi liquids. They all look similar to the corresponding formulae for an ordinary Fermi gas, but contain the effective mass $m^*$ (10.16) instead of $m$, and are renormalized by the respective

Landau parameters. Thus, the specific heat is

$$c = \frac{m^*}{m} c_0 , \tag{10.17}$$

magnetic susceptibility

$$\chi = \frac{m^*/m}{1 + F_0^a} \chi_0 , \tag{10.18}$$

compressibility

$$\kappa = \frac{m^*/m}{1 + F_0^s} \kappa_0 , \tag{10.19}$$

etc.

Already from these expressions one can draw certain general conclusions. Thus, from (10.18), (10.19) one gets the criteria of stability of a homogeneous Fermi liquid without any extra ordering (the so-called Pomeranchuk criteria[1]):

$$1 + F_0^a > 0 , \qquad 1 + F_0^s > 0 . \tag{10.20}$$

Indeed, if, e.g. $1 + F_0^s < 0$, this would give negative compressibility, which means an instability of the system and will lead to a certain phase transition (e.g. structural phase transition in metals). One can show that this is actually a general requirement: the system remains stable if

$$1 + \frac{1}{2l + 1} F_l^{s,a} = 1 + \tilde{F}_l^{s,a} > 0 \tag{10.21}$$

for all $l$. For instance, for $l = 1$ the condition (10.21) guarantees that the effective mass (10.16) is positive. Systems close to magnetic instability have $1 + F_0^a \ll 1$. One can experimentally measure this parameter by comparing susceptibility and specific heat: the Wilson ratio (7.31) becomes in this case

$$R_{\mathrm{W}} = \frac{\pi^2 \chi T}{3\mu_{\mathrm{B}}^2 c} = \frac{1}{1 + F_0^a} \gg 1 . \tag{10.22}$$

More interesting are the consequences of quasiparticle interactions, described by Landau parameters, for collective modes and for transport properties. Thus, e.g. by considering the nonequilibrium states one can show that there can exist in a Fermi liquid (with short-range interaction) a collective mode which is called *zero sound*. Usually there exists in liquids a sound mode in which the equilibrium is restored due to collisions – this is the ordinary hydrodynamic sound which exists for

[1] Very often one speaks of Pomeranchuk instability in a narrow sense, as in the magnetic instability of the normal Fermi liquid, when $F_0^a < -1$.

$\omega\tau \ll 1$, where $\omega$ is the sound frequency, and $\tau$ is the collision time. With increasing frequency, when $\omega\tau \sim 1$, this mode becomes strongly dissipative, and sound ceases to exist. In an interacting system, such as a Fermi liquid, however, another mechanism of equilibration exists: collective interactions. As a result the sound mode exists also in the opposite limit $\omega\tau \gg 1$, and its velocity $u_0$ or dimensionless velocity $s = u_0/v_F$ (where $v_F$ is the Fermi velocity) is given by the expression

$$\frac{s}{2}\ln\frac{s+1}{s-1} = 1 + \frac{1}{F_0^s} \,. \tag{10.23}$$

For $F_0^s \to 0$, $s \to 1$, i.e. the velocity of zero sound approaches the Fermi velocity, $u_0 \to v_F$. This value differs from the velocity of classical hydrodynamic sound modes, which is $v = \frac{1}{3}v_F$ for a classical electron liquid, where the equilibrium is established by electron–electron collisions and where every elementary volume is in thermal equilibrium. Physically zero sound may be viewed as oscillations of the shape of the Fermi surface, and not just a density wave.

There may also exist in a Fermi liquid analogous oscillations connected with fermion spins. They may be called spin collective modes, although formally there is no magnetic ordering in the usual sense. The velocities of such spin modes are in general different from the 'sound' velocity (10.23) because they contain other Landau interaction parameters. Such 'spin waves' were indeed observed experimentally in some simple nonmagnetic metals.

One may ask the following question: we have learned before and stressed several times that the presence of gapless collective modes, in particular sound modes (and maybe of the electron and hole quasiparticles themselves?) is usually connected with some ordering of the system and with broken continuous symmetry. And we see that we have such gapless modes in the normal Landau Fermi liquid. Can we, then, view the normal Fermi liquid as such an ordered system too, and if so, what kind of order does exist in it?

This question is now under hot discussion, and it is not finally solved, or at least the solution is not universally accepted, but one can indeed give some arguments that this is really the case. The very ground state of the Fermi liquid – the filled Fermi sea – is a unique well-defined state with zero entropy, and that is what we usually have in an ordered state! What is then the 'order parameter', and which symmetry is broken here? The possible answer is that the 'order parameter' is *the Fermi surface itself.* It is a well-defined object, a sharp surface in momentum space, and the continuous symmetry corresponding to the possibility of arbitrary shift of all *momenta* $\boldsymbol{k}$ is thus broken by the presence of this Fermi surface. As the mathematical quantity characterizing this 'ordered' state we can for example take the $Z$-factor (8.54), (8.55), which gives the jump in the electron distribution function $n(p)$ at the Fermi surface, see (8.55); it can play a role similar to that of the order parameter in the Landau theory of phase transitions. At high temperatures the Fermi distribution is broadened, and a sharp Fermi surface, formally speaking, disappears, but at $T = 0$ we may speak of an 'ordered state'. In this picture the zero sound discussed above – the oscillations of the shape of the Fermi surface – may be naturally treated as the corresponding collective excitation (the 'Goldstone mode').

This general point of view, the notion of Fermi liquid as a special type of ordered state, does not give us directly any new results, but it is rather important conceptually. It puts the normal Fermi liquid state in the same row, same category as other types of ordered states that usually exist in nature when the temperature goes to zero.

## 10.2 Non-Fermi-liquid states

There exist, besides normal Fermi liquids, several other possible ground states of interacting electron systems. The system may become superconducting; different types of magnetic ordering are possible; there may occur states with inhomogeneous charge distributions. Some of these states may become insulating. We will discuss several such possibilities later, but now we ask the following question: is the conventional Landau Fermi liquid the only possible state of a normal *metal* without long-range order of some kind?

This problem has been actively discussed in recent years, in the beginning mostly in connection with some unusual normal state properties of high-temperature superconductors, but now in a much broader context. There are several indications in the experiment that these and possibly some other materials do not obey the usual rules and behave in a way that is much different from the normal Fermi liquid.

### *10.2.1 Marginal Fermi liquid*

There exist at present several suggestions or 'scenarios' which can in principle give a non-Fermi-liquid ground state. One of the first was the concept of the so-called *marginal Fermi liquid*, formulated phenomenologically mainly by C. M. Varma. In this theory it is assumed that the polarization operator $\Pi(\boldsymbol{q}, \omega)$, see Sections 8.7.1 and 9.1, does not strongly depend on $\boldsymbol{q}$, and its $\omega$-dependence is such that

$$\operatorname{Im} \Pi(\boldsymbol{q}, \omega) = \begin{cases} -\rho(\varepsilon_{\mathrm{F}})\, \omega/T & \text{for } \omega \ll T \\ -\rho(\varepsilon_{\mathrm{F}}) & \text{for } T \ll \omega \ll \omega_{\mathrm{c}}\,. \end{cases} \tag{10.24}$$

Here $\omega_{\mathrm{c}}$ is a certain cut-off frequency, which is supposed to be $\omega_{\mathrm{c}} \ll \varepsilon_{\mathrm{F}}$. From the dispersion relation for $\Pi$ (similar to the Kramers–Kronig relations well known in optics) one then gets

$$\mathrm{Re}\,\Pi(\boldsymbol{q},\omega) \sim \rho(\varepsilon_{\mathrm{F}}) \ln\left(\frac{\omega}{T}\right) . \tag{10.25}$$

Putting these expressions into the self-energy of electrons, we get

$\Sigma =$ 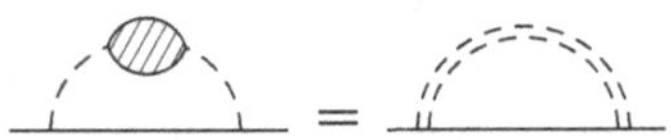 (10.26)

$$\Sigma(\omega) = \lambda\omega \left[ \ln \frac{x}{\omega_{\mathrm{c}}} + i\frac{\pi}{2} x \,\mathrm{sign}(\omega) \right] \tag{10.27}$$

where $x = \max(\omega, T)$. This gives, using equation (8.87), the quasiparticle weight $Z$ (residue of the pole of the Green function)

$$Z = \frac{1}{1 - \partial\,\mathrm{Re}\,\Sigma/\partial\omega} \sim \frac{1}{1 - \lambda \ln(y/\omega_{\mathrm{c}})} , \tag{10.28}$$

where $y = \max\big[(\varepsilon_p - \mu), T\big]$. Thus we see that the assumption (10.24) alone tells us that the quasiparticle weight goes to zero at the Fermi surface, which means that the jump in the electron distribution function $n(p)$ at the Fermi surface, proportional to $Z$, disappears, and that the single-particle density of states tends to zero for $\varepsilon \rightarrow \varepsilon_{\mathrm{F}}$. Simultaneously the quasiparticle lifetime, given by the imaginary part of $\Sigma$ (10.27), is linear in energy, $\tau^{-1} \sim (\varepsilon_p - \mu)$. As a result the quasiparticles are only 'marginally' defined at the Fermi surface: in ordinary Fermi liquids it is required (and assumed) that the lifetime is long enough, $\tau^{-1} \sim (\varepsilon_{\mathrm{F}} - \mu)^2$, see (10.3). If $\tau^{-1}$ were to remain constant at $\varepsilon_{\mathrm{F}}$, the Fermi surface would be completely smeared out; here it is still defined, but the properties of quasiparticles are much different, and the quasiparticles themselves are not so well-defined as in an ordinary Fermi liquid.

### *10.2.2 Non-Fermi-liquid close to a quantum critical point*

The marginal Fermi liquid discussed above is only one of several possible types of non-Fermi liquids. In general there can exist many other situations in which a metal behaves as a non-Fermi liquid. The most widely discussed situation is that with quantum critical points (QCP) in metallic systems, cf. Section 2.6. As explained in that section, this is the situation in which the critical temperature of some ordering is suppressed (goes to zero) by some external parameter (pressure, magnetic field, etc.). When $T_{\mathrm{c}} \rightarrow 0$, besides thermal fluctuations always present in the vicinity of $T_{\mathrm{c}}$, quantum fluctuations start to play an important role. They can significantly modify the properties of the system in the vicinity of such QCP. In particular, if we are dealing with a metallic system, which is, say, a normal Fermi liquid

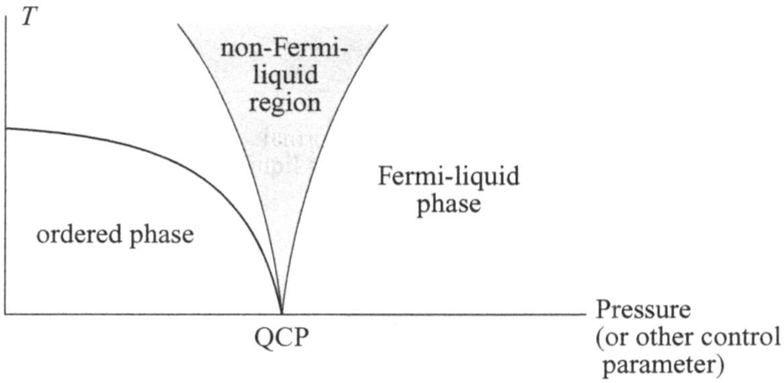


Fig. 10.2

at high pressures, see Fig. 10.2, often the conventional Fermi-liquid behaviour is violated in the vicinity of QCP, see e.g. von Löhneisen *et al.* (2007). Notably, different thermodynamic and transport properties behave in an abnormal way: e.g. the specific heat, instead of the standard linear temperature dependence $c = \gamma T$, behaves as $c \sim T^{\alpha}$; the resistivity, instead of the usual Fermi liquid asymptotic behaviour $\rho \sim AT^2$, becomes linear in temperature, $\rho \sim T$, or $\rho \sim T^{4/3}$, etc.

Qualitatively the general explanation of these deviations from the normal Fermi-liquid behaviour is that close to QCP the electrons strongly interact with collective modes specific for a particular type of ordering which is especially strong when $T_c \to 0$ and when these modes have strong quantum character. Specific features of corresponding couplings in general depend on the particular ordering in question, so that here, similarly to second-order phase transitions, cf. Section 2.5, we may have in different situations different laws, with different exponents. A microscopic theory of most of these phenomena is still absent.

Yet another very surprising phenomenon was discovered during the study of metallic systems close to a quantum critical point. It was found that in many such cases not only do we have non-Fermi-liquid behaviour, but there may appear in the vicinity of QCP a novel phase – a superconducting phase, see Fig. 10.3. Such is, e.g. the situation in some itinerant magnets, e.g. $UGe_2$, URhGe, where the magnetic ordering can be suppressed by pressure. A similar phenomenon is observed in several heavy-fermion compounds (see Chapter 13), e.g. in $CePd_2Si_2$, and in some organic compounds, for example in $(TMTTF)_2PF_6$. There are also ideas that this is what can happen in high-$T_c$ cuprates, although this question is still very controversial. In any case, there are strong reasons to believe that the superconductivity which appears in this situation may be not of the conventional $s$-wave BCS (Bardeen–Cooper–Schrieffer) type, but is unconventional, e.g. with the singlet $d$-wave or triplet $p$-wave pairing.

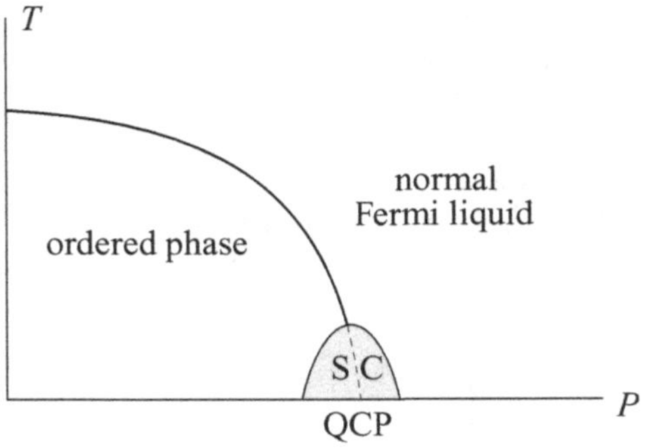


SC – superconducting phase

Fig. 10.3

### 10.2.3 Microscopic mechanisms of non-Fermi-liquid behaviour; Luttinger liquid

It is not really clear what are the different possible microscopic mechanisms of non-Fermi-liquid behaviour. Several microscopic models have been suggested: singular electron–electron interaction originating from scattering on very soft excitons; an interaction with quadrupolar Kondo-centres; etc. None of these mechanisms has been really shown to result in the desired behaviour and to work, e.g. in high-$T_c$ cuprates. Probably the most reliable case is the one-dimensional electron system, where one can indeed show that there exists a state resembling a Fermi liquid, but different from it. One can show that in 1d systems of interacting electrons formally there exists a 'Fermi surface', but it is not a Fermi surface of ordinary electrons, but rather of *spinons*, cf. Section 6.4.2. The real electron distribution function does not have any jump at $\varepsilon_F$ (recall that the existence of such a jump is a prerequisite for ordinary Fermi liquids), but behaves instead as $n(p) - n(p_F) \sim |p - p_F|^\delta \operatorname{sign}(p - p_F)$, i.e. the momentum distribution and the single-particle density of states have a power-law singularity at $\varepsilon_F$. The exponent $\delta$ depends on the electron–electron interaction; in typical cases $\delta \sim \frac{1}{8}$. Physically this behaviour is connected with the *spin–charge separation*: there exist in this case two types of elementary excitations, spinons carrying spin but no charge, and holons which have charge but no spin (the situation here is similar to the RVB picture discussed in Chapter 6). Such a state is often called a *Luttinger liquid*. Simple arguments which show that there should be spin–charge separation in the one-dimensional case may be given using the example of one-dimensional Hubbard model with strong interactions (more detailed discussion of the Hubbard model in general will be given later, in Chapter 12). For one electron per site the ground state with localized electrons is 'almost antiferromagnetic', $\uparrow\downarrow\uparrow\downarrow\uparrow\downarrow\uparrow\downarrow$. If we now

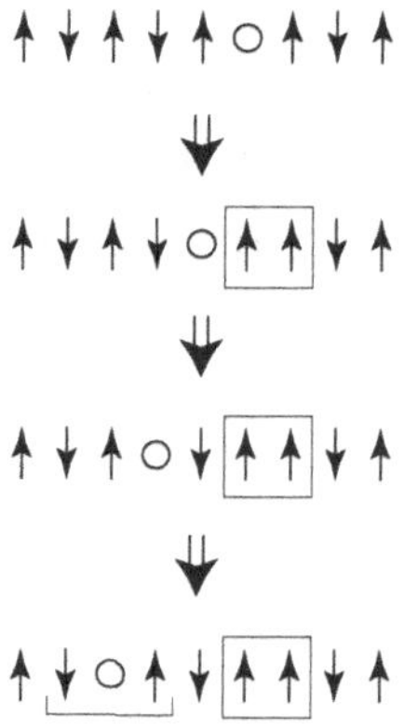

Fig. 10.4

create one hole, then due to the exchange process the hole could move, say, to the left, see Fig. 10.4, so that as a result the hole, which initially was surrounded by two spins ↑, i.e. ↑ ∘ ↑, finally is surrounded by the 'correct' antiferromagnetic spins, ↓ ∘ ↑, so that if we 'contract' the system by removing the hole, the remaining spin structure close to the location of the hole would be undisturbed. However, at the same time in another part of the chain a pair of wrong spins, ↑↑, would remain. Thus the hole would live without disturbing the spin order of the remaining electrons in its vicinity, but another purely spin excitation – a spinon – would be created; this is what is meant by spin–charge separation. This is the typical situation in Luttinger liquids in general. This is the best-proven case of non-Fermi-liquid behaviour – unfortunately only in the 1d case!

Whether the behaviour of 2d or 3d systems may be in any respect similar to the one-dimensional case, is not clear at present, although one may think that the RVB state described above, Section 6.4.2, would lead to a similar picture. However, one can easily see that in the 2d case the same process as illustrated in Fig. 10.4 would not lead to simple spin–charge separation; there will be a 'trace' of wrong spins connecting the hole and the remaining spin defect (the original location of the hole). This will be discussed in more detail in Section 12.4. Thus, unfortunately, even for the 2d case, not to speak of real 3d systems, there is no rigorous proof that the state with spin–charge separation and with the properties similar to a Luttinger liquid can be realized in practice.

# 11

# Instabilities and phase transitions in electronic systems

We have already mentioned several times before (e.g. while discussing the giant Kohn anomalies in Section 9.2 or the Pomeranchuk criteria of stability of the Fermi liquid (10.20) in Chapter 10) that there may exist situations when the usual Fermi-liquid state of electrons in metals becomes unstable, even for weak interactions. In this case a transition to some new state, often an insulating state, can take place. In this chapter we will discuss several such cases, using different approaches to illustrate how the general theoretical methods described above really work. We start with the Peierls transition.

## 11.1 Peierls structural transition

One of the first historically, and conceptually simplest and best-known examples of possible instabilities of the usual metallic state even for weak interactions is the Peierls instability. We first give a simple qualitative picture, and then present several equivalent ways to treat it, thus illustrating some of the approaches described above.

### *11.1.1 Qualitative considerations*

We start with the simplest case of a one-dimensional chain with one electron per site, treating electrons in the tight-binding approximation. The energy spectrum in this case is

$$\varepsilon(k) = -2t \cos ka \,, \tag{11.1}$$

where $t$ is the electron–electron hopping between neighbouring sites, and $a$ is the lattice period (see Chapter 12 for more details). This spectrum is shown in Fig. 11.1($a$), and for one electron per site the corresponding band will be exactly

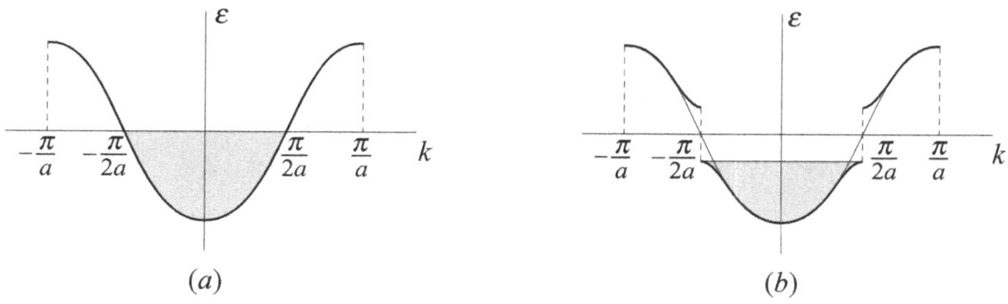


Fig. 11.1

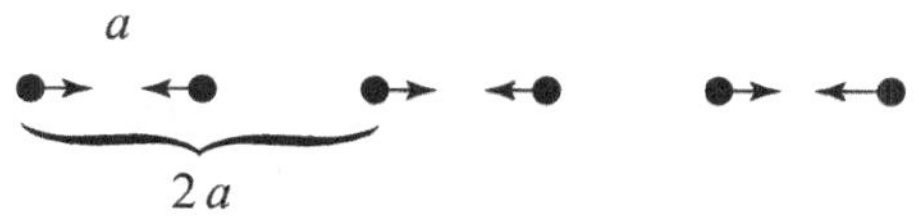


Fig. 11.2

half-filled (Fermi momentum $k_F = \pm\pi/2a$).[1] If we now double the unit cell, shifting every second atom and forming dimers, Fig. 11.2, the new period will be $2a$, and the new Brillouin zone boundaries, instead of $\pm\pi/a$, will be $\pm\pi/2a$, i.e. they will coincide with the original Fermi wavevectors. As always, there appear energy gaps at the Brillouin zone boundaries, thus the spectrum would change from that of Fig. 11.1(*a*) to that of Fig. 11.1(*b*). Thus we see that in effect the energies of all occupied electronic states decrease, i.e. we gain electron energy in this process. Of course we have to deform the lattice, which costs us elastic energy $\sim u^2$, where $u$ is the lattice distortion in going from the regular to the dimerized arrangement of atoms or ions. But if we gain more than we lose, this process will be energetically favourable, and the homogeneous chain will be unstable with respect to dimerization.

These were the arguments first given by R. Peierls (2001) in a short section of his book (first published in 1955). He had indeed found that the electron energy gain always exceeds the elastic energy loss (the electronic energy goes as $\sim u^2 \ln u$ whereas the elastic energy is quadratic, $\sim u^2$). However, in that book Peierls had not presented a mathematical proof, thinking that it is either rather self-evident or too simple (the mathematics is presented in a very nice short book by Peierls (1991), devoted to some problems he had encountered in the course of his long life in physics). We will give a mathematical treatment of this situation below, using several different approaches. This problem, conceptually rather clear, can

[1] In the physics of one-dimensional systems it is more common to denote momentum by $k$, and to speak about $2k_F$- or $4k_F$-instabilities, etc.; we follow this convention in this chapter.

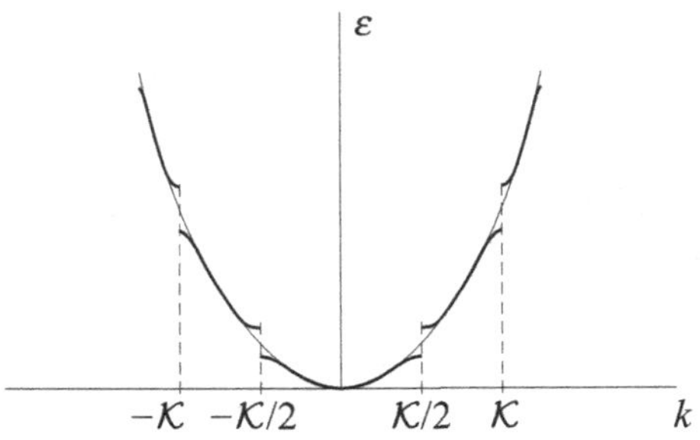


Fig. 11.3

give us a good opportunity to illustrate different theoretical approaches which we have discussed in general above and which are often used to treat this and similar problems. This is why some of the conclusions of this section are repeated several times below – just to see how one can obtain them using different methods.

First, however, we note that the conclusion about the Peierls instability of one-dimensional metallic systems does not apply only to the case of one electron per site and to tight binding (where it gives Peierls dimerization): one would get the same instability also for free electrons in a weak periodic potential, and for arbitrary band fillings.

### *11.1.2 Peierls instability in the general case*

Let us consider in more detail a one-dimensional metal in a nearly free-electron approximation. As is well known, in the presence of a weak periodic lattice potential with Fourier component $U_{\mathcal{K}}$ the electron wavefunction $\Psi(x)$ obeys the Schrödinger equation

$$\left[\frac{k^2}{2m} + U_{\mathcal{K}}\left(e^{i\mathcal{K}x} + e^{-i\mathcal{K}x}\right)\right]\Psi(x) = \varepsilon\Psi(x) . \tag{11.2}$$

Here $\mathcal{K}$ are the Umklapp wavevectors; we have also used the fact that due to symmetry $U_{\mathcal{K}} = U_{-\mathcal{K}}$. Equation (11.2) has solutions in the form of energy bands, with energy gaps at $k = \pm\mathcal{K}n/2$, Fig. 11.3, which appear due to mixing of states with $k$ and $k \pm n\mathcal{K}$ (later on we put the lattice constant $a = 1$).

For weak interactions, when $U_{\mathcal{K}}$ is treated as a perturbation, the energy spectrum $E(k)$ close to the points $\pm\mathcal{K}$ (Brillouin zone boundaries) has the form

$$E(k) = \frac{\varepsilon_k + \varepsilon_{k-\mathcal{K}}}{2} \pm \sqrt{\left(\frac{\varepsilon_k - \varepsilon_{k-\mathcal{K}}}{2}\right)^2 + \Delta^2} , \tag{11.3}$$

where $\varepsilon_k = k^2/2m$ and the energy gap is $\Delta = U_{\mathcal{K}}$.

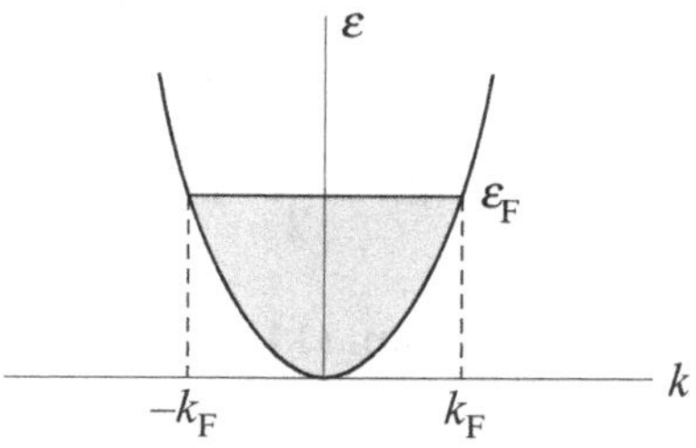


Fig. 11.4

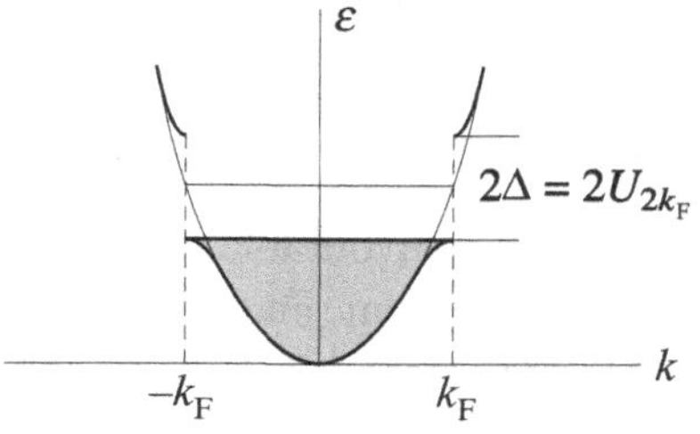


Fig. 11.5

Let us start now with free electrons with the energy band filled up to the Fermi momentum $k_F$, Fig. 11.4. If we now create a periodic potential with period $\tilde{a} = 2\pi a/2k_F$, i.e. with Fourier harmonic $\mathcal{K} = 2k_F$, e.g. if we shift all the atoms of our one-dimensional chain according to the rule

$$u_x = R_x - R_x^0 = \tilde{u}\cos 2k_F x \ , \tag{11.4}$$

then we create energy gaps at $\pm k_F$, i.e. exactly at the Fermi surface, see Fig. 11.5. After such a distortion our system will become an insulator with the gap $2\Delta = 2U_{2k_F}$. The strength of the extra periodic potential $U_{2k_F}$ thus created will be proportional to the amplitude of the distortion $\tilde{u}$.

We see that, similar to the case discussed in Section 11.1.1, after distortion the energies of all occupied states *decrease*. Thus we gain some energy in this process. But by making the lattice deformation (11.4) we also lose elastic energy $\frac{1}{2}B\tilde{u}^2$, where $B$ is the bulk modulus. Therefore the energy balance between the electronic energy gain and the elastic energy loss determines whether such distortion will be energetically favourable, and if it is, what will be the amplitude of corresponding distortion.

We can easily calculate the change in the electronic energy using equation (11.3):

$$\delta\mathcal{E}_{\rm el} = \int_{-k_F}^{k_F} \big[E(k) - \varepsilon(k)\big]\, dk \ . \tag{11.5}$$

This straightforward calculation shows that the dominant term in $\delta\mathcal{E}_{\rm el}$ for weak distortion $\tilde{u}$ or small energy gap $\Delta = U_{2k_{\rm F}} \sim \tilde{u}$ has the form

$$\delta\mathcal{E}_{\rm el} = -\Delta^2 \frac{\rho(\varepsilon_{\rm F})}{2} \ln \frac{\varepsilon_{\rm F}}{\Delta} . \tag{11.6}$$

**Problem:** Check this, using the spectrum (11.2).

**Solution:** The easiest way to get this result is to linearize the spectrum $\varepsilon(k)$ in the vicinity of the Fermi points $\pm k_{\rm F}$, $\varepsilon(k) = v_{\rm F}(|k| \mp k_{\rm F})$, or to transform the integral (11.5) to an integral in $\epsilon$ by the usual rules $\int dk \to \int \rho(\varepsilon)\, d\varepsilon$; for simplicity we consider here the case corresponding to the doubling of the period.

### *11.1.3 Different theoretical ways to treat Peierls distortion*

We see from (11.6) that it is always favourable in 1d metals to make a distortion with the wavevector $q = 2k_{\rm F}$: the electronic energy gain $\sim \Delta^2 \ln(\varepsilon_{\rm F}/\Delta) \sim \tilde{u}^2 \ln \tilde{u}$ is always bigger than the elastic energy loss $\frac{1}{2} B\tilde{u}^2$. To find the value of this distortion, we have to minimize the total energy

$$\mathcal{E}(\Delta) = -\frac{\Delta^2 \rho(\varepsilon_{\rm F})}{2} \ln \frac{\varepsilon_{\rm F}}{\Delta} + \frac{B\tilde{u}^2}{2} , \tag{11.7}$$

but for that we first have to find the connection between $\tilde{u}$ and $\Delta$. We consider below the simplest case of one electron per site, i.e. half-filled tight-binding bands, in which case $2k_{\rm F} = \pi$ and the Peierls transition corresponds to *dimerization* – often one uses the term 'Peierls transition' just for this case. (We put here the original lattice constant $a = 1$.)

The exact relation between $U_{2k_{\rm F}} = \Delta$ and the distortion $\tilde{u}$ can be found from the following considerations. Usually one writes the electron–lattice interaction in the form

$$\mathcal{H}_{\rm e-ph} = \sum_{k,q,\sigma} g(q)\, c^\dagger_{k-q,\sigma}\, c_{k,\sigma}\, (b^\dagger_q + b_{-q}) \tag{11.8}$$

(*Fröhlich Hamiltonian*, see Section 8.6). The classical treatment corresponds to treating lattice displacements with period $q = 2k_{\rm F}$ as static, i.e. substituting $b^\dagger_q + b_{-q} = 2b\,\delta(q - 2k_{\rm F})$. Separating terms with this $q$ and denoting $g(2k_{\rm F}) = \tilde{g}$, $\omega(2k_{\rm F}) = \tilde{\omega}$, we see that the Fourier component of the extra potential $U_{2k_{\rm F}}$, equal to the energy gap $\Delta$, is

$$U_{2k_{\rm F}} = \Delta = 2\tilde{g}b . \tag{11.9}$$

Remembering now that the distortion $u$ is connected to the phonon operators by $u = \frac{1}{\sqrt{2M\omega}}(b^\dagger + b)$ (see (4.10)), we have the connection between the energy gap

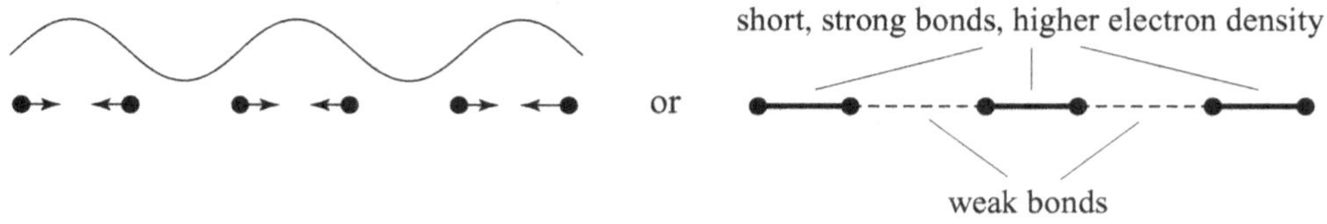


Fig. 11.6

and the distortion $u(2k_{\mathrm{F}}) = \tilde{u}$ in the form

$$\Delta = \tilde{g}\sqrt{2M\tilde{\omega}} \cdot \tilde{u} \,. \qquad (11.10)$$

Putting (11.10) into (11.7) and minimizing $\mathcal{E}(\Delta)$ with respect to $\Delta$, we obtain finally in the weak coupling case ($\Delta \ll \varepsilon_{\mathrm{F}}$, $\tilde{g} \to 0$)

$$\Delta = \varepsilon_{\mathrm{F}}\, e^{-1/2\lambda} \,, \qquad (11.11)$$

where the dimensionless electron–phonon coupling constant $\lambda$ is

$$\lambda = \frac{\tilde{g}^2 \rho(\varepsilon_{\mathrm{F}})}{\tilde{\omega}} \qquad (11.12)$$

(here we have taken into account that $\tilde{\omega}^2 = B/M$).

For arbitrary band filling ($2k_{\mathrm{F}} \neq \pi$) the period of the superstructure in general will be incommensurate with the underlying lattice period; in this case there will be not $1/2\lambda$, but $1/\lambda$ in the exponent in (11.11).[2] The instability of a one-dimensional metal towards lattice distortion with wavefactor $2k_{\mathrm{F}}$ – the Peierls instability – simultaneously means the creation of *charge-density wave* (the short notation, which is now widely used, is CDW). For instance when the period is doubled, the electron density alternates at consecutive bonds, see Fig. 11.6; this will be mathematically shown below.

The fact that our one-dimensional metallic system is unstable with respect to the formation of CDW or to Peierls distortion is connected with the logarithmic divergence of the response function or the polarization operator (giant Kohn anomaly), cf. (9.20), (9.21), (9.30):

$$\Pi(q, \omega = 0) \sim \frac{1}{\pi v_{\mathrm{F}}} \frac{2k_{\mathrm{F}}}{q} \ln\left|\frac{1 + q/2k_{\mathrm{F}}}{1 - q/2k_{\mathrm{F}}}\right| \,. \qquad (11.13)$$

One can see this instability if one considers phonon renormalization: as discussed in Chapter 9 (see (9.34)) the phonon frequency $\omega(q)$ becomes imaginary at $q = 2k_{\mathrm{F}}$, which tells us that if we start from a homogeneous undistorted metallic chain, the

[2] The factor of 2 in equation (11.11) is characteristic of a half-filled band (Peierls dimerization) and is due to Umklapp processes (in this case the processes with momentum transfer $2k_{\mathrm{F}} = \pi$ are equivalent to the Umklapp process with $\mathcal{K} - 2k_{\mathrm{F}} = 2\pi - \pi = \pi$, which doubles the coupling constant $\lambda$ in equation (11.11)).

increment of the vibrations with $q = 2k_F$ will be positive, and a macroscopic standing wave with this period will be created, i.e. Peierls distortion will occur.

There are different theoretical methods to consider this situation. One of them is the 'classical' treatment described above. We can also proceed directly from the electron–phonon Hamiltonian

$$\mathcal{H} = \sum_{k,\sigma} \varepsilon(k)\, c^\dagger_{k,\sigma}\, c_{k,\sigma} + \sum_{k,q,\sigma} g(q)\, c^\dagger_{k-q,\sigma}\, c_{k,\sigma}\, (b^\dagger_q + b_{-q}) + \sum_q \omega(q) b^\dagger_q b_q\ . \tag{11.14}$$

We expect that the phonon mode with $q = 2k_F$ will be macroscopically occupied, i.e. it will behave classically. Accordingly we replace operators $b_{q=\pm 2k_F}$, $b^\dagger_{\pm 2k_F}$ by c-numbers, $\tilde{b} = \langle b_{\pm 2k_F}\rangle$. Keeping only these main terms, we have

$$\mathcal{H} = \sum_{k,\sigma} \varepsilon(k)\, c^\dagger_{k,\sigma}\, c_{k,\sigma} + \sum_{k,\sigma} \tilde{g}\, c^\dagger_{k\pm 2k_F,\sigma}\, c_{k,\sigma} \cdot 2\tilde{b} + \tilde{\omega}\tilde{b}^2\ . \tag{11.15}$$

Here $\tilde{g} = g(2k_F)$, $\tilde{\omega} = \omega(2k_F)$; for simplicity we take $b$ to be real (it is quite easy to include also the phase of $\tilde{b}$). We also omit for simplicity $\pm$ signs at $2k_F$, remembering only that we should take symmetrized sums when necessary, and take care of momentum conservation.

We can proceed in several formally different but equivalent ways. One is to use the Bogolyubov canonical transformation: the Hamiltonian (11.15) is already quadratic, but not yet diagonal; it contains terms $c^\dagger_{k\pm 2k_F} c_k$. This is done below, see equations (11.24)–(11.25). Another possible way is to write down the average energy $\mathcal{E} = \langle \mathcal{H} \rangle$ which will have terms of the type

$$\mathcal{E} = \tilde{g} \left\langle \sum_{k,\sigma} c^\dagger_{k-2k_F,\sigma}\, c_{k,\sigma} \right\rangle \cdot 2\tilde{b} + \tilde{\omega}\tilde{b}^2 + \mathcal{E}_0 \tag{11.16}$$

(and similar terms with $k + 2k_F$). Minimizing $\mathcal{E}$ with respect to $\tilde{b}$, we obtain

$$\tilde{b} = -\frac{\tilde{g}}{\tilde{\omega}} \left\langle \sum_{k,\sigma} c^\dagger_{k-2k_F,\sigma}\, c_{k,\sigma} \right\rangle . \tag{11.17}$$

We see from (11.17) that if, as we have assumed, $\tilde{b} = \langle b_{q=2k_F}\rangle \neq 0$, then the average $\langle \sum_{k,\sigma} c^\dagger_{k-2k_F,\sigma} c_{k,\sigma}\rangle$ is also nonzero. One can show that this implies modulation of the electron density with the same wavevector $2k_F$, i.e. the formation of a charge-density wave mentioned above. Indeed the electron density is

$$\hat{\rho}(x) = \sum_q \Psi^\dagger_\sigma(x)\, \Psi_\sigma(x) = \sum_q \rho_q\, e^{iqx}\ . \tag{11.18}$$

Its Fourier transform is

$$\rho = \sum_{k,\sigma} c^{\dagger}_{k-q,\sigma}\, c_{k,\sigma} \,. \tag{11.19}$$

(The product in coordinate space, here $\Psi^{\dagger}(x)\,\Psi(x)$, goes over to a convolution in momentum space.) Thus the average density $\langle \rho_q \rangle$ is equal to $\left\langle \sum_{k,\sigma} c^{\dagger}_{k-q,\sigma}\, c_{k,\sigma} \right\rangle$.

If we were to have, as in the normal case, that only $\langle c^{\dagger}_k c_k \rangle \neq 0$, i.e. $\langle c^{\dagger}_{k-q} c_k \rangle = \rho_0 \delta(q)$, then, from (11.18), $\big\langle \rho(x) \big\rangle = \rho_0 = \text{const}$. In our case, however, we have $\langle b_{q=2k_{\mathrm{F}}} \rangle \neq 0$, and, from (11.17), also $\langle \rho_{2k_{\mathrm{F}}} \rangle \neq 0$ (11.19). In other words

$$\langle \rho_q \rangle = \rho_0 \delta(q) + \rho_1 \delta(q - 2k_{\mathrm{F}}) \,, \qquad \rho_1 = \left\langle \sum_{k,\sigma} c^{\dagger}_{k-2k_{\mathrm{F}},\sigma}\, c_{k,\sigma} \right\rangle . \tag{11.20}$$

From (11.17) we see then that there appears modulation of the electron density with period $(2k_{\mathrm{F}})^{-1}$:

$$\big\langle \rho(x) \big\rangle = \rho_0 + \rho_1 e^{i 2k_{\mathrm{F}} x} \Longrightarrow \rho_0 + \rho_1 \cos(2k_{\mathrm{F}} x) \,. \tag{11.21}$$

(In a more accurate treatment of course $\big\langle \rho(x) \big\rangle = \mathrm{Re}\left[\rho_0 + \rho_1 e^{i 2k_{\mathrm{F}} x}\right]$, or, in other words, if the Fourier harmonic with $q = 2k_{\mathrm{F}}$ enters in (11.20), so also does the harmonic with $q = -2k_{\mathrm{F}}$, which gives $\delta\rho(x) \sim \cos(2k_{\mathrm{F}} x)$ as in (11.21).)

From (11.20), (11.21) and (11.17) we also see that the electron density wave, or CDW, is proportional to the lattice distortion. Thus indeed the Peierls transition goes hand in hand with charge-density wave formation; these are different sides of the same phenomenon. Usually people speak about a Peierls transition when dealing with one-dimensional systems, and use the terminology 'CDW' for similar phenomena in 2d or 3d materials (such transitions may occur in these cases when special conditions – nesting of the parts of the Fermi surface – are fulfilled, see equation (9.37) and discussions thereof).

There is yet another approach to our problem. We can exclude phonon variables and go over to an effective electron–electron interaction. The correct way to do this would be, e.g. by writing down the electron–electron vertex due to the exchange of a phonon,

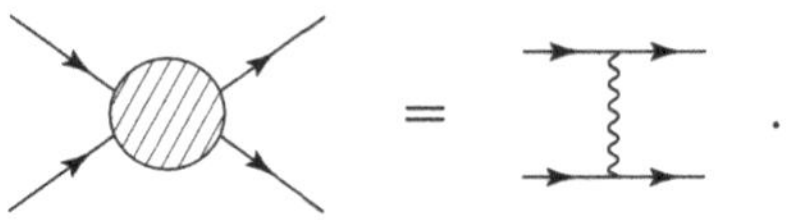

One can also proceed in a less rigorous way, starting from equation (11.15). Putting the expression for $\tilde{b}$ (11.17) back into (11.15) we would get

$$\mathcal{H} = \sum_{k,\sigma} \varepsilon(k)\, c^{\dagger}_{k,\sigma}\, c_{k,\sigma} - \frac{q^2}{\tilde{\omega}} \sum_{kk',\sigma\sigma'} c^{\dagger}_{k-2k_{\rm F},\sigma}\, c_{k,\sigma} \langle c^{\dagger}_{k'+2k_{\rm F},\sigma'}\, c_{k',\sigma'} \rangle$$
$$+ \Big[ (-2k_{\rm F}) \rightarrow (+2k_{\rm F}) \Big] . \qquad (11.22)$$

One may treat this as a mean field decoupling of the four-fermion interaction

$$\mathcal{H}_{\rm int} = -\frac{\tilde{g}^2}{\tilde{\omega}} \sum_{kk',\sigma\sigma'} c^{\dagger}_{k-q,\sigma}\, c_{k,\sigma}\, c^{\dagger}_{k'+q,\sigma'}\, c_{k',\sigma'} \,, \qquad (11.23)$$

with $q = \pm 2k_{\rm F}$; this is exactly the interaction we would obtain by excluding phonons as mentioned above.

We see that the exchange of phonons leads to an effective electron–electron attraction. (Actually such interactions will be retarded, or frequency-dependent; our treatment is carried out in the static limit, $\omega \rightarrow 0$. This attraction is in fact the mechanism of Cooper pair formation and of superconductivity in ordinary superconductors.) We thus see that we can proceed in two ways; either treat electron–phonon interactions in an apparent way, or first exclude phonons and then consider electrons with an effective attractive interaction. We would get essentially the same instability, which in the first method would be ascribed to lattice distortions – here we use the term 'Peierls transition'; in the second approach we would speak of a CDW transition. If there exists an electron–electron attraction of any other origin besides the electron–phonon interaction, it will also lead in the 1d case to CDW instability; the lattice will of course then follow. Note right away that if there exists not an attraction but an effective *repulsion* between electrons, 1d systems would still be unstable, however not with respect to *charge*-density wave, but to *spin*-density wave (SDW) formation – see below, Section 11.6.

The treatment given above is not yet the complete solution. We have to make only one more step. Actually we have already obtained the Hamiltonian (equation (11.15) or (11.22)) which is quadratic in the electron operators. It can be written as

$$\mathcal{H} = \sum_{k,\sigma} \Big[ \varepsilon(k)\, c^{\dagger}_{k,\sigma}\, c_{k,\sigma} + \Delta c^{\dagger}_{k-2k_{\rm F},\sigma}\, c_{k,\sigma} + \text{h.c.} \Big] \qquad (11.24)$$

with (as yet undetermined) $\Delta = \tilde{g}\tilde{b} = -\frac{\tilde{g}^2}{\tilde{\omega}} \big\langle \sum_{k,\sigma} c^{\dagger}_{k-2k_{\rm F},\sigma}\, c_{k,\sigma} \big\rangle$, which should be found later self-consistently.

We see that the situation now resembles the case of Bose condensation, Chapter 5, or the treatment of antiferromagnetic magnons, Chapter 7: the Hamiltonian (11.24) is quadratic, but still nondiagonal; it contains terms $c^\dagger_{k-2k_F,\sigma}\, c_{k,\sigma}$. Thus we again have to use the Bogolyubov canonical transformation to diagonalize it:

$$\begin{aligned} \alpha_{k,\sigma} &= u_k c_{k,\sigma} + v_k c_{k-2k_F,\sigma} \\ \beta_{k,\sigma} &= v_k c_{k,\sigma} - u_k c_{k-2k_F,\sigma} \; . \end{aligned} \tag{11.25}$$

With a proper choice of coefficients $u_k$, $v_k$ the Hamiltonian (11.24) now takes the form

$$\mathcal{H} = \sum_{k,\sigma} E(k)(\alpha^\dagger_{k,\sigma}\alpha_{k,\sigma} + \beta^\dagger_{k,\sigma}\beta_{k,\sigma}) + \text{const.} \; , \tag{11.26}$$

with the spectrum (11.3) (with $\mathcal{K} = 2k_F$), as of course it should. This transformation is also very similar to the well-known transformation in the theory of superconductivity, see below, Section 11.5, equation (11.50). The difference is that whereas in the latter case we mix electrons and holes, $u_k c_{k,\sigma} + v_k c^\dagger_{-k,-\sigma}$, here we mix two electrons with different wavevectors. Canonical transformations in the theory of superconductivity imply nonzero averages of the type $\langle c_{k,\sigma} c_{-k,-\sigma}\rangle$, the famous Cooper pairs. Analogously, the transformation (11.25) corresponds to nonzero averages $\langle c^\dagger_{k-2k_F,\sigma}\, c_{k,\sigma}\rangle$, see (11.17). We may call them electron–hole pairs, or excitons. Thus we can speak here about an electron–hole, or exciton condensate.

Note also that this procedure (mixing electron states with momenta $k$ and $k \pm 2k_F$) is essentially the same procedure as is usually done in treatments of the energy spectrum and band formation in a periodic potential, e.g. in the weak coupling approximation, see (11.2), (11.3) and any textbook on solid state physics. Similar to this case, we have obtained here that after this transformation the spectrum acquires a gap at $k = \pm 2k_F$, i.e. exactly at the position of the initial Fermi surface, in agreement with equation (11.3) and Fig. 11.5.

Thus we see that there are several, technically somewhat different, but conceptually very close ways to treat Peierls distortion: a purely classical treatment of equations (11.7)–(11.12); a treatment using the electron–phonon (Fröhlich) Hamiltonian with mean field decoupling; the transition to the effective electron–electron interaction by excluding phonons, with similar mean field decoupling at this level. At this (mean field) level all these methods are equivalent and give the same results. However, if one would want to go beyond a self-consistent treatment and consider fluctuation effects, both classical and quantum, the different starting points (electron–phonon model, or the effective electron–electron one) would become more, or less appropriate, depending on the effects studied.

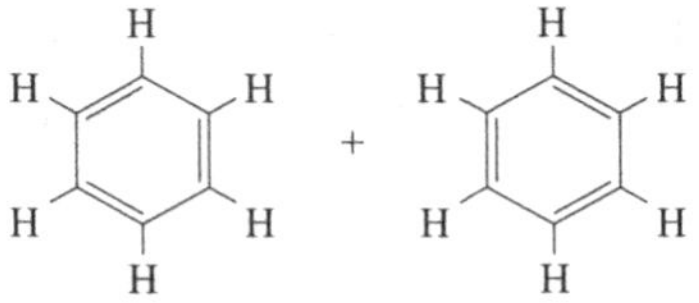

Fig. 11.7

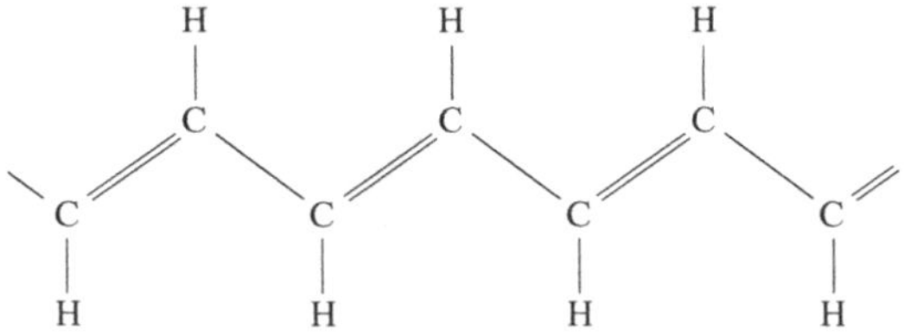

Fig. 11.8

### *11.1.4 Peierls distortion and some of its physical consequences in real systems*

Peierls distortion is observed experimentally in a number of quasi-one-dimensional compounds. Although formally in a pure 1d system there should be no phase transition, in systems consisting of many chains, with weak interchain interaction, the Peierls transition actually occurs as a real phase transition.

Probably the best-known and the clearest example of a system whose properties are described using this picture is met in organic materials with conjugate bonds. Carbon is usually four-valent, but often its s- and p-electrons, due to $sp^2$ hybridization, form three strong chemical bonds (called $\sigma$-bonds) lying in one plane at 120° to one another, and the remaining electron occupies the perpendicular $p_z$-orbital. This 'extra' electron (one per carbon ion) is called a $\pi$-electron. These extra electrons can either localize at certain bonds, giving what is called in chemistry double bonds, or $\pi$-bonds; or these electrons may be delocalized, forming broad bands. They are in fact responsible for electric conduction in graphite, or in (very popular nowadays) graphene, single-layer sheets of graphite. In small organic molecules these double bonds can give rise to 'resonating' structures, e.g. in the benzene molecule $C_6H_6$, see Fig. 11.7 (this actually gave rise to the notion of resonating valence bonds, discussed above, in Chapter 6). And in some cases these half-filled $\pi$-bands (one $\pi$-electron per site!) may show the phenomenon of Peierls distortion. A good example is polyacetylene $C_nH_n$, which is usually depicted as shown in Fig. 11.8. In chemists' notation double bonds correspond to shorter distances between carbon atoms. This structure can be visualized as originating from the homogeneous one  as a result of Peierls distortion with the doubling of the period (which actually would have occurred at temperatures much above the temperature of decomposition of this substance).

There is one very interesting aspect of the physics of polyacetylene and other similar systems. In this particular case (doubled period) there exist two equivalent configurations:  and . One can form a defect – a domain wall . Similarly to one-dimensional magnets, this defect may propagate along the chain, i.e. it forms an elementary excitation, a soliton. One can see that there are different possible states of such an object. It may be neutral: each carbon usually has four electrons (four bonds in Fig. 11.8). When we create a defect of the type , an unpaired electron may remain at the appropriate carbon atom (actually it will be delocalized over a certain distance $\xi$). In this case the object, the soliton, will be neutral: there will again be four electrons at this site. However, in contrast to the initial case, when all electrons participate in valence bonds and form singlets, such a neutral soliton will have an *unpaired spin* $\frac{1}{2}$. Thus it is an elementary excitation which carries *spin but no charge*.

But we can also create charged excitations. Let us start from the neutral soliton described above, and let us remove an electron (take it to infinity). The remaining object will have uncompensated positive charge $+1$, but no spin. Similarly, we can put onto a neutral soliton an extra electron which will form a singlet pair with the existing unpaired one. The resulting object will have charge $-1$ and again no spin. Thus we see that, in contrast to the usual electrons in metals or semiconductors, which carry both charge and spin, there is here a spin–charge separation: neutral solitons carry spin $\frac{1}{2}$, and charged solitons carry charge $\pm 1$ but no spin. Such solitons were indeed observed in polyacetylene. (This picture was developed by Su, Schrieffer and Heeger, and these solitons are sometimes called SSH solitons.)

As we see, there exists a close analogy between the properties of one-dimensional systems with Peierls distortion (neutral solitons carrying spin $\frac{1}{2}$, charged soliton with spin 0) and the properties of elementary excitations (spinons and holons) in the RVB state discussed in Chapter 6. This is not accidental: actually the dimerization in one-dimensional materials leads in the strong coupling case to the formation of the usual valence bonds (singlet states of two electrons $\frac{1}{\sqrt{2}}(1\uparrow 2\downarrow - 1\downarrow 2\uparrow)$) well known in chemistry; we have actually already used this when we depicted possible states in polyacetylene. Thus one can say that the Peierls dimerization of a half-filled one-dimensional metal is a first step towards the formation of a molecular crystal, like molecular hydrogen, consisting of $H_2$ molecules. Actually, this property of spin–charge separation can be really proven only in 1d systems, whereas its existence in higher dimensions (e.g. in the 2d case discussed in Chapter 6) is still hypothetical. And, as mentioned in Chapter 10, such spin–charge separation is responsible for the formation of the Luttinger liquid state, one of the very few cases in which the existence of a non-Fermi-liquid state is really established theoretically.

(Note that the treatment presented above, which gave an insulating state due to a Peierls or CDW transition, was actually a mean field treatment implicitly relying on the interchain coupling in a quasi-one-dimensional system. Formally in the purely 1d case quantum fluctuations would be so strong that they would destroy this mean field long-range order, and consequently there will appear no gap in the energy spectrum. But the spectrum itself would be very different from the usual electron spectrum; there will be spinon and holon excitations. This resulting state is just the Luttinger liquid state mentioned above.)

One more interesting consequence of the picture described above becomes clear if we consider Peierls transitions for a different band filling, e.g. corresponding not to dimerization, but to tripling of the period. It is clear that there will now be *three* possible degenerate ground states, e.g.

$$\begin{array}{cccccccc} = & - & - & = & - & - & & (1) \\ - & = & - & - & = & - & & (2) \\ - & - & = & - & - & = & & (3) \end{array} \tag{11.27}$$

and correspondingly more (essentially three independent) types of domain walls, or solitons. Arguments similar to those given above lead to a striking conclusion: the charge of these solitons will not be $\pm e$, but will be fractional ($\pm\frac{2}{3}e$), like quarks in elementary particle theory![3]

Returning to the case of double periodicity one can show that energetically the solitons are indeed the most favourable excitations. The spectrum of one-particle excitations (electrons and holes) is given by equation (11.3) which in the case of a half-filled tight-binding band looks like the one shown in Fig. 11.9, i.e. it consists of two subbands, Fig. 11.10. Consequently, the lowest electron–hole excitations (for a fixed lattice) have minimum energy $2\Delta$. It turns out that the soliton level lies exactly in the middle of the energy gap, Fig. 11.11, thus it costs less energy to create such an excitation. In the symmetric case the chemical potential $\mu$ also lies in the middle of the gap, so it seems that formally the creation of a soliton would cost *no* energy. However, to make the soliton we also have to change the lattice configuration, e.g. create two neighbouring long bonds; this also costs some

[3] Indeed the simplest excitation which keeps the state the same at the right and left ends of the molecule would be the one in which we remove one double bond from one of the states, e.g. from the state (1) in (11.27). But then we can move the double bonds so as to create a domain wall of the type $= \boxed{- - -} = - - =$ (three single bonds in a row). We go back to the original domain (1) after three such steps, such domain walls, i.e. the missing charge $-2e$ (two removed electrons) will now be equally split into three solitons, or domain walls, and thus the charge of each of them will be $-\frac{2}{3}e$.

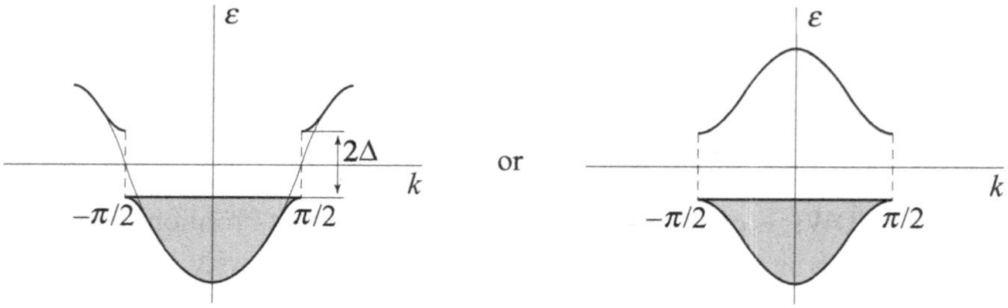


Fig. 11.9

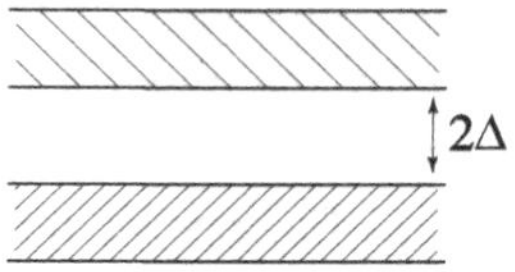


Fig. 11.10

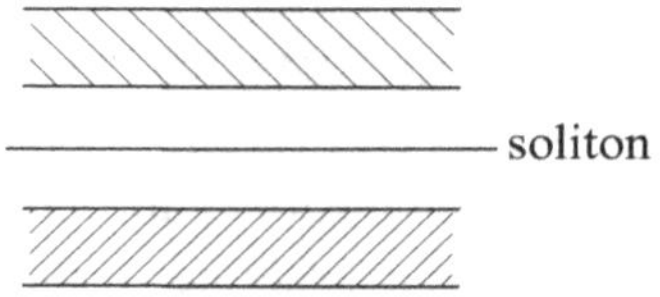


Fig. 11.11

energy, so that the resulting energy of the soliton is not $\Delta$, but $\frac{2}{\pi}\Delta$. Nevertheless it is still more favourable to create two charged solitons (or soliton and antisoliton) instead of an independent electron–hole pair; it costs $\frac{4}{\pi}\Delta$ instead of $2\Delta$. Similarly charge carriers introduced by doping also create solitons and are bound to them.

An interesting situation may occur if such a CDW superstructure is incommensurate with the underlying lattice period. In this case the density wave which has the form $\rho(x) = \rho_0 \cos(2k_{\mathrm{F}}x + \varphi)$ may have an arbitrary phase, or may be located at an arbitrary position relative to the lattice. There exists a collective mode – oscillations of the phase $\varphi$, the so-called phasons, with the gapless spectrum

$$\Omega(q) = a\, v_{\mathrm{F}}\, q \;, \qquad a^{-1} = 1 + \frac{4\Delta^2}{\lambda\omega_0^2} \;. \tag{11.28}$$

Physically such a mode (it is also called a Fröhlich collective mode) for $q \to 0$ corresponds to a CDW which *slides* along the chain, Fig. 11.12. Such a sliding mode contributes to the conductivity of the system, because it carries with itself an extra charge; at finite frequency it is optically active. This mechanism of conductivity,

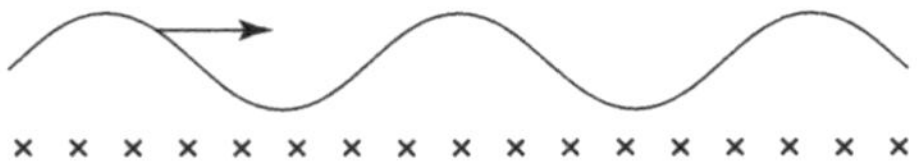

Fig. 11.12

due to sliding CDWs, is called *Fröhlich conductivity*; in fact Fröhlich suggested this picture in the 1950s as a possible explanation of superconductivity. It turned out later that the real superconductivity is explained differently, but Fröhlich conductivity was observed recently in quasi-one-dimensional system $NbSe_3$ and in several other materials. (Usually a CDW is pinned to the lattice both by commensurability effects and, e.g. by impurities, defects, etc., thus one needs a certain critical electric field to set it in motion. Thus experimentally the fingerprint of conductivity due to sliding CDW is the existence of such a threshold, or nonlinear $I$–$V$ characteristics. Another experimental check is the presence of a specific noise spectrum in conductivity.)

## 11.2 Spin-Peierls transition

There exists another class of systems for which Peierls instability may be important: these are (quasi-)one-dimensional antiferromagnets. We have already discussed in Chapter 6 the properties of one-dimensional antiferromagnets and have shown that the valence bond wavefunction, the structure of the type $\times\!\!-\!\!\!-\!\!\times \quad \times\!\!-\!\!\!-\!\!\times$, gives a lower energy than the usual two-sublattice Néel structure even without resonance, see (6.114)–(6.118). In chemical language this corresponds to the formation of valence bonds, $\frac{1}{\sqrt{2}}(1\!\uparrow\!2\!\downarrow - 1\!\downarrow\!2\!\uparrow)$, as in molecular hydrogen $H_2$. Actually in the spin chain with a fixed lattice and equal distances between spins there is an even better state in which there exists *resonance* between these bonds. However, if we 'release' the lattice and permit it to adjust to the spin structure, it may turn out that the state of the type drawn above may become preferential: there may occur a *dimerization* of the lattice, leading to an alternation of the exchange constants,

$$J_{n,n+1} = J_0 + \delta J \cdot (-1)^n = J_0 \pm \delta J \ . \tag{11.29}$$

One can easily see that this is indeed the case in the one-dimensional $xy$ model (6.109). We have seen in Chapter 6 that this model can be mapped by a Jordan–Wigner transformation onto the model of noninteracting spinless fermions with the spectrum $\varepsilon_k = 2J\cos k$, see (6.111). As discussed there, the ground state in this case corresponds to a half-filled band, Fig. 11.13. It is evident that when we now take into account interaction with the lattice and allow the lattice to distort, there will occur the same Peierls transition, which in this case is called a *spin-Peierls transition*. (Of course, here again we have to include coupling to neighbouring spin

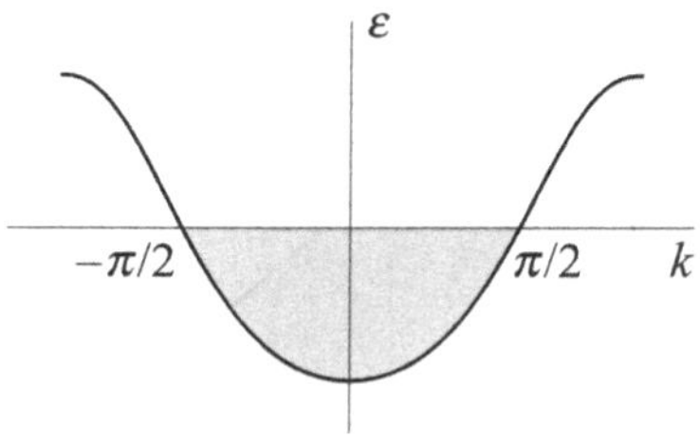


Fig. 11.13

chains, otherwise quantum fluctuations would suppress such long-range order. Usually one invokes here an elastic interaction between distortions on different chains; lattice distortion is here treated classically.)

An effective spin–lattice interaction (or fermion–lattice interaction) originates from the distance dependence of the exchange interaction

$$J(R)\boldsymbol{S}_i \cdot \boldsymbol{S}_{i+1} = J(R_0 + u)\boldsymbol{S}_i \cdot \boldsymbol{S}_{i+1} = \left[J_0 + \frac{\partial J}{\partial R}\,u\right]\boldsymbol{S}_i \cdot \boldsymbol{S}_j$$
$$= J_0\boldsymbol{S}_i \cdot \boldsymbol{S}_{i+1} + g\boldsymbol{S}_i \cdot \boldsymbol{S}_{i+1}\,u \qquad (11.30)$$

i.e. the spin–lattice coupling constant is $g \sim \partial J/\partial R$. Similarly to (11.11), there will be a distortion and a gap in the spectrum of spin excitations

$$\Delta \sim J\,e^{-1/\lambda}\,, \qquad \lambda \sim \frac{g^2}{\omega_0 J}\,. \qquad (11.31)$$

Consequently, e.g. the magnetic susceptibility of such a system will behave as $\chi \sim e^{-\Delta/T}$ at temperatures below the spin-Peierls transition temperature. The typical behaviour of susceptibility in this case is shown schematically in Fig. 11.14. The susceptibility follows the Curie–Weiss law at high temperatures, then passes through the maximum at about $T^* \sim 0.7J$ (called the Bonner–Fisher maximum), and at lower temperature $T_c \sim \Delta$ (11.31) the system has a spin-Peierls phase transition, below which the susceptibility decreases exponentially. One can show (this was done by Cross and Fisher) that the Heisenberg model is even more unstable with respect to the spin-Peierls transition than the $xy$ model, so that the energy gap is not proportional to the lattice displacement $u$ as in equation (11.10), but to $\Delta \sim u^{4/3}$.

An interesting situation occurs when we put a spin-Peierls system in an external magnetic field $H$. As discussed in Chapter 6, such a field plays the role of the chemical potential of Jordan–Wigner fermions,

$$\mathcal{H} = \sum_n 2J\cos k\, c_k^\dagger c_k - H\sum_n c_k^\dagger c_k\,. \qquad (11.32)$$

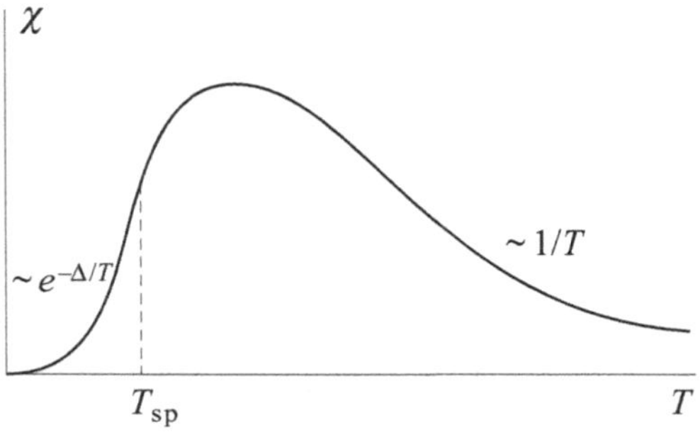


Fig. 11.14

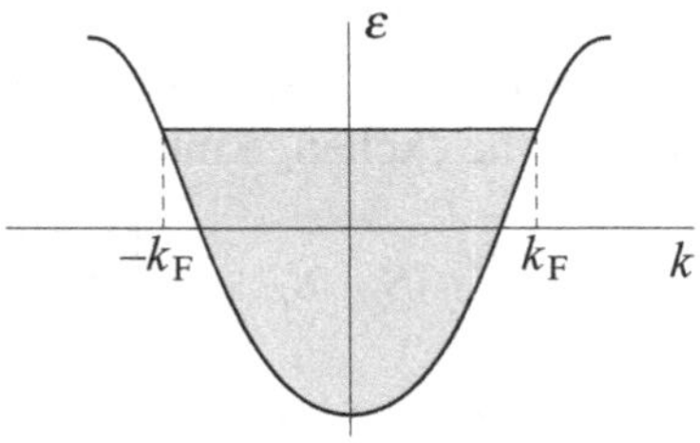


Fig. 11.15

Consequently, for $H \neq 0$ the situation would correspond to a fermion band filled by more (or less) than one fermion per site, Fig. 11.15. This means that the wavevector of the instability $Q = 2k_F$ will be different from $\pi$, i.e. there will occur not a simple dimerization, but there will appear instead a superstructure with $Q = 2k_F$, in general incommensurate with the initial crystal lattice, and the period will depend on the magnetic field. Actually due to the fact that for the dimerization the effective coupling constant is twice as large as that for other periods (see (11.11) and discussions there), the energy gain for the dimerization is larger than for other periods. Therefore, for small enough fields (band close to being half-filled) the dimerized state will still be the most favourable one (pinning of the distortion in the commensurate structure), but away from this 'region of attraction to dimerization' the period will indeed change and become in general incommensurate. This problem may be solved at least close to $T_c$, and the resulting phase diagram has the form shown in Fig. 11.16, where $*$ denotes the point of cross-over to the incommensurate structure, which is an example of the Lifshits point discussed in Section 2.4.

The phenomenon of the spin-Peierls transition was experimentally observed in several quasi-one-dimensional organic compounds and recently in the inorganic compound $CuGeO_3$ which contains chains of ions $Cu^{2+}$ with spin $\frac{1}{2}$. The phase

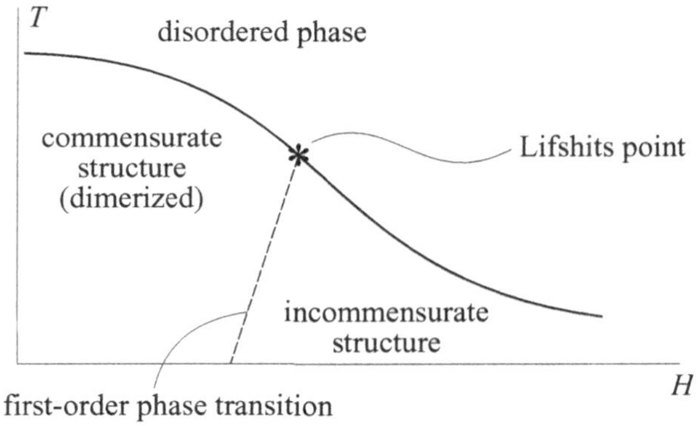


Fig. 11.16

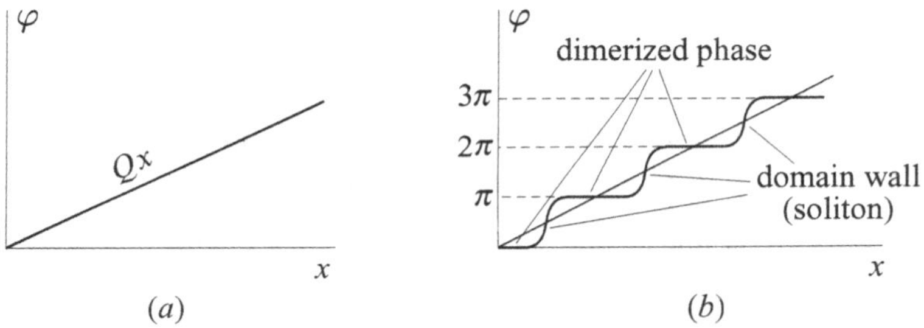


Fig. 11.17

diagram of the type shown in Fig. 11.16 (which was already presented in the general discussion in Section 2.4) has indeed been observed experimentally.

There is yet another very interesting aspect of this story. In the first approximation the distortion in the incommensurate phase is indeed simply $u(x) = \tilde{u}\, e^{iQx}$, with the period $l = 2\pi/Q$ in general incommensurate with the underlying lattice period. We can also write this down as $u(x) = \tilde{u}\, e^{i\varphi(x)}$, with the phase $\varphi(x) = Qx$, see Fig. 11.17($a$). However, in a more detailed treatment the picture is different. On average the distortion is similar to the previous case, but actually there appears inhomogeneous distortion, with the domains of the dimerized phase divided by domain walls. This is illustrated schematically in Fig. 11.17($b$).[4] These domain walls are actually the same solitons discussed above, with the difference that here they are not excitations, but exist in the ground state and form a soliton lattice. Inside the commensurate domains, spins are paired into singlets, but at each domain wall there is an unpaired spin. All these spins are parallel to the external magnetic

[4] Interestingly, such a curve appeared in 1904 on the first pages of the science fiction novel *The Food of the Gods* by H. G. Wells, and with a rather similar scientific content!

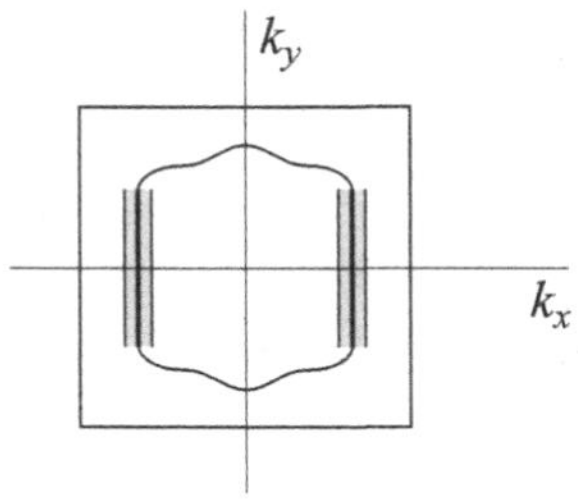


Fig. 11.18

field, and it is this energy gain which in fact leads to the formation of such a state. Interestingly enough, there may exist also new excitations – oscillations of this new 'superlattice', similar to ordinary phonons in crystals, etc. Such a soliton lattice seems to be observed experimentally in $CuGeO_3$.

## 11.3 Charge-density waves and structural transitions, higher-dimensional systems

The idea of the Peierls transition was first put forth for one-dimensional systems. It was realized later that in principle one may have similar situations also in 2d and 3d systems with special features of the electron energy spectrum. As already mentioned in Chapter 9, there may exist situations with *nesting* of the whole or parts of the Fermi surface, see equation (9.37). If the energy spectrum satisfies the condition $\varepsilon(\boldsymbol{k} + \boldsymbol{Q}) = -\varepsilon(\boldsymbol{k})$ (with energies counted from the Fermi energy), the response functions and polarization operator will behave essentially similarly to the one-dimensional case, see (11.13). Consequently, such systems will be unstable with respect to CDW formation and lattice distortion with the wavevector $\boldsymbol{Q}$ (if the dominant interaction is the electron–phonon interaction; however if the dominant electron–electron interaction is repulsive, we will have similar instability in the spin channel, and a spin-density wave – SDW – will be formed instead, see Section 11.6 below).

If the condition $\varepsilon(\boldsymbol{k} + \boldsymbol{Q}) = -\varepsilon(\boldsymbol{Q})$ is not fulfilled over the whole Fermi surface, but only parts of it are nested, such transitions may still occur; this will lead to the creation of an energy gap at appropriate parts of the Fermi surface, leaving the remaining parts intact; see the schematic picture in Fig. 11.18. This phenomenon is observed in some layered transition metal dichalcogenides ($NbSe_2$; $TaS_2$). In some of them (e.g. $1T$–$TaS_2$) when there is strong nesting and the gap covers nearly the whole Fermi surface, the material becomes practically insulating. In others, e.g. in $2H$–$NbSe_2$, there occurs partial gapping of the Fermi surface, but the remaining parts of the Fermi surface make these materials still metallic below $T_{\mathrm{CDW}}$ ($NbSe_2$

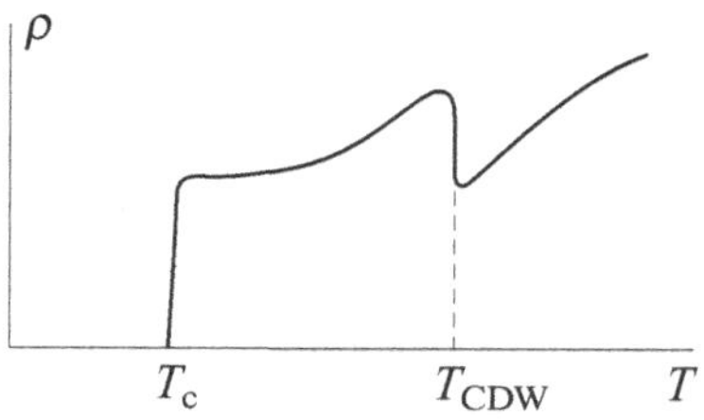


Fig. 11.19

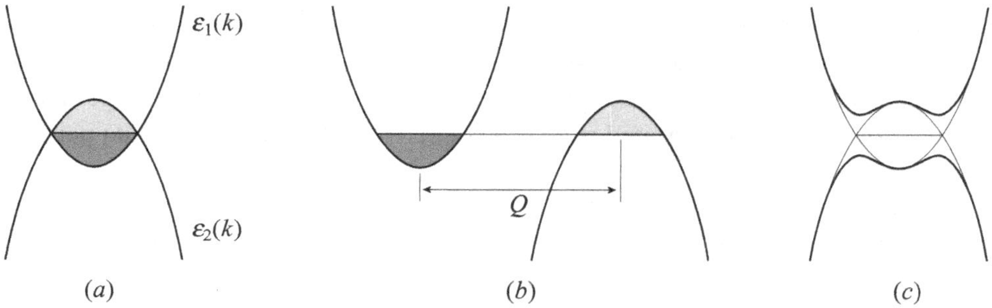


Fig. 11.20

even becomes superconducting at still lower temperatures). One may expect that the typical behaviour of resistivity in this case may look as shown in Fig. 11.19: the removal of a part of the Fermi surface (reduction of the number of carriers) can increase resistivity at $T_{CDW}$ (and the system may still become superconducting at $T_c < T_{CDW}$). However, the results may be even opposite and the resistivity may decrease below CDW formation, if in this process the scattering of electrons decreases; this seems to be the case in $NbSe_2$.

## 11.4 Excitonic insulators

(The following section has a somewhat more technical character. But the methods of the theoretical treatment of the situation discussed below are widely used now in different situations, and I think this is a good opportunity to learn how these methods work.)

Phenomena very similar to those considered in the previous sections may occur in another situation – in the so-called *excitonic insulators*. These are systems in which the initial band structure consists of two overlapping bands, one of which is filled by electrons and the other by holes, see Fig. 11.20. (If the band overlap is small, such materials are called semimetals.) There exist in this case electron and hole pockets of the Fermi surface. If the electron and hole Fermi surfaces coincide, as in Fig. 11.20($a$), or will coincide after a shift by a certain wavevector $Q$ as in Fig. 11.20($b$), then this will be exactly equivalent to the nesting situation (9.37), and there will exist an instability resembling the Peierls instability in one-dimensional systems: the attraction between electrons and holes will lead to the formation of 'excitons' or rather of an *excitonic condensate*. As a result the material will become insulating, with the new band structure shown in Fig. 11.20($c$) by thick lines. (I put 'excitons' in quotation marks because it turns out that for weak coupling the radius of such 'excitons' is much larger than the average distance between them, so that such an 'excitonic insulator' should be visualized as a collective state with excitonic correlations, but not actually consisting of real excitons. The situation here very much resembles that in weak coupling BCS superconductors, where Cooper pairs strongly overlap and the ground state is a collective state with appropriate pairing correlations.)

The mathematical treatment of this problem is also very similar to that of the Peierls transition and of the BCS theory of superconductivity. First of all one can check that there indeed exists an instability of the normal state. One can see this by looking at the two-particle (electron–hole) Green function

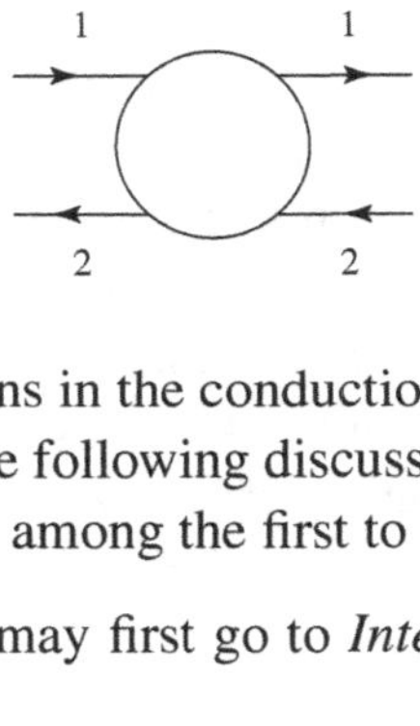


where $\xrightarrow{\;1\;}$ denotes electrons in the conduction band, and $\xleftarrow{\;2\;}$ denotes holes in the valence band (in the following discussion I mostly follow the work by Keldysh and Kopaev who were among the first to study this problem).

(An option: from here on you may first go to *Intermezzo*, Section 11.5, and then come back to this page.)

Consider the situation shown in Fig. 11.20($a$) and take the spectra

$$\varepsilon_{1,2}(\boldsymbol{k}) = \pm\left(\frac{-\boldsymbol{k}^2}{2m_{1,2}} + \frac{k_F^2}{2m_{1,2}}\right) \tag{11.33}$$

(below, for simplicity, we will consider the case of equal masses, $m_1 = m_2$). The calculation of the electron–hole two-particle Green function, or, equivalently, the scattering amplitude for electron–hole scattering $\Gamma$ in the usual RPA approximation (we use here the alternative form of Feynman diagrams for the electron–electron interaction, rather than the one used in Chapter 9, because it is more convenient, but of course equivalent, for four-fermion interactions (11.37))

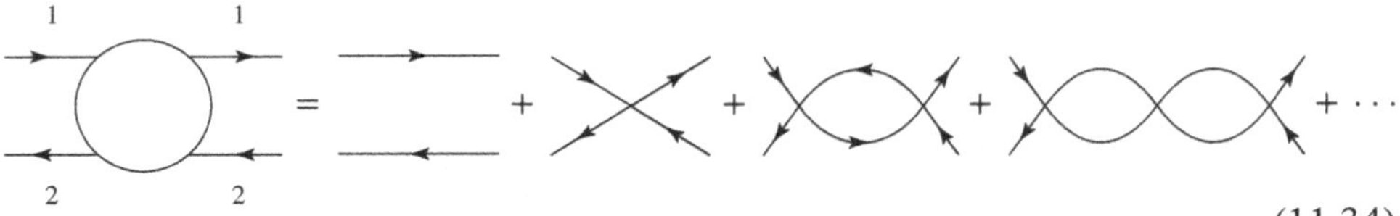


(11.34)

gives the following result:

$$\Gamma(\boldsymbol{k}=\boldsymbol{0}, \omega) = \frac{\lambda}{1+\lambda\left(\ln\left|\frac{\omega}{2\omega_0}\right| - i\frac{\pi}{2}\right)} . \tag{11.35}$$

Here $\omega_0$ is a cut-off of the order of $\varepsilon_F$, $\boldsymbol{k}$ and $\omega$ are the total momentum and energy, and $\lambda$ is the electron–hole coupling constant which depends on the detailed interaction involved. The origin of the logarithm in the denominator of (11.35) is actually the same as in the case of the one-dimensional Peierls instability or its three-dimensional analogue, the giant Kohn anomaly in the case of nesting, cf. Section 9.2. Here the corresponding expression is written accurately, and one sees that the electron–hole Green function, or the scattering amplitude, has an imaginary pole at $\omega = i\Omega$, if $\lambda < 0$ (which is the case here, as we have electron–hole attraction). The position of this pole is given by the expression

$$\Omega = 2\omega_0 e^{-1/|\lambda|} . \tag{11.36}$$

The existence of this pole shows that our system is unstable towards the formation of electron–hole pairs (excitons) from different bands close to the Fermi surface. The position of the pole $\Omega = 2\omega_0 e^{-1/|\lambda|}$ gives the binding energy of the pair, i.e. it determines the magnitude of the energy gap in the spectrum, see Fig. 11.20(*c*). To find this spectrum and to discuss the properties of the resulting state one can proceed in different ways. One of the methods is similar to the one which we have used, e.g. in the treatment of the Peierls transition: one writes the interband interaction as

$$\mathcal{H}_{\text{int}} = \sum V(\boldsymbol{k}, \boldsymbol{k}', \boldsymbol{q})\, c^{\dagger}_{1\sigma}(\boldsymbol{k}+\boldsymbol{q})\, c^{\dagger}_{2\sigma'}(\boldsymbol{k}'-\boldsymbol{q})\, c_{2\sigma'}(\boldsymbol{k}')\, c_{1\sigma}(\boldsymbol{k}) \tag{11.37}$$

and keeps the most divergent terms which correspond to the scattering of electrons from band 1 and holes from band 2 with the same wavevector, i.e. the terms with $\boldsymbol{k}+\boldsymbol{q} = \boldsymbol{k}'$. Then we make a decoupling

$$\mathcal{H}_{\text{int}} \sim c^{\dagger}_{1\sigma}(\boldsymbol{k}')\, c^{\dagger}_{2\sigma'}(\boldsymbol{k})\, c_{2\sigma'}(\boldsymbol{k}')\, c_{1\sigma}(\boldsymbol{k}) \Longrightarrow \tilde{\mathcal{H}}_{\text{int}} \sim c^{\dagger}_{1\sigma}(\boldsymbol{k}')\, c_{2\sigma'}(\boldsymbol{k}') \left\langle c^{\dagger}_{2\sigma'}(\boldsymbol{k})\, c_{1\sigma}(\boldsymbol{k})\right\rangle \tag{11.38}$$

and diagonalize the resulting quadratic Hamiltonian $\mathcal{H} = \mathcal{H}_0 + \tilde{\mathcal{H}}_{\rm int}$ by the Bogolyubov canonical transformation. The spin structure of the average

$$\langle c^{\dagger}_{2\sigma'}(\boldsymbol{k})\, c_{1\sigma}(\boldsymbol{k})\rangle \sim \Delta_{\sigma\sigma'}(\boldsymbol{k})\ , \qquad (11.39)$$

which plays the role of the order parameter, may be in principle different, i.e. exciton correlations may occur either in a singlet or in a triplet channel. We consider below the singlet case $\sigma = \sigma'$ and omit the spin indices.[5]

The simplified interaction term $\tilde{\mathcal{H}}_{\rm int}$ (11.38) is still nondiagonal in the band indices 1, 2, and we have to use the transformation practically identical to the one used for treating Bose condensation in Section 5.2 or for studying antiferromagnetic spin waves in Section 6.3.2, with the only difference that because of the Fermi statistics here the signs in corresponding linear transformations should be different. Thus we go from the operators $c_1$, $c_2$ to new operators $\alpha$, $\beta$, defined as

$$c_1(\boldsymbol{k}) = u_{\boldsymbol{k}}\alpha(\boldsymbol{k}) + v_{\boldsymbol{k}}\beta(\boldsymbol{k})\ , \qquad (11.40)$$

etc. and require that in terms of these new operators the Hamiltonian should be diagonal. We leave the details to the reader (the treatment coincides with that of Sections 5.2 and 6.3.2; see also the next section). Finally we get the following spectrum of one-particle excitations:

$$\omega_{1,2}(\boldsymbol{k}) = \frac{\varepsilon_1(\boldsymbol{k}) + \varepsilon_2(\boldsymbol{k})}{2} \pm \sqrt{\frac{\left[\varepsilon_1(\boldsymbol{k}) - \varepsilon_2(\boldsymbol{k})\right]^2}{4} + \Delta^2}\ , \qquad (11.41)$$

where $\Delta$ is equal to $\Omega$ (11.36). This is the anticipated result shown in Fig. 11.20(*c*).

Note that the ground state thus obtained is an eigenstate of the new operators $\alpha$, $\beta$, which are linear combinations of the original electrons $c_1$, $c_2$. As a result in the ground state we have a nonzero *anomalous average* $\langle c_1^{\dagger}(\boldsymbol{k})c_2(\boldsymbol{k})\rangle \neq 0$, which describes the excitonic correlations (electron–hole pairing) discussed above. This anomalous average serves as the order parameter of the excitonic insulator.

One can use this model to illustrate yet another method widely used for treating similar problems – the method of *anomalous Green functions*. It was introduced by Gor'kov for treating superconductivity, and the corresponding equations are called *Gor'kov equations*. In addition to the usual Green functions $G_1(\boldsymbol{k}, \omega)$, $G_2(\boldsymbol{k}, \omega)$ for the two bands, one introduces new 'mixed' Green functions $F$, $F^{\dagger}$: in the coordinate representation they are defined as (cf. (8.31))

$$\begin{aligned} F(\boldsymbol{r}_1, t_1; \boldsymbol{r}_2, t_2) &= -\,i\big\langle \mathrm{T}\big\{\Psi_2(\boldsymbol{r}_2, t_2)\, \Psi_1^{\dagger}(\boldsymbol{r}_1, t_1)\big\}\big\rangle \\ F^{\dagger}(\boldsymbol{r}_1, t_1; \boldsymbol{r}_2, t_2) &= i\big\langle \mathrm{T}\big\{\Psi_1(\boldsymbol{r}_1, t_1)\, \Psi_2^{\dagger}(\boldsymbol{r}_2, t_2)\big\}\big\rangle\ . \end{aligned} \qquad (11.42)$$

[5] This order parameter can also be imaginary, in which case the resulting state would correspond to that with orbital currents (B. Halperin and T. M. Rice), see Section 11.7 below.

These functions are necessary because, according to our expectations and to the physical picture, there will occur some mixing between the two bands 1 and 2 (proportional to the excitonic average (11.39)). The new Green functions $F$, $F^{\dagger}$ introduced above describe just this effect. Writing down the equations of motion for the ordinary Green function, e.g. $G_1$, we obtain on the right-hand side terms of higher order which should be decoupled in some way. Similar to (11.38) we keep the interband terms, which will give the anomalous Green functions $F$ (11.42). The resulting equations have the form

$$\begin{aligned} \Big(\omega - \varepsilon_1(\boldsymbol{k})\Big)\, G_1(\boldsymbol{k},\omega) + \Delta\, F(\boldsymbol{k},\omega) &= 1 \\ \Big(\omega - \varepsilon_2(\boldsymbol{k})\Big)\, F(\boldsymbol{k},\omega) - \Delta\, G_1(\boldsymbol{k},\omega) &= 0\,, \end{aligned} \tag{11.43}$$

where the quantity (c-number) $\Delta$ is

$$\Delta(\boldsymbol{k}) = i\int F(\boldsymbol{p},\varepsilon)\, V(\boldsymbol{p}-\boldsymbol{k})\, \frac{d^3\boldsymbol{p}\, d\varepsilon}{(2\pi)^4}\,. \tag{11.44}$$

The solution of equations (11.43) has the form

$$\begin{aligned} G_1(\boldsymbol{k},\omega) &= \frac{\omega - \varepsilon_2(\boldsymbol{k})}{\omega^2 - \big(\varepsilon_1(\boldsymbol{k}) + \varepsilon_2(\boldsymbol{k})\big)\omega + \varepsilon_1(\boldsymbol{k})\,\varepsilon_2(\boldsymbol{k}) - \big|\Delta(\boldsymbol{k})\big|^2} \\ F(\boldsymbol{k},\omega) &= \frac{\Delta(\boldsymbol{k})}{\omega^2 - \big(\varepsilon_1(\boldsymbol{k}) + \varepsilon_2(\boldsymbol{k})\big)\omega + \varepsilon_1(\boldsymbol{k})\,\varepsilon_2(\boldsymbol{k}) - \big|\Delta(\boldsymbol{k})\big|^2}\,, \end{aligned} \tag{11.45}$$

from which we again find the spectrum (11.41), and the solution of the self-consistent equation (11.45) gives the same energy gap $\Delta = \Omega$ (11.36).

We have used this model to illustrate how the different methods discussed above work in a particular case. By studying the scattering amplitude in a normal state we have detected an incipient instability of this system. Then, guided by the kind of instability, we devised the method(s) to treat the new state which will appear. There are different ways to describe this new state (the Bogolyubov canonical transformation, or the introduction of anomalous Green functions and solutions of the Gor'kov equations) which give coinciding results and are actually rather close physically, although technically different.

Returning to the excitonic insulators themselves, one should say that we have obtained an insulating state which at first glance is rather different from conventional insulators: we treated it in a rather sophisticated way, and used such terms as 'electron–hole bound states', 'exciton condensate', etc. The question arises, are the physical properties of this state very special?

The model we have used is theoretically indeed very nice and rich. One can use here all the machinery of modern theoretical physics; there are many different possible types of orderings accompanying this transition. We did not specify them above, but in general one can get here structural phase transitions, or spin-density waves – see the next section – or even orbital antiferromagnetism. However, it turns out that in most respects such a state is rather similar to the usual insulator. Thus, there are no 'supercurrents' connected with an 'excitonic condensate', although there were attempts to obtain some 'superproperties'. Probably the most interesting feature of this state is its close proximity to a 'prototype' metallic phase, so that there may occur here an insulator–metal transition at not too high temperature. The influence of impurities on such states may also have some specific features. Nevertheless, most of the properties of excitonic insulators are (unfortunately) rather similar to ordinary insulators, which nevertheless does not make this model less attractive and beautiful.

## 11.5 Intermezzo: BCS theory of superconductivity

It is instructive at this point to compare the treatment presented above with the corresponding treatment of superconductivity (Bardeen–Cooper–Schrieffer, or BCS theory). Superconductivity is a big field in itself, which is well covered in many textbooks and monographs – see e.g. Schrieffer (1999); we partially follow in this section the presentation by Abrikosov *et al.* (1975). Without going into much detail, we present here only some of the results to illustrate how the general methods described above work in this case. (Actually some of these methods, like the Gor'kov equations, were introduced first in the theory of superconductivity, and later on 'borrowed' by other fields.)

We start with the electron–electron interaction and first consider the lowest-order terms in the electron–electron scattering (as in Section 11.4, we use here the form of Feynman diagrams convenient for the interaction (11.49)):

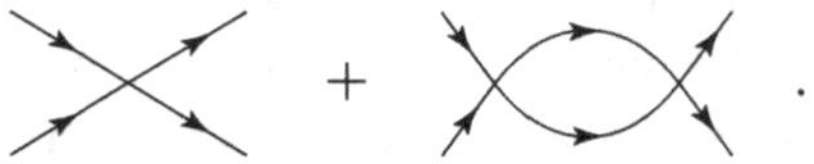

The second diagram can be shown to give the term

$$-\lambda^2 \frac{mk_{\rm F}}{2\pi^2} \ln \frac{2\omega_0}{\omega} \tag{11.46}$$

where $\lambda$ is the coupling constant (the vertex in these diagrams, see (11.49)), and $\omega_0$ is a cut-off frequency (since in the BCS theory, devised for conventional superconductors, the actual electron–electron attraction is due to the exchange of phonons, this cut-off is usually taken equal to the average phonon energy, the Debye frequency $\omega_{\rm D}$, although it may be different for other situations, with other microscopic mechanisms of pairing). As this lowest-order loop gives a logarithmically large contribution, we have to sum the whole series:

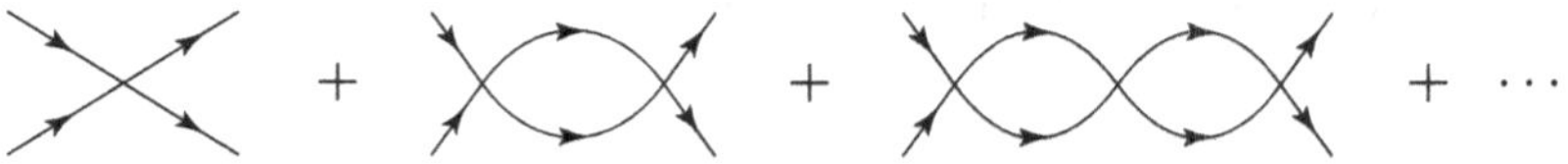

which will give us, analogously to (11.35), the effective scattering amplitude, or electron–electron (two-particle) Green function

$$\Gamma(\boldsymbol{q}=\mathbf{0},\omega)=\frac{\lambda}{1+\left(\frac{\lambda m k_{\mathrm{F}}}{2\pi^2}\right)\left[\ln\left|\frac{2\omega_0}{\omega}\right|+i\frac{\pi}{2}\right]}\,. \tag{11.47}$$

Thus for an attraction between electrons (coupling constant $\lambda < 0$) $\Gamma(\omega)$ has a pole at $\omega = i\Omega$,

$$\Omega = 2\omega_0 e^{-2\pi^2/|\lambda| m k_{\mathrm{F}}} \tag{11.48}$$

(cf. (11.35), (11.36)), where $\omega$ is the cut-off frequency. These results are quite similar to those presented above for the excitonic insulators, the difference being that in the case of excitonic insulators this pole corresponded to an electron–hole bound state (exciton) whereas here it is a *Cooper pair* (electron–electron pair) which exists when the electron–electron interaction is attractive ($\lambda < 0$).

Further treatment of these two problems is also similar. One of the methods often used is the Bogolyubov canonical transformation. To take into account the electron–electron correlations (the formation of Cooper pairs – pairs of electrons with opposite momenta $\boldsymbol{k}$, $-\boldsymbol{k}$, and opposite spins,[6] so that the total momentum $\boldsymbol{q}$ in (11.47) is zero) one has to keep the averages $\langle c^\dagger_{\boldsymbol{k},\sigma}\, c^\dagger_{-\boldsymbol{k},-\sigma}\rangle$. Making a corresponding decoupling in the electron–electron interaction term

$$\underset{(\boldsymbol{k}_1+\boldsymbol{k}_2=\boldsymbol{k}_3+\boldsymbol{k}_4)}{\mathcal{H}_{\text{int}} \sim \lambda\, c^\dagger_{\boldsymbol{k}_1,\sigma}\, c^\dagger_{\boldsymbol{k}_2,\sigma'}\, c_{\boldsymbol{k}_3,\sigma'}\, c_{\boldsymbol{k}_4,\sigma}} \quad \underset{\begin{pmatrix}\boldsymbol{k}_1=-\boldsymbol{k}_2=\boldsymbol{k}\\ \boldsymbol{k}_3=-\boldsymbol{k}_4=-\boldsymbol{k}'\\ \sigma=-\sigma'\end{pmatrix}}{\Longrightarrow} \quad \lambda\left\{c^\dagger_{\boldsymbol{k},\sigma}\, c^\dagger_{-\boldsymbol{k},-\sigma}\langle c_{-\boldsymbol{k}',-\sigma}\, c_{\boldsymbol{k}',\sigma}\rangle + \text{h.c.}\right\} \tag{11.49}$$

we diagonalize the resulting Hamiltonian by the transformation (for simplicity we drop spin indices):

$$\begin{aligned} b^\dagger_{\boldsymbol{k}} &= u_{\boldsymbol{k}} c^\dagger_{\boldsymbol{k}} - v_{\boldsymbol{k}} c_{-\boldsymbol{k}}\,, & b_{\boldsymbol{k}} &= u_{\boldsymbol{k}} c_{\boldsymbol{k}} - v_{\boldsymbol{k}} c^\dagger_{-\boldsymbol{k}}\,,\\ b^\dagger_{-\boldsymbol{k}} &= u_{\boldsymbol{k}} c^\dagger_{-\boldsymbol{k}} + v_{\boldsymbol{k}} c_{\boldsymbol{k}}\,, & b_{-\boldsymbol{k}} &= u_{\boldsymbol{k}} c_{-\boldsymbol{k}} + v_{\boldsymbol{k}} c^\dagger_{\boldsymbol{k}}\,. \end{aligned} \tag{11.50}$$

[6] We consider here the simplest case of singlet $s$-wave Cooper pairs. This is the situation in most conventional superconductors. In principle more complicated types of pairing may exist, e.g. singlet pairings in different orbital states (thus, such $d$-wave pairing most probably exists in high-$T_c$ superconductors), or spin-triplet pairing which is realized in $^3$He and probably in superconducting $Sr_2RuO_4$. According to the general rules of quantum mechanics singlet pairs may have orbital momenta $l = 0$ ($s$-wave pairing), $l = 2$ ($d$-wave pairing), $l = 4$, etc. whereas triplet pairing with $S_{\text{tot}} = 1$ may exist with odd values of $l$ ($l = 1$, the $p$-wave pairing, $l = 3$ etc.).

Note that the transformation (11.50) corresponds to the appearance of anomalous averages $\langle c^\dagger_k c^\dagger_{-k}\rangle \neq 0$, similar to the excitonic averages $\langle c^\dagger_{1k} c_{2k}\rangle$ in excitonic insulators, cf. (11.39).

The resulting spectrum is the well-known spectrum of superconductors:

$$E(\boldsymbol{k}) = \sqrt{(\varepsilon_{\boldsymbol{k}} - \mu)^2 + \Delta^2} \equiv \sqrt{\xi_{\boldsymbol{k}}^2 + \Delta^2}\,, \tag{11.51}$$

where $\xi_{\boldsymbol{k}} = \varepsilon_{\boldsymbol{k}} - \mu$ and $\Delta$ is the energy gap equal to $\Omega$ (11.48). The coefficients of the Bogolyubov transformation (11.50) have the form

$$\begin{aligned} u_{\boldsymbol{k}}^2 &= \frac{1}{2}\left[1 + \frac{\xi_{\boldsymbol{k}}}{\sqrt{\xi_{\boldsymbol{k}}^2 + \Delta^2}}\right] = \frac{1}{2}\left[1 + \frac{\xi_{\boldsymbol{k}}}{E_{\boldsymbol{k}}}\right] \\ v_{\boldsymbol{k}}^2 &= \frac{1}{2}\left[1 - \frac{\xi_{\boldsymbol{k}}}{\sqrt{\xi_{\boldsymbol{k}}^2 + \Delta^2}}\right] = \frac{1}{2}\left[1 - \frac{\xi_{\boldsymbol{k}}}{E_{\boldsymbol{k}}}\right]\,; \end{aligned} \tag{11.52}$$

these expressions are very important and often used, e.g. in the treatment of coherence effects in superconductors.

Now we illustrate on the same example the use of anomalous Green functions and the Gor'kov equations. As the anomalous averages $\langle c^\dagger_k c^\dagger_{-k}\rangle$ are nonzero in this case, we introduce the corresponding anomalous Green functions

$$F_{\alpha\beta}(\boldsymbol{r}_1, t_1; \boldsymbol{r}_2, t_2) = \Big\langle \mathrm{T}\big\{\Psi_\alpha(\boldsymbol{r}_1, t_1)\,\Psi_\beta(\boldsymbol{r}_2, t_2)\big\}\Big\rangle\,, \tag{11.53}$$

where $\alpha$, $\beta$ are spin indices. The coupled equations of motion for the normal Green function $G(\boldsymbol{k}, \omega)$ and the anomalous Green function $F(\boldsymbol{k}, \omega)$ have the form (cf. (11.43)):

$$\begin{aligned} (\omega - \xi_{\boldsymbol{k}})\, G(\boldsymbol{k}, \omega) - i\Delta\, F^\dagger(\boldsymbol{k}, \omega) &= 1 \\ (\omega + \xi_{\boldsymbol{k}})\, F^\dagger(\boldsymbol{k}, \omega) + i\Delta\, G(\boldsymbol{k}, \omega) &= 0\,. \end{aligned} \tag{11.54}$$

Just these equations were originally known as the Gor'kov equations (although now this term is used in a broader sense). Their solution has the form

$$G(\boldsymbol{k}, \omega) = \frac{\omega + \xi_{\boldsymbol{k}}}{\omega^2 - \xi_{\boldsymbol{k}}^2 - \Delta^2}\,, \qquad F^\dagger(\boldsymbol{k}, \omega) = -i\,\frac{\Delta}{\omega^2 - \xi_{\boldsymbol{k}}^2 - \Delta^2}\,, \tag{11.55}$$

where, similar to (11.44),

$$\Delta = \int F^\dagger(\boldsymbol{k}, \omega)\frac{d^3\boldsymbol{k}\, d\omega}{(2\pi)^4} \tag{11.56}$$

(note that, as compared to (11.42)–(11.44), here the definition of the functions $F$, $F^\dagger$, and $\Delta$ differs by a factor of $i$). From (11.55) we see that we obtain the same spectrum (11.51) (poles of the Green functions (11.55)). The Green functions themselves can be written in the form

$$\begin{aligned} G(\boldsymbol{k}, \omega) &= \frac{u_k^2}{\omega - E(\boldsymbol{k}) + i\delta} + \frac{v_k^2}{\omega + E(\boldsymbol{k}) - i\delta} \\ F^\dagger(\boldsymbol{k}, \omega) &= -i\frac{\Delta}{\big(\omega - E(\boldsymbol{k}) + i\delta\big)\big(\omega + E(\boldsymbol{k}) - i\delta\big)} \end{aligned} \tag{11.57}$$

where the functions $u_k^2$ and $v_k^2$ are given by the expressions (11.52), i.e. they coincide with the coefficients of the Bogolyubov canonical transformation.

By putting the expression (11.57) into (11.55), and performing integration over $\omega$ with the help of residues, we obtain the self-consistency equation for the energy gap $\Delta$:

$$1 = -\frac{\lambda}{2(2\pi)^3}\int \frac{d^3\boldsymbol{k}}{\sqrt{\xi_k^2 + \Delta^2}}\,, \tag{11.58}$$

from which we finally obtain the energy gap

$$\Delta = 2\omega_0 e^{-1/\tilde{\lambda}}\,, \qquad \tilde{\lambda} = \frac{\lambda m k_{\mathrm{F}}}{2\pi^2}\,. \tag{11.59}$$

This expression coincides with (11.48), as it should. Here we have again introduced a cut-off $\omega_0$ in the integral (11.58), which depends on the physical situation considered. Thus, for the superconductivity induced by the electron–phonon interaction (as in most of the conventional superconductors) $\omega_0$ is the typical phonon frequency (Debye frequency $\omega_{\mathrm{D}}$). If pairing is due to some other mechanism of attraction, $\omega_0$ may be different, coinciding usually with the characteristic energy of the 'intermediate quanta' carrying the interaction (e.g. instead of phonons we can consider electronic excitations – excitons, or magnons, etc.).

One must say that the BCS theory of superconductivity is one of the most successful theories in the modern quantum theory of solids. Many concepts and methods were actually first introduced here and later on transferred to other fields. This is true in particular for the use of Green functions for treating different nontrivial situations, where, depending on the specific problem, one has to introduce different anomalous Green functions and solve the corresponding coupled equations. The treatment of excitonic insulators presented above gives a good illustration of that. But we can also use this method, e.g. for considering Peierls transitions or the formation of CDW and SDW states.

## 11.6 Spin-density waves

We have seen above that in certain cases (specifically when the Fermi surface has nested parts) there may occur a structural phase transition or formation of CDWs which creates energy gaps at parts or at the whole of the Fermi surface and which can sometimes lead to metal–nonmetal transitions. Similar phenomena occur in two-band systems with coinciding electron and hole Fermi surfaces leading to excitonic insulators. In these cases we had an effective attraction between relevant particles: the effective attraction via phonons in one-band models, cf. (11.23), or the usual (Coulomb) attraction between electrons and holes in the two-band case. However, even if the resulting interaction is repulsive, the system satisfying the nesting condition (9.37) will still be unstable, however, not in the charge but in the spin channel, i.e. not CDW, but a spin superstructure with the corresponding wavevector will be created, known as a spin-density wave (SDW). The corresponding instability was first discussed in the one-dimensional case by Overhauser and is sometimes called by his name.

One can see that the same nesting condition leads to an instability in the spin channel if one looks, e.g. at the expression for magnetic susceptibility for local repulsion (9.16):

$$\chi(\boldsymbol{q},\omega)=\frac{\chi_0(\boldsymbol{q},\omega)}{1+U\,\Pi_0(\boldsymbol{q},\omega)}=\frac{\chi_0(\boldsymbol{q},\omega)}{1-\frac{U}{\mu_B^2}\,\chi_0(\boldsymbol{q},\omega)}\,. \tag{11.60}$$

(Remember that with our definitions $\Pi_0<0$, and $\chi_0=-\mu_B^2\Pi_0$.) One sees that in one-dimensional systems, as well as in 2d and 3d systems with nesting, we have the same divergence of magnetic susceptibility in the case of repulsion $U>0$ as we had in the charge response function (dielectric function) for an attraction; this again signals an instability, here towards the formation of SDW. It will occur in this case at arbitrarily small repulsion $U$. In principle the corresponding instability may exist even in the case of incomplete nesting or even without it; in this case, however, we will need the interaction strength to exceed a certain critical value, so that the condition

$$U\left|\Pi_0(\boldsymbol{q},0)\right|>1 \tag{11.61}$$

is satisfied. This condition is well known for ferromagnetism ($\boldsymbol{q}=0$); as $\Pi(\boldsymbol{q}\to 0,0)=-\rho(\varepsilon_F)$, see (9.24), it takes the form

$$U\,\rho(\varepsilon_F)>1\,, \tag{11.62}$$

which is known as the *Stoner criterion* of ferromagnetism. In general $\chi(\boldsymbol{q},0)$ may have a maximum at different values of $\boldsymbol{q}=\boldsymbol{Q}$, and the corresponding instability would first appear at the corresponding wavevector $\boldsymbol{Q}$. In the case of nesting, where there is a real divergence of $\Pi_0(\boldsymbol{q}, 0)$, the 'Stoner'-like criterion (11.61) will be definitely fulfilled at the wavevector $\boldsymbol{q}$ equal to the nesting wavevector, even for an arbitrary weak interaction.

One can consider the resulting problem similarly to the treatment of the Peierls transition or CDW (as an example we look below at the 1d case). We start from the Hamiltonian of the so-called Hubbard model (see Chapter 12) in which one takes only on-site electron–electron interactions $U n_{i\uparrow} n_{i\downarrow}$. In the momentum representation the model has the form

$$\mathcal{H} = \sum_{k,\sigma} \varepsilon_k \, c^\dagger_{k,\sigma} \, c_{k,\sigma} + \frac{U}{V} \sum_{kk'q} a^\dagger_{k\uparrow} \, a_{k+q\uparrow} \, a^\dagger_{k'\downarrow} \, a_{k'-q\downarrow} \, . \tag{11.63}$$

The spin density at point $x$ is given by the general expression

$$\begin{aligned} S^z(x) &= \frac{1}{2}\left[c^\dagger_\uparrow(x)\, c_\uparrow(x) - c^\dagger_\downarrow(x)\, c_\downarrow(x)\right] \\ &= \frac{1}{2V} \sum_{kk'} \left[c^\dagger_{k\uparrow}\, c_{k'\uparrow} - c^\dagger_{k\downarrow}\, c_{k'\downarrow}\right] e^{-i(k-k')x} \, . \end{aligned} \tag{11.64}$$

As the susceptibility (11.60) diverges at $q = 2k_{\mathrm{F}}$, we keep in (11.64) only terms with $k - k' = \pm 2k_{\mathrm{F}}$. Then we have

$$\langle S^z(x)\rangle = \frac{1}{2V} \sum_k \left[\left\langle c^\dagger_{k\uparrow}\, c_{k+2k_{\mathrm{F}}\uparrow} - c^\dagger_{k\downarrow}\, c_{k+2k_{\mathrm{F}}\downarrow}\right\rangle\right] e^{i2k_{\mathrm{F}}x} + \text{c.c.} \tag{11.65}$$

(The term with $k' = k - 2k_{\mathrm{F}}$ becomes the complex conjugate of the first term in (11.65) after a change of the summation index $k \to k + 2k_{\mathrm{F}}$.) Denoting

$$\frac{1}{V} \sum_k \left\langle c^\dagger_{k\uparrow}\, c_{k+2k_{\mathrm{F}}\uparrow} - c^\dagger_{k\downarrow}\, c_{k+2k_{\mathrm{F}}\downarrow}\right\rangle = \langle s\rangle = |s| e^{i\varphi} \tag{11.66}$$

we obtain

$$\langle S^z(x)\rangle = \tfrac{1}{2}\mathrm{Re}\left(\langle s\rangle e^{i2k_{\mathrm{F}}x}\right) = |s| \cos(2k_{\mathrm{F}}x + \varphi) \, . \tag{11.67}$$

Making the corresponding mean field decoupling in the Hamiltonian (11.63) (keeping again terms with $q = 2k_{\mathrm{F}}$) we obtain

$$\mathcal{H} = \sum_{k,\sigma} \left\{\varepsilon_{k,\sigma}\, c^\dagger_{k,\sigma}\, c_{k,\sigma} + \left(\Delta\, c^\dagger_{k+2k_{\mathrm{F}},\sigma}\, c_{k,\sigma} + \text{h.c.}\right)\right\} , \tag{11.68}$$

where $\Delta = \frac{U}{N}\langle S^z\rangle$.

Diagonalizing (11.68), we obtain the new energy spectrum

$$E_k = \mu \pm \sqrt{(\varepsilon_k - \mu)^2 + |\Delta|^2} \, . \tag{11.69}$$

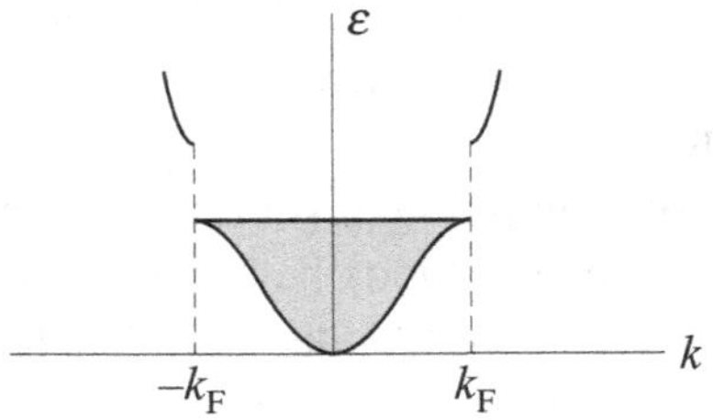


Fig. 11.21

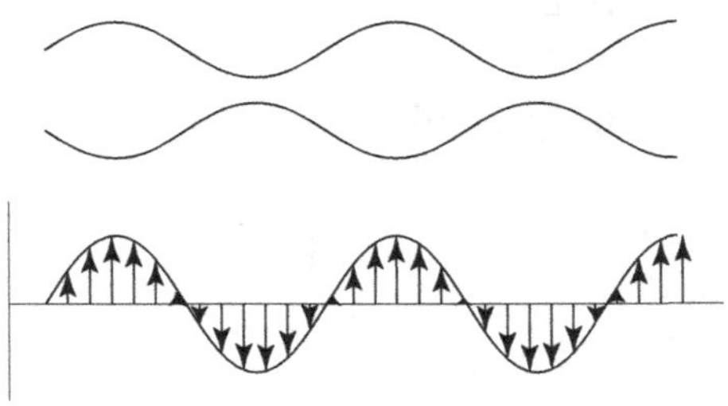


Fig. 11.22

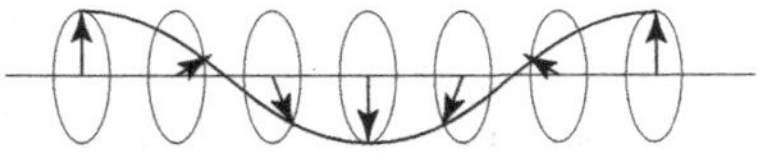

Fig. 11.23

All these results are quite similar to those obtained earlier for the Peierls transition or excitonic insulators; we obtain the spectrum with the gap at $\pm k_F$, Fig. 11.21 (in principle there will also be smaller gaps at $\pm 2k_F$, $\pm 4k_F$, etc.). One can visualize the resulting structure as two density waves, for electrons with spin ↑ and spin ↓, but being in antiphase, so that the total charge density remains uniform, and the spin density oscillates with the period $l = 2\pi/2k_F$, see (11.67) and Fig. 11.22. Such spin-density waves are called sinusoidal SDW.

There exists yet another solution, the so-called helicoidal SDW, in which the magnitude of the spin remains constant, but its direction rotates in the $xy$-plane, $\langle S^x \rangle = |S| \cos(2k_F x + \varphi)$, $\langle S^y \rangle = |S| \sin(2k_F x + \varphi)$, see Fig. 11.23. Usually this structure is energetically more favourable, because the length of spins at each site here remains the same, which, e.g. optimizes the exchange interaction (6.21). Sinusoidal SDW may be stabilized if there is strong uniaxial anisotropy in the system. Note also that the plane of spin rotation may lie in different directions in the crystal and should not be perpendicular to the spiral axis: in the absence of spin–orbit interaction the axes in spin space have nothing to do with the real crystal

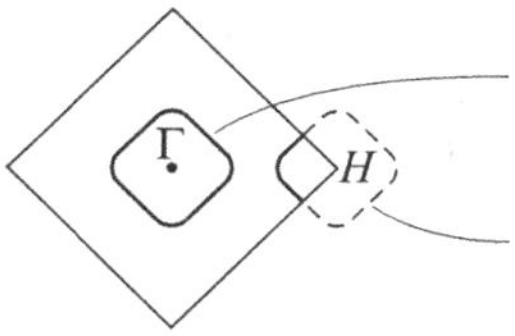


Fig. 11.24

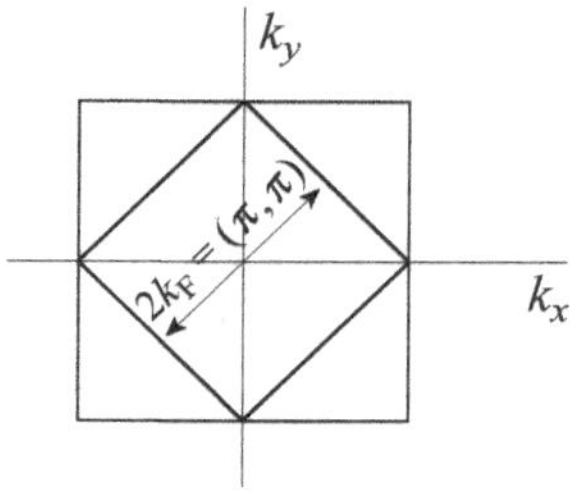


Fig. 11.25

axes! (This is true for all magnetic systems, including those with localized spins, discussed in Chapter 6.)[7]

One can also study thermodynamic properties of this model. It turns out that in the mean field approximation there will be a phase transition in which SDW and the energy gap in the electron spectrum simultaneously disappear. The corresponding formulae are very similar to those in the BCS theory of superconductivity.

A similar treatment can be carried out also for higher dimensions with nesting. Experimentally SDW of this kind were observed in a number of quasi-one-dimensional organic compounds. But probably the most important and the best-studied example is the magnetic structure of the metal Cr. The explanation of the magnetic structure of Cr was provided by this model (or rather by its two-band analogue, similar to the model of excitonic insulators), with the Fermi surface having nearly nested electron and hole pockets, schematically shown in Fig. 11.24. Similar physics seems to work also in many recently discovered FeAs superconductors.

It is also instructive to consider a prototype higher-dimensional system with nesting – the two-dimensional square lattice in the tight-binding approximation. As shown in Fig. 9.4, for the half-filled band ($n = 1$) the Fermi surface is a square, i.e. it is perfectly nested with the wavevector $\boldsymbol{Q} = (\pi, \pi)$, see Fig. 11.25. Consequently in the case of repulsive interactions SDW will appear here with the wavevector $(\pi, \pi)$ which will open a gap in the whole Fermi surface, i.e. the system

[7] Thus the spins can rotate in the plane containing the wavevector of the spiral; the resulting structure may be called a *cycloidal* spiral. Interestingly enough, cycloidal spirals give rise to ferroelectricity (Katsura, Nagaosa and Balatzky; Mostovoy), i.e. the corresponding magnets will be *multiferroic*.

will become insulating. But an SDW with this wavevector is nothing else but the usual two-sublattice antiferromagnetism, $\begin{matrix} \uparrow\downarrow\uparrow\downarrow\uparrow\downarrow \\ \downarrow\uparrow\downarrow\uparrow\downarrow\uparrow \\ \uparrow\downarrow\uparrow\downarrow\uparrow\downarrow \end{matrix}$. Thus one can obtain this state both in the model of localized electrons (localized spins) with antiferromagnetic exchange, or proceeding from the band picture with appropriate conditions (one electron per site, full nesting).

There exist many systems with this type of magnetic ordering. Probably the best-known nowadays is the insulating $La_2CuO_4$, the parent compound for high-$T_c$ superconductors. Usually the insulating and antiferromagnetic structure of $La_2CuO_4$ is explained in the model of localized electrons, starting, e.g. from the Hubbard model with strong coupling, see below, Chapter 12. However, if we did not know that the electrons in $La_2CuO_4$ were localized, and tried to describe the electronic structure of this material in the usual band theory, using the tight-binding approximation, we would obtain exactly the situation described above – a metal with perfect nesting and with the Fermi surface shown in Fig. 11.25. But then this situation would be unstable, SDW would appear, and we would end up in an antiferromagnetic insulator of exactly the same type as in the usual description! Thus these two approaches have much in common, and many features are similar, although there are also important differences, such as the value of sublattice magnetization, which in the weak-coupling approach used here may be (much) less than the nominal spin $\frac{1}{2}$, and also in the behaviour at finite temperatures: the SDW theory predicts an insulator–metal transition when the magnetic ordering disappears, whereas in the localized picture the system remains insulating (Mott insulator) even in the paramagnetic phase, see Chapter 12.

## 11.7 Different types of CDW and SDW

In the previous sections we have discussed several cases of instabilities of normal metals, many of which we could describe as the formation of charge or spin superstructures (Peierls distortion, CDW, SDW, excitonic insulator formation). We discussed these cases separately, but actually in all these cases we are dealing with a very similar situation: due to specific features of the energy spectrum, typically with nesting, the system is unstable towards the formation of a novel state with electron–hole correlations, described by the appearance of anomalous averages of the type $\langle c^\dagger_k c_{k+Q} \rangle$, either describing the modulation of charge density, or spin density; see equations (11.16), (11.39), (11.65). What then is the relation between these seemingly different phenomena?

The situation with CDW seems to be the simplest. Let us consider the situations with density modulation (CDW, Peierls distortion); spin superstructures can be

(a) (b)

Fig. 11.26

discussed similarly. Here we describe a new ordering by the anomalous average

$$\Delta = \langle c_k^\dagger c_{k+Q} \rangle \neq 0 \,, \tag{11.70}$$

where we either assume summation over spins, or, for simplicity, consider spinless fermions. The vector $\boldsymbol{Q}$ is determined by the nesting condition; let us consider, as an example, the half-filled one-dimensional band treated in Section 11.1, for which $Q = \pi$. As discussed in that section, the appearance of the anomalous average (11.70) leads to dimerization of the system and to the opening of the gap at the Fermi surface.

Generally speaking, the average (11.70), for a fixed value of $Q$ (e.g. equal to $\pi$) can still depend on the momentum $k$. It can also have an arbitrary phase, in particular it can be real or imaginary. It turns out that all these possibilities in fact describe different physical properties of the resulting state.

Let us first take $\Delta$ constant and real, $\Delta(k) = \Delta_0$. We can calculate (the modulation of) the electron density at site $n$ in the state with this order parameter:

$$\rho_n = \langle c_n^\dagger c_n \rangle = \frac{1}{N} \sum_{k,p} e^{-ikn}\, e^{ipn} \langle c_k^\dagger c_p \rangle \,. \tag{11.71}$$

With the only nonzero average $\langle c_k^\dagger c_{k+Q} \rangle = \Delta_0$, we immediately get that the electron density will oscillate as

$$\delta\rho_n = \Delta_0\, e^{i\pi n} = \Delta_0\, (-1)^n \,. \tag{11.72}$$

Thus this solution describes a *site-centred CDW* – the electron density alternates from site to site, see Fig. 11.26(*a*). As the order parameter is here a constant, it may be called an *s-wave CDW* (analogous solutions in the 2d and 3d cases would be spherically symmetric).

How then can one describe Peierls distortion and the modulation of the electron density connected with it? It is clear that in this case all ions remain identical and have similar charge, but the *bonds* are no longer identical: some of them become shorter, and some longer. Consequently we expect that the electron density will increase at the short bonds and decrease at the long ones, see Fig. 11.26(*b*). (Actually we may think of the process of Peierls distortion as a first step in forming *molecules*, like forming $H_2$ molecules out of a row of hydrogen atoms.)

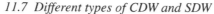

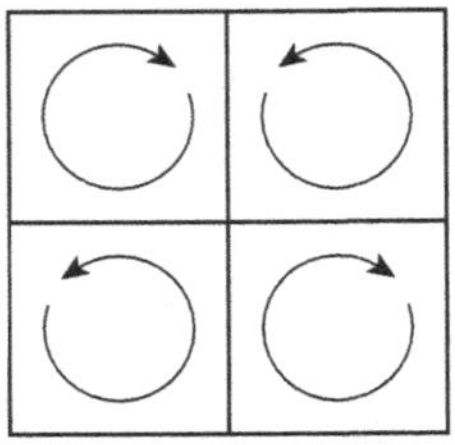

Fig. 11.27

One can show that we can describe this process if we take the anomalous average (11.70) in the form

$$\Delta_k = i\Delta_0 \sin k \ . \tag{11.73}$$

One can easily check that in this case the electron density at each site remains the same, $\delta\rho_n = 0$. But the *bond average* $\delta\rho_{n,n+1} = \langle c_n^\dagger c_{n+1} + c_{n+1}^\dagger c_n \rangle$ is now nonzero.

**Check this** making a Fourier transform of this 'bond density' and taking the anomalous average $\langle c_k^\dagger c_p \rangle = i\Delta_0 \delta(p - k - \pi)$.

As a result we obtain exactly what we wanted: with the order parameter (11.73) we see that the 'bond density' alternates from bond to bond, $\delta\rho_{n,n+1} = \Delta_0(-1)^n$, Fig. 11.26(*b*). Thus such a bond-centred CDW, associated with a Peierls transition, is described by the imaginary $k$-dependent order parameter (11.73), which because of its sine-like $k$-dependence may be called a *p-wave CDW*.

Similarly, one can also consider other possible types of $k$-dependence of the excitonic-like anomalous averages (11.70). Thus, e.g. in the 2d case with a square lattice, the order parameter $\Delta_k \sim \Delta_0(\cos k_x - \cos k_y)$, which may be called a *d-wave CDW*, in fact describes *orbital currents* running around plaquettes, see Fig. 11.27. This state is better known as a *flux phase* – there will be magnetic fluxes alternatingly piercing these plaquettes. Similar states were discussed in connection with high-temperature superconductivity, with currents running on Cu–O–O triangles (C. M. Varma). Analogously, if we consider other spin structures of the averages (11.39), (11.70), we would obtain either states with different spin structures (site-centred CDW of different kinds, see above), or states with *spin currents*.

## 11.8 Weakly and strongly interacting fermions. Wigner crystallization

Yet another situation in which the usual description of electrons in metals (ordinary band theory, the Fermi surface, etc.) breaks down is the case of strong electron–electron interactions. We will mostly discuss this topic in the next chapter, but here we want to give a general consideration of when we should consider an electron system as weakly or strongly interacting and which qualitative consequences we can expect.

It turns out that the answer strongly depends on the character of the interaction we are dealing with. Consider first the case of short-range interactions; this may not be very realistic for electrons which always experience long-range Coulomb interactions, but we can meet this situation in other Fermi systems such as $^3$He or nuclear matter.

In all these cases we cannot simply take the interaction as weak. Thus, e.g. both in $^3$He and for neutrons in nuclear matter or in neutron stars, where particles interact weakly at large enough distances, the interaction definitely becomes very strong when they come close together: there is always what we may call a hard core repulsion. But despite this in certain situations we can consider such a system as effectively weakly interacting, like the so-called nearly ideal Fermi gas. The condition for this is that the radius of the interaction (e.g. the radius of hard cores) $r_0$ is small compared to the average distance between fermions $\tilde{r}_s \sim (V/N)^{1/3} = n^{-1/3}$, see (7.5). This will definitely be the case for low-density fermion systems with short-range interactions.

When the condition

$$r_0 \ll \tilde{r}_s \sim \left(\frac{V}{N}\right)^{1/3} \tag{11.74}$$

is satisfied, we have simultaneously the inequality

$$k_\mathrm{F} r_0/\hbar \ll 1 \tag{11.75}$$

for characteristic values of the momentum $k \sim k_\mathrm{F}$, because $k_\mathrm{F} \sim (\frac{N}{V})^{1/3}\hbar$, see (7.3). This means that we are dealing only with 'slow' fermions. It is known in quantum mechanics that for the interaction (collision) of slow particles we can replace the interaction matrix element by the scattering amplitude, which for small momenta describes only $s$-wave scattering and which tends to a constant $(-a)$, called the scattering length:

$$-a = -\frac{M}{4\pi\hbar^2} U_0 \,, \qquad U_0 = \int U(\boldsymbol{r})\, d^3\boldsymbol{r} \,. \tag{11.76}$$

The actual small parameter which can be used in perturbation theory is

$$\frac{k_\mathrm{F} a}{\hbar} \ll 1 \,. \tag{11.77}$$

In terms of diagrams this corresponds to the summation of the so-called *ladder* diagrams. We want to describe the scattering of two particles,

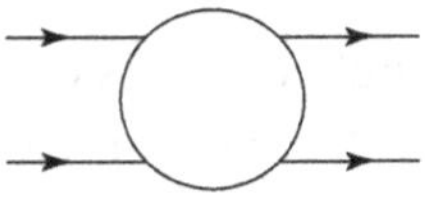 .

The lowest (Born) approximation corresponds to the diagram

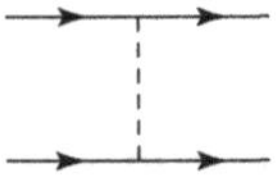

where the dashed line describes the interaction $U$. When we have a low-density system, the main effect is the multiple (repeated) scattering of the same pair of particles, which is described by the 'ladder' diagrams

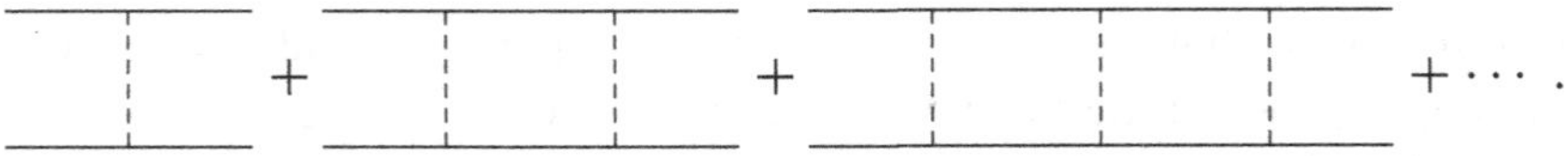

All other processes will be connected with the creation of electron–hole pairs, e.g.

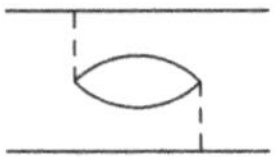 ,

or

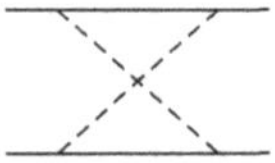

which is equivalent to

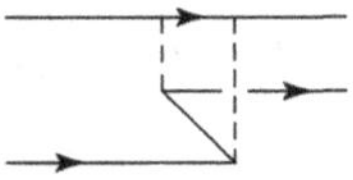 ,

and for low density they can be neglected. Such a ladder summation leads to effective replacement of the interaction $U$ by the scattering amplitude $\Gamma$ which for a contact interaction is

$$\Gamma \sim U_{\text{eff}} = \frac{U}{1 + U\rho(\varepsilon_{\text{F}})} \,. \tag{11.78}$$

Thus even for very strong (hard core) local repulsion the effective vertex 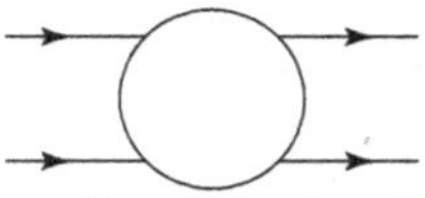 remains finite, equal to a constant $-a$ (the scattering length), and for low density one can carry out all these calculations. It turns out that in this limit the Fermi liquid (or, rather, weakly interacting Fermi gas) picture is valid, even far away from the Fermi surface. I will only give here a couple of results which are obtained in this limit:

- The ground state energy of the system is (K. Huang, C. N. Yang)

$$E_0 = N\frac{3k_F^2}{10m}\left[1 + \frac{10}{9\pi}\frac{k_F a}{\hbar} + \frac{4(11-2\ln 2)}{21\pi^2}\left(\frac{k_F a}{\hbar}\right)^2\right]. \tag{11.79}$$

- The effective mass is (A. A. Abrikosov, I. M. Khalatnikov)

$$\frac{m^*}{m} = 1 + \frac{8}{15\pi^2}(7\ln 2 - 1)\left(\frac{k_F a}{\hbar}\right)^2 \tag{11.80}$$

(cf. the case of Coulomb interaction (9.46)).

We see that these expressions are the first terms of a series in the small parameter $(k_F a/\hbar)$ (11.77). From the treatment given above it is clear, in particular, that one should not expect any instabilities for a low-density system with repulsive interactions. In particular one should not expect magnetic instabilities which could be present in systems with repulsion: e.g. the Stoner criterion (11.62) will not be satisfied, as we should replace there $U$ by $U/\left[1 + U\rho(\varepsilon_F)\right]$. Note that the situation with an attraction is drastically different: in this case, as we know, there may exist two-particle bound states, and as a result the system becomes unstable for arbitrarily weak attraction; this is the superconducting instability.

Let us now turn to the case of long-range Coulomb interactions. It turns out that the situation here is just the opposite. Whereas for short-range interactions the low-density limit corresponds to a weakly interacting regime, for the Coulomb interaction, in contrast to that, the low-density situation corresponds to the strong-coupling case, and the high-density system behaves as weakly interacting. This can be understood as follows (see also Chapter 3):

We should compare the average potential energy

$$E_{\text{pot}} = \frac{e^2}{\tilde{r}_s} \tag{11.81}$$

with the average kinetic energy of the electrons

$$E_{\text{kin}} = \frac{\hbar^2}{m\tilde{r}_s^2}, \tag{11.82}$$

where $\tilde{r}_s$ is the average distance between electrons. The system can be treated as weakly interacting if $E_{\text{pot}} < E_{\text{kin}}$, or

$$\tilde{r}_s < \frac{\hbar^2}{me^2} = a_0\,, \qquad \text{i.e.} \qquad r_s = \frac{\tilde{r}_s}{a_0} < 1\,, \tag{11.83}$$

where $a_0$ is the Bohr radius. In the opposite limit, $r_s > 1$, the interaction becomes comparable to or bigger than the kinetic energy, and one can expect that the standard Fermi liquid description breaks down. This indeed happens: it was shown by Wigner that the better state, the one with lower energy, is reached if electrons, instead of being in a 'liquid' state filling the Fermi surface, were localized in space, forming the so-called *Wigner crystal*. By doing this we increase the average kinetic energy of the electrons (due to the uncertainty relation, when the position of an electron is confined to a certain volume $\sim \tilde{r}_s^3$, its momentum and corresponding kinetic energy will increase). However, by keeping electrons apart, as far away from each other as possible, we gain more in potential energy, if $r_s > 1$.

What is the critical value of $r_s$, beyond which a Wigner crystal becomes stable, is still a matter of debate. Different calculations give for 3d systems different values, from $r_s \sim 10$ to $\sim 170$; the most probable value seems to be $r_s \sim 80$. There were many attempts to observe Wigner crystallization experimentally. Up till now it was observed for ions at the surface of liquid He, and there were reports about observation of Wigner crystallization of electrons in some semiconducting structures with low electron concentration (the conditions for Wigner crystallization are less stringent in the two-dimensional case, to which both these examples actually belong). A phenomenon very similar to Wigner crystallization was also observed in several transition metal and rare earth compounds with mixed valence: magnetite $Fe_3O_4$, $Eu_3S_4$, $La_{0.5}Ca_{0.5}MnO_3$. The best-known example is magnetite, in which the ions $Fe^{2+}$ and $Fe^{3+}$ order below $\sim 119$ K (this transition is known as the Verwey transition). This ordering can also be visualized as ordering of the 'extra' electrons on $Fe^{2+}$ as compared to $Fe^{3+}$. However, it is not really clear whether the main driving force for this ordering is in this case the Coulomb interaction or, e.g. the electron–lattice interaction.

One can easily calculate the energy of a Wigner crystal. Crudely it is given by an expression similar to equation (11.81) with the coefficient depending on the detailed type of the lattice formed. The calculations give for the best of such lattices the value (in Rydberg)

$$E_{\mathrm{W}} = -\frac{1.8}{r_s}\,\mathrm{Ry}\,. \tag{11.84}$$

By using two limits, that of a high density (9.47), and the low-density limit giving Wigner crystals, one can attempt to make an interpolation for the total energy or for the correlation energy $E_c$ defined (see the definition of $E_c$ in Chapter 9, after equation (9.47)) as the difference between the exact and the Hartree–Fock values. There are several interpolation formulae of this type suggested, e.g. the formula of

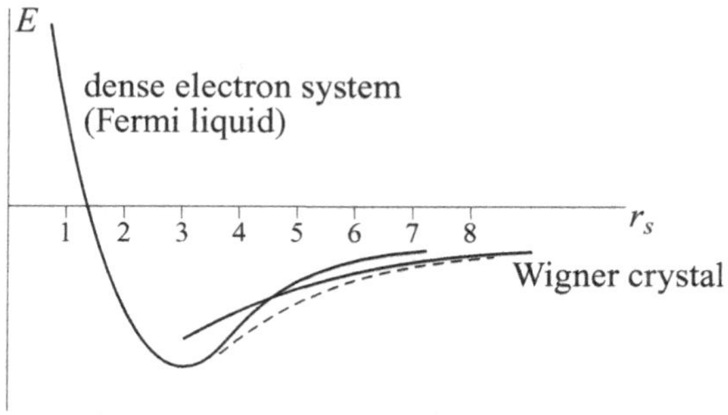


Fig. 11.28

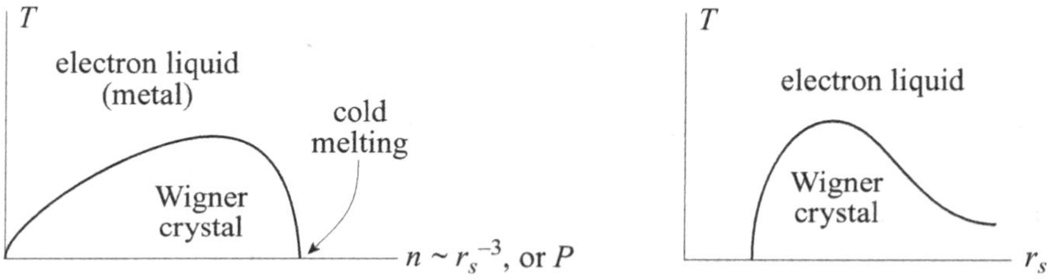


Fig. 11.29

Wigner himself:

$$E_c = -\frac{0.88}{r_s + 7.8} \text{Ry}. \tag{11.85}$$

The total energy as a function of $r_s$ thus looks as shown in Fig. 11.28 (see (9.47) and (11.84)) where the dashed line represents the interpolation line (11.85). There is a certain danger in this interpolation, stressed by P. W. Anderson: actually high- and low-density curves cross, which in the general thermodynamic treatment implies a first-order phase transition. This is consistent with the general statement made in Chapter 2, that the transition between crystal and liquid should always be first order. Melting of the electronic Wigner crystal is no exception.

The melting of the Wigner crystal presents a rather interesting situation. As always we can induce melting by increasing the temperature. But there exists yet another possibility: the Wigner crystal becomes unstable and 'melts' even at $T = 0$ with increasing density of electrons, for example under pressure (so-called cold melting). The resulting phase diagram looks as shown in Fig. 11.29. (This 'cold melting' of the Wigner crystal presents yet another example of a quantum critical point, cf. Sections 2.6 and 10.3.)

One can understand the general shape of the phase diagram from the following simple arguments: at very small density (large $r_s$) the Wigner crystal is stable at

$T = 0$, but its energy is $\sim 1/r_s$, see (11.84). Consequently the temperature needed to melt it will scale as $T_m \sim 1/r_s$. However, for high density (small $r_s$) the Wigner crystal will become unstable even at $T = 0$, so that $T_m$ should go down. One can see a certain analogy with the phase diagram of the insulator–metal transition, see the case of $V_2O_3$, Fig. 2.15. One can also note the similarity of this phenomenon with the question of the existence of liquid due to quantum effects as compared to a crystal state, cf. the discussion in Section 4.4.3 (the discussion of the quantum de Boer parameter, (4.73), and nearby). In this sense we see that the low-density Coulomb system behaves more or less as a classical one, whereas the high-density limit (Fermi liquid) is indeed a quantum liquid.

Three extra remarks are in order here. First, when considering the criteria ($r_{s\ \mathrm{crit}}$) for Wigner crystallization, one should be aware of other possible instabilities as well. Thus, a simple treatment shows that a 3d electron gas may develop magnetic instabilities at the values of $r_s$ smaller than those necessary for Wigner crystallization; one has thus to consider these different possibilities simultaneously.

The second problem concerns possible magnetic states of the Wigner crystal. When we form a crystal made of electrons, there will be localized spins $\frac{1}{2}$ at each lattice site which should order in some fashion. Most probably this ordering will be antiferromagnetic (cf. the next chapter). However, this question has not really been investigated well enough.

And the third point concerns a possible regular description of Wigner crystallization, or at least the instability of the normal Fermi liquid towards it. When describing different possible instabilities in the first parts of this chapter, in most cases we could use (and did use) the standard Feynman diagram technique, looking for some signatures of instabilities, e.g. in the appearance of imaginary poles in certain Green functions or response functions (imaginary frequencies of certain collective excitations). As far as I know there is no similar treatment for the Wigner crystallization, and it is not clear which response function and which subset of Feynman diagrams could show the instability towards it. It may well be that this instability is beyond the standard treatment of most of this chapter: the instabilities discussed above existed already at weak coupling and were related to the special features of the energy spectrum such as nesting, and with the corresponding divergences of certain response functions, or susceptibilities. At the same time Wigner crystallization appears only when the effective coupling reaches a certain critical strength, i.e. it is actually a signature of *strong electron correlations*, the main topic of the next chapter.

# 12

# Strongly correlated electrons

In the previous section we have already seen that in the case of strong electron–electron interactions, when the average interaction energy becomes larger than the corresponding kinetic energy, one can expect drastic changes of the properties of the system. Notably, the electrons will have a tendency to localize, so as to minimize their repulsion at the expense of a certain increase in kinetic energy. Materials and phenomena for which this factor plays an important role are now at the centre of activity of both experimentalists and theoreticians; this interest was especially stimulated by the discovery of high-$T_c$ superconductivity in which electron correlations play a very important role. But even irrespective of the high-$T_c$ problem, there are a lot of other interesting phenomena which are connected with strong electron–electron interactions. These phenomena include electron localization, orbital ordering and certain structural phase transitions, insulator–metal transitions, mixed valence and heavy fermion behaviour. The very existence of localized magnetic moments in solids, both in insulators and in metals, is actually determined by these correlations. That is why this is one of the most actively studied classes of phenomena at present.

Real materials to which one applies the models and the treatment presented in this chapter are mostly transition metal and rare earth compounds, although general ideas developed in this context are now applied to many other systems, including organic materials, nanoparticles or supercooled atoms. The typical situation in transition metal compounds is the one with partially filled d-shells. The corresponding wavefunctions are rather localized; their spatial extension may be smaller than the distance between these atoms or ions, especially if we are dealing with transition metal compounds such as oxides, in which typically there are oxygen ions between transition metal ions. As the effective hopping of electrons between sites determines the electron bandwidth and their kinetic energy, in these cases we can have a situation discussed in Section 11.8, in which the kinetic energy is smaller than the electron interaction. This situation is thus equivalent to a low-density electron system, and we can expect that the standard description of electrons as a weakly interacting Fermi liquid breaks down. What will be the outcome in this case and what are the properties of corresponding systems constitutes the field of *strong electron correlations*.

## 12.1 Hubbard model

The very idea that in the case of a strong electron–electron interaction one can expect drastic changes in the properties of the system has already been explained in the previous section. After the first treatment by Wigner (1937), similar approaches were applied to this situation by Landau and Zeldovich (1943) and by Mott (1948), see below. In applications to most of the relevant cases one now usually uses a somewhat different approach and starts not from free electrons with long-range Coulomb interactions, but rather from the tight-binding model with only on-site interactions, the so-called *Hubbard model*. We consider the system with a fixed lattice and nondegenerate band. In coordinate space the model has the form

$$\mathcal{H} = -t \sum_{\langle ij\rangle,\sigma} c_{i\sigma}^{\dagger} c_{j\sigma} + U \sum_{i} n_{i\uparrow} n_{i\downarrow} , \tag{12.1}$$

where $n_{i\sigma} = c_{i\sigma}^{\dagger} c_{i\sigma}$ and the summation in the first term goes over nearest neighbours $\langle ij\rangle$.[1] The negative sign in equation (12.1) is here chosen simply for convenience, so that the bottom of the corresponding tight-binding band (12.2) would be at $k = 0$ (although in more complicated cases the signs of different hopping matrix elements sometimes have to be fixed and can modify the results). The limitations of this model (omission of longer-range interaction, etc.) are of course rather severe; however it turns out that even this seemingly so simple model describes very rich physics and is at present far from being completely understood.

## 12.2 Mott insulators

There are essentially two parameters in the Hubbard model (12.1). These are the interaction strength (dimensionless parameter $U/t$) and the electron concentration, or band filling $n = N_{\mathrm{el}}/N_{\mathrm{site}}$. Thus, e.g. the often studied case of one electron per site $n = 1$ corresponds to a half-filled band.

The main virtue of the Hubbard model is that it permits us quite naturally to describe two opposite limits: that of weakly interacting electrons $U \ll t$, and the case of strongly interacting, or strongly correlated electrons, $U \gg t$.

[1] More generally we can write down the first term in equation (12.1) as $\sum_{ij,\sigma} t_{ij} c_{i\sigma}^{\dagger} c_{j\sigma}$, where $t_{ij}$ describes hopping between sites $i$ and $j$ which are not necessarily nearest neighbours. For nearest-neighbour hopping we usually take $t_{ij} = -t$, as in equation (12.1).

Fig. 12.1

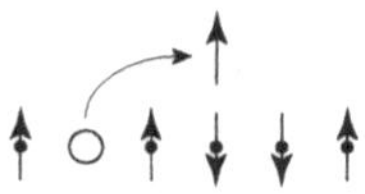

Fig. 12.2

In the first case one expects that the standard Fermi-liquid picture will be valid, at least in the three-dimensional case. In this situation we can treat the interaction term in (12.1) as a small perturbation and use all the techniques developed in Chapter 9.

Consider now the opposite case of strong interactions, $U \gg t$, for a half-filled band, $n = 1$. Formally the standard band theory would give a metal even in this case: the first term in (12.1) (kinetic energy) will in the momentum representation have the form

$$\mathcal{H}_{\text{kin}} = \sum_{\boldsymbol{k},\sigma} \varepsilon_{\boldsymbol{k}}\, c^{\dagger}_{\boldsymbol{k},\sigma}\, c_{\boldsymbol{k},\sigma} \,, \qquad \varepsilon_{\boldsymbol{k}} = \tilde{t}(\boldsymbol{k}) = -2t(\cos k_x + \cos k_y + \cos k_z) \,, \tag{12.2}$$

where $\tilde{t}(\boldsymbol{k})$ is the Fourier transform of the hopping matrix elements $t_{ij}$, which for nearest-neighbour hopping and for a simple cubic lattice gives the spectrum (12.2). Thus we have here a simple tight-binding band, and for $n = 1$ this band will be half-filled, i.e. we would have a metal, irrespective of the distance between the ions and of the value of the hopping matrix element $t$ and of the corresponding bandwidths $W = 2zt$ ($z$ is the number of nearest neighbours), which may be very small! However, this is a rather unphysical conclusion: one can argue that for a sufficiently narrow band $t \ll U$ (which is definitely the case if the lattice parameter, i.e. the distance between sites, is much larger than the radius of corresponding orbitals) the electrons will be localized at each site, and there will be no metallic conductivity. Indeed, such a localization of electrons helps to minimize the Coulomb repulsion $U$ at the expense of the kinetic energy $t$, and for $n = 1$ and $U \gg t$ such a state is clearly preferable. (The situation here is very similar to the case of the Wigner crystal, cf. Section 11.4.)

One can depict such a state as in Fig. 12.1, where we have chosen certain but arbitrary directions of the spins of the electrons ($\uparrow$ or $\downarrow$). We see that if we start from this state, then to create charge-carrying excitations we have to transfer one electron from its site to another, Fig. 12.2. After such an excitation is made, the electron (doubly occupied site) and the hole (empty site) will already propagate freely (if we ignore initially the background magnetic structure). This will give an energy gain $\sim t$ (both electron and hole will now be at the bottom of respective bands (12.2)). However to create such a pair one has to spend the energy $U$, the repulsion of two electrons at the doubly occupied site. And if $U \gg t$, this process costs energy $\sim U \gg t$ (or rather $U \gg W = 2zt$, where $z$ is the number of nearest neighbours), i.e. such a system will behave as an insulator with the energy gap $E_g \simeq U - 2zt$. Thus the resulting state should be an insulator – this is the famous *Mott insulator*.

Actually in his original publication in 1948 Mott used different arguments and proceeded from the picture of electrons with long-range Coulomb interactions. He argued that if we start from an insulator, and create electron and hole excitations, they would be bound to excitons, so that the system would remain insulating. And only if the concentration of these electrons and holes exceeds a certain critical value, then the screening of the Coulomb attraction would be sufficiently strong so that the excitonic bound state would disappear. (Recall that in quantum mechanics one needs a certain minimum strength of short-range attraction for the bound state to appear in the three-dimensional case.) The condition of that is qualitatively that the Debye screening length $r_D$, given by equation (9.22), $r_D^{-2} = \kappa_D^2 = 6\pi ne^2/\varepsilon_F \simeq 4\pi e^2 n^{1/3}/\hbar^2$ becomes smaller than the Bohr radius $a_0 = \hbar^2/me^2$. From this condition Mott obtained his famous criterion for the transition from an insulator to a metal:

$$n^{1/3} a_0 \geq 0.25 \; . \tag{12.3}$$

(Note here a certain relation of this picture to that of the excitonic insulators, Section 11.4, in which, however, due to the assumed nesting, an excitonic instability and the transition to an insulating state with a gap appeared at arbitrary small band overlap and respective electron and hole concentration.) As we see from this estimate, very narrow bands (large values of the effective mass $m^*$ and small Bohr radius $a_0$) always favour the formation of such bound states and prevent the transition to a metallic state, so that such narrow-band materials would remain insulating.[2] And only somewhat later, when dealing with transition metal

[2] In fact already five years earlier, in a paper published in 1943 (and which largely remained unknown) practically the same arguments were presented by Landau and Zeldovich [ZhETF **32**, 1944 (1943); Acta Phys.-chem. URSS **18**, 194 (1943)] who, in turn, referred to an earlier remark by Peierls. They wrote: 'A dielectric differs from a metal by the presence of an energy gap in the electronic spectrum. Can, however, this gap tend to zero when the transition point into a metal is approached (on the side of the dielectric)? In this case we should have to do with a transition without latent heat, without change of volume and of other properties. Peierls has pointed out that a continuous transition – in this sense – is impossible. Let us consider the excited state of the dielectric in which it is capable of conducting an electric current: an electron has left its place, leaving a positive charge in a certain place of the lattice and is moving throughout the latter. At large distances from the positive charge, the electron must certainly suffer a Coulomb attraction tending to bring it back. In a Coulomb attraction field there always exist discrete levels of negative energy, corresponding to a binding of the electron; the excited conducting state of the dielectric must therefore always be separated from the fundamental one, in which the electron is bound, by a gap of a finite width.' This is the same picture as the one first put forth by Mott in 1948, only the estimate (12.3) had not been made by Landau and Zeldovich.

compounds, Mott himself used arguments of the type which we started with, which became standard after the very important contribution of Hubbard in 1964, who formulated in a clear way and solved in certain approximations (see Section 12.4 below) the model (12.1) which now carries his name.

Actually the situation described by the Hubbard model with $t \ll U$ is met in many transition metal and rare earth compounds, typical examples being oxides such as NiO, CoO. They have partially filled 3d levels, and according to the band theory they should be metals. Experimentally, however, they are good insulators. This is explained by the mechanism described above: the radius of 3d orbitals ($\sim 0.6\,\text{Å}$) is much smaller than the metal–metal (e.g. Ni–Ni) distance ($\sim 3\,\text{Å}$), so that the effective *d–d* hopping $t$ (which actually goes via an intermediate oxygen) is much smaller than the on-site Coulomb repulsion $U$.[3]

Thus the ground state of the Hubbard model for $U \gg t$ and $n = 1$ is an insulating state with electrons localized one at each site. Note that this state is an insulator of a completely different type from those described by the standard band theory: in the usual cases the conventional band scheme is applicable, and the material is insulating, or semiconducting, if the valence band(s) is completely full, and conduction band(s) empty. The energy gap in these cases is determined by the interaction of electrons with the periodic lattice potential, and interactions between electrons do not play a crucial role there, i.e. this conventional gap is obtained already in the one-electron picture. Here, however, the very insulating character is determined *by the interaction between electrons*. This is completely different physics, and consequently many properties of such states are quite different from those of ordinary band insulators.

In the situation considered ($U \gg t$) the theoretical description should be reversed as compared to the standard one: whereas usually one takes the first term in the Hamiltonian (12.1) as the zero-order Hamiltonian and treats the interaction as a perturbation, here we should invert the description and treat the second term as the zero-order Hamiltonian, the kinetic energy playing the role of the perturbation:

$$\begin{aligned} \mathcal{H}_0 &= U \sum_i n_{i\uparrow} n_{i\downarrow} \,, \\ \mathcal{H}' &= -t \sum_{\langle ij \rangle, \sigma} c^\dagger_{i\sigma} c_{j\sigma} \,. \end{aligned} \tag{12.4}$$

[3] It turns out that in the oxides of heavier transition metals such as NiO and CoO another process is important: charge transfer between metal and oxygen, so that just these oxides which are usually cited as typical examples of Mott, or Mott–Hubbard insulators, are actually insulators of a somewhat different type – they are called charge-transfer insulators (Zaanen, Sawatzky and Allen), see Section 12.10 below. However, many of their properties are similar to those of Mott insulators, and very often one also uses for their description the Hubbard model (12.1) (although there are situations for which this reduction is not valid).

$\delta E = 0$ $\qquad\qquad$ $\delta E = -\dfrac{2t^2}{U}$

(*a*) $\qquad\qquad$ (*b*)

Fig. 12.3

## 12.3 Magnetic ordering in Mott insulators

The state with localized electrons (Mott, or Mott–Hubbard insulator) is a ground state of our zero Hamiltonian $\mathcal{H}_0$. However, it is not yet a unique state. There is still spin degeneracy left: each localized electron (localized spin) can have two orientations, $\uparrow$ or $\downarrow$, thus the degeneracy is $2^N$ where $N$ is the number of sites. This degeneracy is lifted by electron hopping (the term $\mathcal{H}'$ in (12.4)). This term, treated as a perturbation, in the first approximation creates a nearby electron–hole pair (or a 'doublon'–hole pair), with energy $U$, i.e. it creates a polar state which lies outside the degenerate ground-state manifold. However, a second application of $\mathcal{H}'$ can return us to the nonpolar state, so that the second-order terms in the perturbation expansion in $t/U$ give nonzero average in the subspace of localized states. It is these terms which lift the spin degeneracy and give magnetic (here antiferromagnetic) order.

One can explain the tendency to antiferromagnetism very simply. Consider two neighbouring sites with one electron at each. There are two possibilities: their spins may be parallel or antiparallel. The processes of virtual hopping of electrons (the second-order contributions in $\mathcal{H}'$) giving the energy change $\delta E$ may be illustrated as shown in Fig. 12.3. There we show the second-order contribution to the ground state energy for each case. Here $t^2$ comes because of applying the hopping term $\mathcal{H}'$ twice, and $U$ in the denominator is the energy of the intermediate state, as always in perturbation theory. In the case of parallel spins such hopping is forbidden by the Pauli principle, thus the corresponding contribution is zero. But for antiparallel spins it is allowed, and this extra delocalization of each electron (virtual 'excursions' to neighbouring sites) leads to a decrease of kinetic energy, according to the Heisenberg uncertainty relation. (Recall that in quantum mechanics the second-order contribution to the ground state energy is always negative.)

Mathematically one has to calculate the second-order term in perturbation theory $\langle\beta|\mathcal{H}'\frac{1}{E_0-\mathcal{H}_0}\mathcal{H}'|\alpha\rangle$, where $|\alpha\rangle$, $|\beta\rangle$ are states of the $2^N$-degenerate ground state manifold. In this subspace we obtain the secular equation with the effective Hamiltonian

$$\mathcal{H}_{\text{eff}} = \mathcal{H}'\frac{1}{E_0 - \mathcal{H}_0}\mathcal{H}' = -\frac{t^2}{U}\sum_{\langle ij\rangle,\sigma,\sigma'} c^{\dagger}_{i\sigma}\, c_{j\sigma}\, c^{\dagger}_{j\sigma'}\, c_{j\sigma'}\,. \tag{12.5}$$

In our subspace with one electron per site ($n_{i\uparrow} + n_{i\downarrow} = 1$) the electron operators can be expressed via the spin operators:

$$c^\dagger_{i\uparrow} c_{i\uparrow} = n_{i\uparrow} = \tfrac{1}{2} + S^z_i, \quad c^\dagger_{i\downarrow} c_{i\downarrow} = n_{i\downarrow} = \tfrac{1}{2} - S^z_i, \quad c^\dagger_{i\uparrow} c_{i\downarrow} = S^+_i, \quad c^\dagger_{i\downarrow} c_{i\uparrow} = S^-_i . \tag{12.6}$$

Putting these into (12.5) we obtain that the effective Hamiltonian is[4]

$$\mathcal{H}_{\text{eff}} = \text{const.} + \frac{2t^2}{U} \sum_{\langle ij \rangle} \boldsymbol{S}_i \cdot \boldsymbol{S}_j . \tag{12.7}$$

**Check this** using commutation relations and the expressions (12.6). Obtain the value of the constant in (12.7). Express the resulting exchange Hamiltonian through the projection operators onto the singlet state of the pair of spins $i$, $j$, $P_{S_{\text{tot}}=0} = \frac{1}{4} - \langle \boldsymbol{S}_i \cdot \boldsymbol{S}_j \rangle$, and onto the triplet state $P_{S_{\text{tot}}=1} = \frac{3}{4} + \langle \boldsymbol{S}_i \cdot \boldsymbol{S}_j \rangle$, where $\langle \boldsymbol{S}_i \cdot \boldsymbol{S}_j \rangle$ is the spin correlation function, $\langle \boldsymbol{S}_i \cdot \boldsymbol{S}_j \rangle = -\frac{3}{4}$ for the singlet state (total spin of the pair $S_{\text{tot}} = 0$), and $\langle \boldsymbol{S}_i \cdot \boldsymbol{S}_j \rangle = \frac{1}{4}$ for the triplet state ($S_{\text{tot}} = 1$).

The result (12.7) indeed confirms our qualitative considerations and shows that the ground state of our system is antiferromagnetic. Such an exchange mechanism is called *superexchange* (sometimes one also uses the term 'kinetic exchange'). It is the main mechanism of antiferromagnetism in insulators. One can calculate from (12.7) all standard properties such as spin-wave spectrum, etc. as in Chapter 6.

The Hamiltonian (12.7) also describes thermodynamic properties of our system when $T \ll U$. At $T = T_{\text{N}} \sim t^2/U$ the antiferromagnetic order disappears, and the material becomes paramagnetic. Nevertheless the material remains insulating, because strong electron correlations are still present and still prevent the formation of charge carriers. This is an essential difference from the case of spin-density waves (SDW) which exist in the case of weak interaction due to nesting, see Chapter 11; in this latter case the material can also be insulating in the SDW state, but above $T_{\text{SDW}}$, when the SDW disappears, it would become a metal. Here, due to strong correlations, the situation is qualitatively different (although the description of the ground state itself may be rather similar in both cases).

## 12.4 One-particle spectrum of strongly correlated systems

The structure and properties of one-particle excitations in the strongly interacting Hubbard model is highly nontrivial. This is connected with the electron–electron

[4] Here, according to our convention of Chapter 6, the summation goes over all site indices $i$, $j$ independently, i.e. every bond is counted twice. If one should count every bond only once, the exchange constant in (12.7) would be $4t^2/U$.

interaction, which manifests itself in the strong interaction of electron and hole excitations with the underlying magnetic structure, or, in other words, interaction between charge and spin degrees of freedom.

### 12.4.1 Aproximate treatment (Hubbard I decoupling)

Before discussing these particular questions, we present here the general treatment of the electron spectrum, using the method of decoupling of the equations of motion (cf. Section 6.3.1), first used for this problem by Hubbard (it is often called the Hubbard I decoupling scheme). It will also give us a good opportunity to illustrate how the method of equations of motion works in general.

Hubbard himself used the so-called double-time Green functions which are slightly different from the usual Green functions introduced in Chapter 8. For our purposes it is sufficient to write down the equations of motion for operators.

Let us write down the standard equation of motion for the operators in the Heisenberg representation. For the Hubbard model (12.1) (with arbitrary hopping $t_{ij}$) we have

$$\omega c_{i,\sigma} = [c_{i,\sigma}, \mathcal{H}] = \sum_j t_{ij} c_{j,\sigma} + U n_{i,-\sigma} c_{i,\sigma} \,. \tag{12.8}$$

**Check this** using the standard anticommutation relation for fermions $c^\dagger_{i,\sigma} c_{j,\sigma'} + c_{j,\sigma'} c^\dagger_{i,\sigma} = \delta_{ij}\delta_{\sigma\sigma'}$.

On the right-hand side of this equation we have a new operator $n_{i,-\sigma}c_{i,\sigma}$. If we make a decoupling at this stage, $n_{i,-\sigma}c_{i,\sigma} \to \langle n_{i,-\sigma}\rangle c_{i,\sigma}$, we would have a closed equation for $c_{i,\sigma}$ which can be solved by Fourier transform. This would correspond to the Hartree–Fock (mean field) approximation with the bare electron energies $\varepsilon_{\boldsymbol{k},\sigma} \to \varepsilon_{\boldsymbol{k},\sigma} + U\langle n_{-\sigma}\rangle = \varepsilon_{\boldsymbol{k}} + \frac{1}{2}U\langle n\rangle$ (for the nonmagnetic case, $\langle n_{i,\sigma}\rangle = \frac{1}{2}\langle n\rangle$). However this approximation does not account for the correlation effects; the system would remain, in this approximation, a metal.

To correct for this deficiency, we have to keep at least the terms describing on-site correlations, i.e. we keep the term $n_{i,-\sigma}c_{i,\sigma}$ and write down the equation of motion for this term; we will make a decoupling in similar terms containing operators at different sites. Thus we write

$$\begin{aligned}\omega n_{i,-\sigma} c_{i,\sigma} &= [n_{i,-\sigma} c_{i,\sigma}, \mathcal{H}] \\ &= \sum_j t_{ij} n_{i,-\sigma} c_{j,\sigma} + U n_{i,-\sigma} c_{i,\sigma} + \sum_j t_{ij} \left\{ c^\dagger_{i,-\sigma} c_{j,-\sigma} c_{i,\sigma} - c^\dagger_{j,-\sigma} c_{i,-\sigma} c_{i,\sigma} \right\} .\end{aligned} \tag{12.9}$$

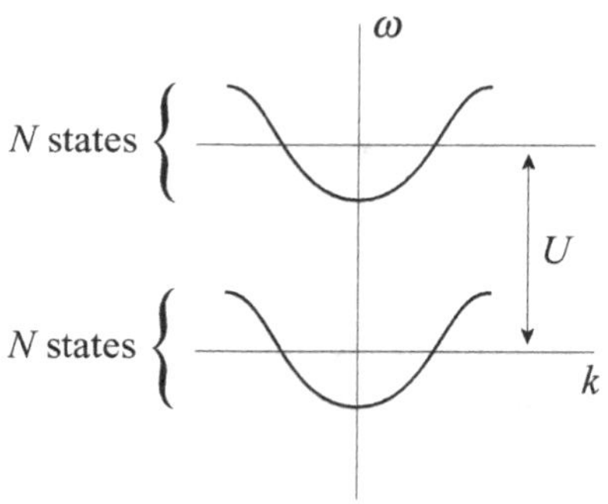


Fig. 12.4

The second term in (12.9) takes this form when we take into account that $n^2_{i,-\sigma} = n_{i,-\sigma}$ ($n_{i,-\sigma}$ is equal to 1 or 0). Now, we make a decoupling in the first term, $n_{i,-\sigma}c_{j,\sigma} = \langle n_{i,-\sigma}\rangle c_{j,\sigma}$ ($i \neq j$!). The last term drops out after a similar decoupling, and instead of (12.9) we have the equation

$$\omega n_{i,-\sigma}c_{i,\sigma} = \sum_j t_{ij}\langle n_{i,-\sigma}\rangle c_{j,\sigma} + U n_{i,-\sigma}c_{i,\sigma} \,. \tag{12.10}$$

Equations (12.8) and (12.10) form a closed set of two equations for $c_{i,\sigma}$ and $n_{i,-\sigma}c_{i,\sigma}$. Solving them (e.g. finding $n_{i,-\sigma}c_{i,\sigma}$ from (12.10) and putting it into (12.8), then performing a Fourier transform) we find the energy spectrum consisting of two branches (we consider here the paramagnetic state with $\langle n_{i,-\sigma}\rangle = \langle n_{i,\sigma}\rangle$ independent of the site index $i$):

$$\omega_\pm = \frac{U + t(\mathbf{k})}{2} \pm \frac{1}{2}\sqrt{\left(U - t(\mathbf{k})\right)^2 + 4U\,t(\mathbf{k})\,\langle n_{-\sigma}\rangle} \tag{12.11}$$

which for the half-filled band $\langle n_{-\sigma}\rangle = \frac{1}{2}\langle n\rangle = \frac{1}{2}$ takes the simple form

$$\omega_\pm = \frac{U + t(\mathbf{k})}{2} \pm \frac{1}{2}\sqrt{U^2 + t^2(\mathbf{k})} \,. \tag{12.12}$$

In contrast to the initial spectrum with the single half-filled band, the spectrum (12.12) describes two bands, one centred around $\omega = 0$ and another at $\omega = U$, see Fig. 12.4. One can show that whereas the original band contained $2N$ places (two per site, with spins ↑ and ↓), these new so-called *upper and lower Hubbard (sub)bands* each contain (for the half-filled case $n = 1$) only $N$ states ($N$ places), so that the lower band will be completely filled, and the upper one empty, see Fig. 12.4. Thus this solution gives an insulating state, which is the Mott–Hubbard insulator. One can check that the solution obtained is exact for the case of an isolated atom: indeed it gives two energy levels, $\omega = 0$ (one electron in the atom) and $\omega = U$ (two electrons).

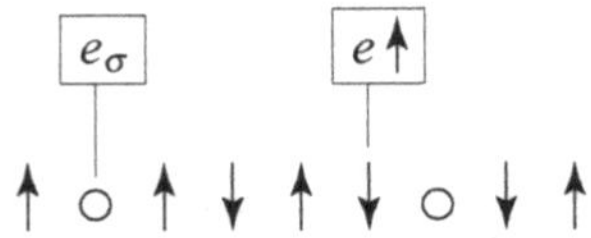


Fig. 12.5

One can also show that in general the one-electron Green function for electrons with spin $\sigma$ in this approximation has the form

$$G_\sigma(\boldsymbol{k}, \omega) = \frac{1 - \langle n_{-\sigma} \rangle}{\omega - \omega_-(\boldsymbol{k})} + \frac{\langle n_{-\sigma} \rangle}{\omega - \omega_+(\boldsymbol{k})} \,. \tag{12.13}$$

For $n = 1$ in the paramagnetic phase $\langle n_{-\sigma} \rangle = \langle n_\sigma \rangle = \frac{1}{2}\langle n \rangle = \frac{1}{2}$, and we indeed obtain that there are two poles in the Green function, the strength of each being $\frac{1}{2}$, which corresponds to the number of possible electron states in each subband being $N$ instead of $2N$. Physically the form of the Green function (12.13) is quite transparent: in accordance with the general considerations of Chapter 8, it describes the spectrum of strongly correlated systems, and the probability to add an electron with spin $\sigma$ at site $i$ with energy $\omega_- \sim 0$ is equal to the probability that there is *no* electron with spin $-\sigma$ at this site (the term with $1 - \langle n_{-\sigma} \rangle$ in (12.13)). If, on the other hand, there is an electron with spin $-\sigma$ at site $i$, $\langle n_{i,-\sigma} \rangle = 1$, then when we add an electron with $\sigma$ at this site, it will have energy $\omega_+ \sim U$ (the term with $\langle n_{-\sigma} \rangle$ in (12.13)), see Fig. 12.5.

There are several appealing features in the Hubbard I solution; it indeed describes the insulating state, which appears due to strong electron correlations, and qualitatively its properties are reasonable. However, there are also several drawbacks: the metallic state for $n \neq 1$ does not obey the Luttinger theorem, so it does not describe an ordinary Fermi liquid; for $n = 1$ there is a gap in the spectrum (12.12) for any $U$, including $U \ll t$, whereas we expect a transition to a metallic state in this case. Thus, this approximation does not describe the insulator–metal transition (Mott transition). It is also noteworthy that this is not a variational state: the energy of the ground state in this approximation is *lower* than the exact value.

### *12.4.2 Dealing with Hubbard bands. Spectral weight transfer*

One has to be very careful in using the picture of upper and lower Hubbard bands and the semiconducting-type analogy of Fig. 12.4: these bands should not be treated as ordinary bands. Thus, as we see from the expression (12.13), the residues of corresponding parts of the Green function, which describe the 'capacity' of each subband, i.e. the number of states in them, depend on the occupation of corresponding states, in contrast to the usual bands, each of which has $2N$ places.

(a) (b)

Fig. 12.6

Here, when we add electrons to our system, the number of states in each subband changes, or there occurs a *transfer of spectral weight* between subbands. This is illustrated in Fig. 12.6. Suppose that there are $N_e$ electrons in a lattice with $N$ sites. In this case there will be $N_e$ occupied sites with energy $\omega \sim 0$ (below the chemical potential), i.e. there will be $N_e$ ways to extract an electron from the system, e.g. by photoemission. On the other hand, there will be $(N - N_e)$ empty sites. At each empty site we can add two electrons with energy $\omega \sim 0$, with spins $\uparrow$ and $\downarrow$, i.e. there will be $2(N - N_e)$ empty states *in the lower Hubbard band*, with energies above the chemical potential. Thus altogether in the lower Hubbard band there will be $N_e + 2(N - N_e) = 2N - N_e$ states. And if we add an electron to an already occupied site, we can do it on $N_e$ sites, each time adding an electron with the opposite sign. Thus there will be $N_e$ empty places in the upper Hubbard band with the energy $\sim U$. In effect, the total structure of the density of states would look schematically as shown in Fig. 12.6(*b*), where we mark occupied states below the chemical potential $\mu$ by the grey region, and unoccupied states are empty. In total both these bands contain $2N$ places, as they should. But we see that, as we change the number of electrons $N_e$, the 'capacity' of each Hubbard band, i.e. its spectral weight, changes; only at $N_e = N$, i.e. $n = N_e/N = 1$, would the total weights of the lower and upper Hubbard bands be equal. This spectral weight redistribution is a very characteristic feature of systems with strongly correlated electrons. It shows, in particular, that it may be very difficult in this case to isolate only low-energy degrees of freedom, as one does, e.g. in the Fermi-liquid theory; there may be a significant admixture to them of the higher-energy states with energies $\sim U$. Such effects are indeed seen in many systems with strongly correlated electrons, for example in their optical properties.

### *12.4.3 Motion of electrons and holes in an antiferromagnetic background*

One of the most important drawbacks in the original treatment of Hubbard presented in Section 12.4.1 is the assumption of a nonmagnetic ground state, whereas we have seen that actually for $n = 1$ and $U \gg t$ the ground state is antiferromagnetically

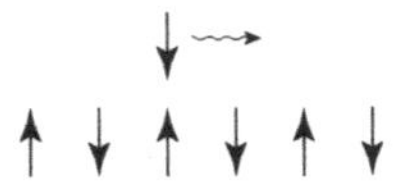

Fig. 12.7

ordered. This factor can significantly influence the motion of electrons and holes and it modifies the one-particle excitation spectrum.

Qualitatively one can understand this from Fig. 12.7: if we start from the simple Néel configuration and add one electron to it, e.g. with spin $\downarrow$, then this extra electron cannot hop onto nearest neighbouring sites because there are already electrons with the same spin on these sites, and the Pauli principle forbids this. This hopping would be allowed if there are spin deviations at neighbouring sites of a thermal or quantum nature. Thus we can expect that the presence of an antiferromagnetic background strongly hinders the motion of current-carrying excitations – electrons or holes.

The simplest way to treat this problem is the following (L. N. Bulaevskii and D. Khomskii, 1967). Let us start from the antiferromagnetic ground state $|\Phi_0\rangle$ and consider one extra electron (or hole) at site $i$, $|i\rangle = c^\dagger_{i,\sigma}|\Phi_0\rangle$. Such states are degenerate, and we seek a solution in the form

$$|\Phi\rangle = \sum_i a_i c^\dagger_{i,\sigma}|\Phi_0\rangle \ . \tag{12.14}$$

From the usual Schrödinger equation

$$\mathcal{H}|\Psi\rangle = E|\Psi\rangle \tag{12.15}$$

we find, with (12.4),

$$\sum_i a_i \langle j|\mathcal{H}'|i\rangle = (E - E_0)\langle j|j\rangle a_j \tag{12.16}$$

(the wavefunction $|i\rangle$ is an eigenfunction of $\mathcal{H}_0$ with the energy $E_0 = U$). The matrix elements of $\mathcal{H}'$ entering (12.16) may be expressed through spin operators via (12.6), and we get

$$\langle j|\mathcal{H}'|i\rangle = t\,\langle 0|\ c_{j,\sigma} \sum_{\langle lm\rangle,\sigma'} c^\dagger_{l,\sigma'} c_{m,\sigma'} c^\dagger_{i,\sigma}\ |0\rangle = t\left(\tfrac{1}{4} + \langle 0|\boldsymbol{S}_i \cdot \boldsymbol{S}_j|0\rangle\right) \ . \tag{12.17}$$

Taking into account also the normalization of the wavefunctions $|i\rangle$, given by

$$\langle i|i\rangle = \langle 0|c_{i,\sigma} c^\dagger_{i,\sigma}|0\rangle = \tfrac{1}{2} \mp \langle 0|S^z_i|0\rangle \tag{12.18}$$

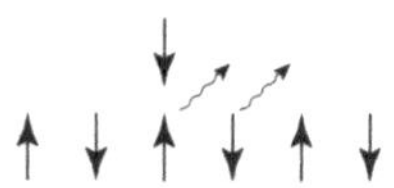

Fig. 12.8

and putting (12.17), (12.18) into equation (12.16), we obtain finally the spectrum of the extra electron added into the antiferromagnet with $n = 1$:

$$E(\boldsymbol{k}) = U + t(\boldsymbol{k})\, \frac{\frac{1}{4} + \langle \boldsymbol{S}_0 \cdot \boldsymbol{S}_1 \rangle}{\sqrt{\frac{1}{4} - S^2}} \,. \tag{12.19}$$

Here we have denoted the nearest-neighbour correlation function $\langle 0|\boldsymbol{S}_i \cdot \boldsymbol{S}_j|0\rangle$ as $\langle \boldsymbol{S}_0 \cdot \boldsymbol{S}_1 \rangle$, and the sublattice magnetization $\left|\langle 0|S_i^z|0\rangle\right| = S$. In the mean field approximation $\langle \boldsymbol{S}_0 \cdot \boldsymbol{S}_1 \rangle = -S^2$, and the spectrum of electrons (12.19) takes the form

$$E(\boldsymbol{k}) = U + t(\boldsymbol{k})\sqrt{\tfrac{1}{4} - S^2} \,. \tag{12.20}$$

We can obtain a similar equation also for a hole in the otherwise half-filled lower Hubbard band, which will have the same form as (12.20) without the constant term $U$.

We see that as a result of antiferromagnetic ordering the motion of charge carriers is indeed severely hindered: the electron and hole bands significantly narrow, $\varepsilon_{\boldsymbol{k}} = t(\boldsymbol{k}) \longrightarrow t(\boldsymbol{k})\sqrt{\frac{1}{4} - S^2}$ where $S \gtrsim \frac{1}{2}$ (only slightly less in the 3d case due to quantum fluctuations, cf. Chapter 6). Thus we see that there is very strong interplay between electron or hole motion and the underlying magnetic structure, so that charge transport may be strongly reduced or completely suppressed. One sees also that above the Néel temperature, where the sublattice magnetization $S$ in (12.20) disappears or the spin correlator $\langle \boldsymbol{S}_0 \cdot \boldsymbol{S}_1 \rangle$ in (12.19) is strongly reduced, electron and hole bandwidths may strongly increase, and conductivity may be significantly enhanced. This factor may play a certain role in the insulator–metal transition from an insulating antiferromagnet to a paramagnetic metal (L. N. Bulaevskii and D. Khomskii, 1970); such transitions occur in many transition metal compounds, e.g. $V_2O_3$.

There is one drawback in the treatment given above. In the picture described above, see Fig. 12.7, we have considered the motion of an *extra* electron added to the system. However, there is essentially no difference between 'extra' and 'own' electrons. Thus, instead of hopping of the extra electron which is forbidden by the Pauli principle, an 'own' electron may hop instead at each new step, see Fig. 12.8. It may seem that this process would lead to a free charge transfer on the

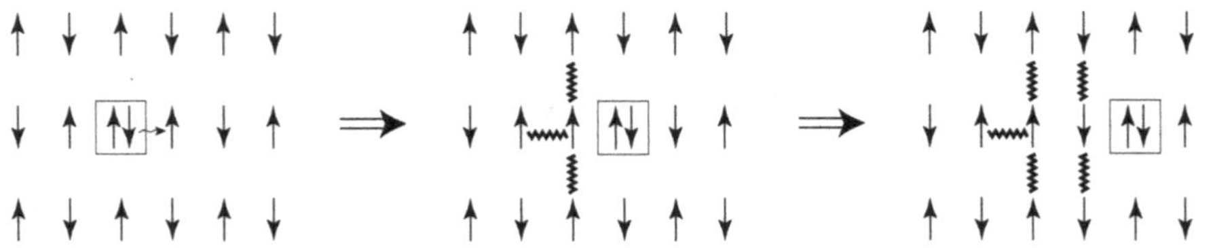


Fig. 12.9

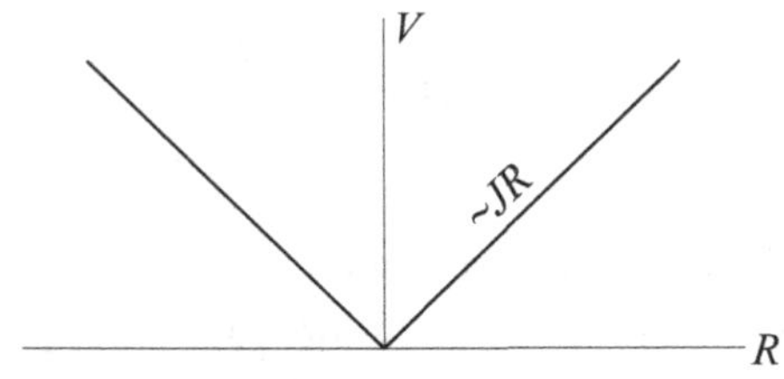


Fig. 12.10

antiferromagnetic background. However, the actual situation is more complicated and much more interesting.

In the one-dimensional case this is what does indeed happen, and we have a separate spinon (a pair of reversed spins at the point where we started this process) and a holon (or, in this case, a 'doublon' – an extra electron, forming one doubly occupied spin singlet site), which could propagate along the chain. But if we were to carry out this process in two- or three-dimensional cases, we would see that the charge carriers could indeed move by this process, but they would leave a 'tail' along their trajectory, a trace of wrong spins: each of the sites through which the hole or electron has passed will have its spin reversed; each such reversal would cost exchange energy $\sim J = 2t^2/U$, see Fig. 12.9 where the 'wrong' bonds are marked by thick wavy lines. Thus the farther away from the initial position the electron moves (nowadays, in connection with high-$T_c$ superconductivity, one speaks more often of holes), the bigger is the energy loss: it will be proportional to $Jl$, where $l$ is the length of the trajectory (L. N. Bulaevskii, E. Nagaev and D. Khomskii, 1970). As a result it looks as though the hole moves in a potential which *increases at least linearly* with the distance $R$ between the sites, see Fig. 12.10 (if we chose the shortest – straight – trajectory leading from point 0 to $R$). Thus there will be a constant force pulling the hole back to the origin, which leads to a *confinement* of the hole (the hole is localized in the vicinity of its initial position). The actual motion of the hole may be visualized as a 'wandering' with predominant return to the origin, , and at each step the spin direction is reversed, so that as a result in this local region the antiferromagnetic order will be strongly disordered or

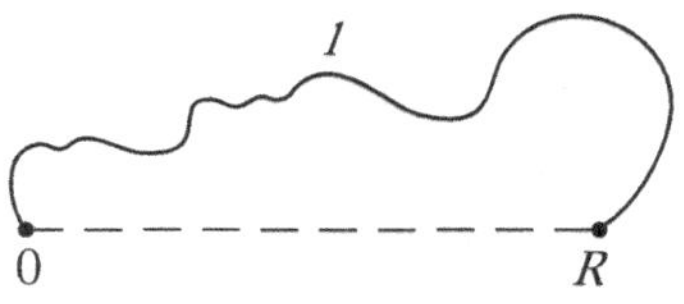


Fig. 12.11

modified. The situation somewhat resembles a polaron state, and it may be called a magnetic polaron: the electron or hole distorts the antiferromagnetic background and is itself localized in this distorted region. The detailed form of the magnetic state inside this region is rather complicated and is actually not well known.

In fact the extra energy of the electron which moved distance $R$ from the origin is not simply a function of $R$, as with an ordinary potential: it depends on the whole *trajectory* along which the electron has travelled, and it is proportional to $Jl$, where $l$ is the length of the trajectory, see Fig. 12.11. However, if we take into account only the least costly, the shortest, i.e. straight trajectory (dashed line in Fig. 12.11), then we indeed reduce our problem to the motion of a particle in an ordinary potential, linearly increasing with distance, $V(R) \sim JR$, as in Fig. 12.10. One can solve the resulting Schrödinger equation, which for the $s$-wave is reduced to the Airy equation. The energy of the ground state is (in a 3d cubic lattice)

$$\mathcal{E}_0 = -6t + 9.32\, t^{1/3} J^{2/3} \, . \tag{12.21}$$

Thus only for $J = 2t^2/U \to 0$ does the extra electron or hole reach the bottom of the free-electron conduction band $-zt$ (which it would do in the ferromagnetic case); for a finite antiferromagnetic exchange there is a certain loss of kinetic energy[5] $\sim t^{1/3} J^{2/3} \simeq t^{5/3}/U^{-2/3}$.

This notion of hole confinement is very important in the modern theory of strongly correlated electron systems. The term 'confinement' here is not accidental: one can indeed establish a close mathematical correspondence with quark confinement in quantum chromodynamics (in particular there the energy also depends on the trajectory – the famous Wilson loops). There are several important implications of this picture. One of the most spectacular is probably that it gives a mechanism of electron pairing which may be relevant for high-$T_c$ superconductors. The picture is qualitatively the following: one electron moving through an antiferromagnet leaves a trace of wrong spins and becomes localized, which decreases its kinetic energy. However, if we have *two electrons* or holes, the second one can move after the first

[5] The description presented above is somewhat simplified. This would be more or less true for the case of Ising interactions between localized spins. In reality, however, in the case of the Heisenberg interaction the 'damage', the trace of wrong spins left along the trajectory of the electron or hole, can be 'repaired' due to terms of the type $S^+S^-$ in the Heisenberg Hamiltonian. Also one can shift a hole along the diagonal of a plaquette by moving it around this plaquette 1.5 times (Trugman's trajectories). Both these effects, however, are weak and do not change the main qualitative conclusions.

one, 'repairing' the damage done, so that such a motion *of the pair* would not cost this exchange energy and the pair would not be confined. As a result the pair would move freely, and the corresponding gain in kinetic energy may well be the physical source of pairing in high-$T_c$ superconductors.

## 12.5 Ferromagnetism in the Hubbard model?

The same physical factors described above were used as an argument that for large enough $U/t$ one can get ferromagnetic ordering in the Hubbard model. Indeed, the motion of electrons or holes is completely unhindered if the background magnetic ordering were ferromagnetic instead of antiferromagnetic. In this case the energy spectrum is equal to the 'bare' spectrum $\varepsilon_{\boldsymbol{k}} = t(\boldsymbol{k})$, and the electron can gain energy $\sim W$, where $W$ is the bandwidth, $W = 2zt$ (the electron occupies the state at the bottom of the band, which for a ferromagnet has the full width $2zt$, without narrowing due to spin correlations). Nagaoka has proven that this is indeed the case if we consider the system with $U = \infty$ and with *one extra electron or hole* ($N_e = N \pm 1$) in a bipartite lattice (which can be subdivided into two sublattices). This rigorous proof is, however, valid only in this somewhat artificial case; it is not known whether such a state would exist in the thermodynamic limit, with a small but finite *concentration* of holes when the total volume of the system or the total number of sites goes to infinity.

After the work of Nagaoka the question of the possible existence of ferromagnetism in the simple nondegenerate Hubbard model (12.1) attracted considerable attention. Very detailed numerical calculations have shown that there may indeed be a region in the $(n, U/t)$ phase diagram in which the ferromagnetic state is a ground state. However, the exact boundaries of this region are not really known. They can also depend on the type of underlying lattice, so that for some lattices, such as simple cubic or bcc, ferromagnetism may even be absent altogether, whereas for other types of lattices, e.g. for the fcc lattice, it can exist for one type of doping, e.g. for electrons, but be absent for holes, or vice versa.

There exist special cases (e.g. with flat bands) for which one can rigorously prove that the ferromagnetic state is the ground state. But these cases are somewhat artificial. Thus the question of the existence of ferromagnetism in the nondegenerate Hubbard model in different situations is still not completely solved (see also the next section).

## 12.6 Phase diagram of the Hubbard model

Let us discuss the possible states of electronic systems described by the Hubbard model (12.1) in the whole $(n, U/t)$-plane. We have seen that for $n = 1$ the system is insulating and antiferromagnetic for $U/t \gg 1$. However this state can extend down to small $U$ for bipartite lattices and for nearest-neighbour hopping: in this case we have the band structure with the nested Fermi surface (condition (9.37) is satisfied), and according to the treatment in the previous chapter, an SDW state will exist even for weak interactions. (In this case the deficiency of the Hubbard I solution mentioned above – the fact that it gives an insulator even for $U \ll t$ – is not really a drawback.) However, if we break the nesting, including, e.g. further neighbour hopping, the insulating antiferromagnetic state will exist only above a certain critical value of $U/t$. An open question here is whether the magnetic order and the energy gap would disappear simultaneously; there are some arguments that this need not be the case, and an intermediate metallic antiferromagnetic phase may exist.

When we go away from $n = 1$, there will be competition between the kinetic energy of extra electrons and holes, which 'does not like' antiferromagnetism and tends to destroy it, and the exchange interaction of remaining localized spins (this was partially discussed in previous sections). In this situation several possibilities exist: antiferromagnetism may exist up to a certain doping $\delta = |1 - n|$, after which kinetic energy starts to dominate. Antiferromagnetic order is stabilized by the exchange interaction $J = 2t^2/U$. However, as we have seen, in the antiferromagnetic state we lose the kinetic energy of extra electrons. If we change the order, e.g. to ferromagnetic order, we gain kinetic energy $\sim t|1 - n| = t\delta$. Thus one can expect a cross-over to ferromagnetic ordering at

$$\delta = |1 - n| \geq \frac{t}{U} . \tag{12.22}$$

For $U \to \infty$ this critical value of doping tends to zero, $\delta \to 0$, which is consistent with the Nagaoka theorem. Nagaoka himself obtained the corresponding criterion in the form

$$0.246\,\delta > \frac{t}{U} . \tag{12.23}$$

However, there exist other possibilities as well. One can show that if one considers only homogeneous solutions, this cross-over from antiferromagnetic to ferromagnetic behaviour occurs not abruptly, but proceeds via intermediate phases. One such phase may be canted antiferromagnetism, Fig. 12.12 (D. Khomskii, 1970). In this state two antiferromagnetic sublattices are preserved, but their moments are canted so that a ferromagnetic component appears. The canting angle $\alpha$ increases with doping $\delta$, and at a certain critical $\delta$ (close to the Nagaoka estimate (12.23)) $\alpha \to \pi/2$, i.e. the ordering becomes ferromagnetic. The resulting phase diagram has the form shown in Fig. 12.13. Yet another possibility is that the intermediate phase, instead of a canted two-sublattice structure, would correspond to spiral

Fig. 12.12

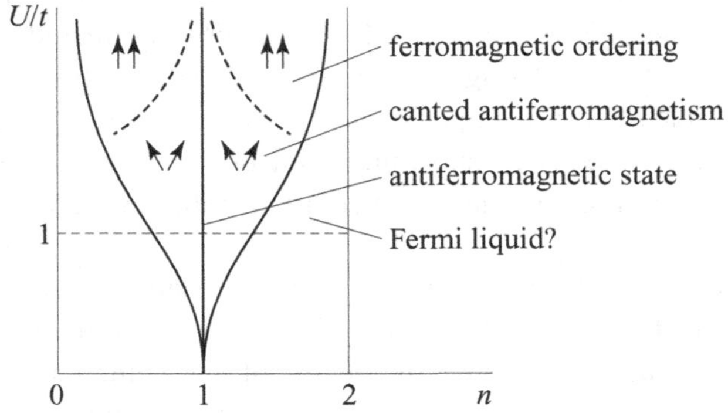


Fig. 12.13

ordering with the wavevector $\boldsymbol{Q}$ decreasing with increasing $\delta$; at $\delta = \delta_c$, $\boldsymbol{Q} \to \boldsymbol{0}$, i.e. the spiral state becomes ferromagnetic. This possibility may be understood using the arguments given in the previous chapter in treating charge and spin density waves: for $\delta \neq 0$ the Fermi momentum of noninteracting electrons deviates from the value $\boldsymbol{k}_{F_0} = \left(\frac{\pi}{2}, \frac{\pi}{2}, \ldots\right)$ which it had for $n = 1$ (or $\delta = 0$), and the SDW with the period determined by $2k_F(\delta)$ may be the most stable solution.

A lot of effort was devoted to the study of the opposite case – the limits of stability of ferromagnetism when the doping $\delta$ becomes large, or the concentration of electrons $n$ becomes small. From the arguments presented in Section 11.7, cf. equation (11.78), one can conclude that at least in the low-density limit $n \to 0$ our system should be in the nonmagnetic Fermi liquid state even for very strong interactions. If this is true there should be a certain critical concentration $n_c$ depending on $U/t$, below which ferromagnetism should disappear. Different approximate schemes give for $n_c(U/t \to \infty)$ the values ~0.2–0.4. However, this question is still open: as already discussed in the previous section, first of all it is not yet rigorously proven that ferromagnetism ever exists in the thermodynamic limit (the Nagaoka state may be just a singular point). There are some arguments that ferromagnetism may be stabilized only in the presence of orbital degeneracy or of the other bands. Second, the applicability of the low-density approximation (ladder summation) of Chapter 11 in the case of very strong interaction can also be questioned: it may turn out that the results actually depend on whether we first take the limit $n \to 0$ and then $U \to \infty$, or whether we take these limits in the opposite order. Thus the real form of the phase diagram of the Hubbard model is actually still unknown. The situation is even more complicated because of the possibility of the existence of spatially inhomogeneous solutions; see the next section.

## 12.7 Phase separation

In the previous treatment we have considered only spatially homogeneous solutions. However, this is not the only possibility. We have already seen that due to competition between kinetic energy and antiferromagnetic exchange an electron may become localized in a magnetically distorted region. For example, this microregion may have ferromagnetic ordering. Such a ferromagnetic microregion is called a ferromagnetic polaron, or ferron (E. L. Nagaev). Recently the same object, introduced in the context of high-temperature superconductivity, was called a spin-bag (J. R. Schrieffer).

One can easily estimate the size of such a ferromagnetic polaron. If we create a ferromagnetic region of radius $R$, the loss of exchange energy is $\frac{4\pi}{3}R^3 J$. The lowest electron energy level in such a potential well is approximately $-tz + t/R^2$, so that the total ferron energy is

$$E(R) = \frac{4\pi}{3}R^3 J - tz + \frac{t}{R^2} . \tag{12.24}$$

Minimizing (12.24) in $R$, we find

$$R_0 \simeq \left(\frac{U}{t}\right)^{1/5} , \qquad E_0 = E(R_0) \simeq -tz + t^{7/5}\, U^{-2/5} = -tz + J^{3/5}\, t^{2/5} . \tag{12.25}$$

Thus the size of such a ferromagnetic region becomes infinite when $U \to \infty$, so that it will occupy the whole sample, in agreement with the Nagaoka theorem. The general situation can then be visualized as an antiferromagnetic background, with doped electrons or holes localized in ferromagnetic 'bubbles' created by themselves.[6]

Now we can take one step further. If *one* electron is localized in a ferromagnetic region, why not two, or three, etc.? In other words, would it not be more favourable to create one big ferromagnetic region, containing *all* the doped electrons, leaving the remaining part of the crystal doping-free and antiferromagnetic? This is the

[6] Comparing (12.25) with (12.21) we see that for $J \ll t$ (or $t \ll U$) the energy of the ferromagnetic microregion is slightly higher than that of the 'string' solution (12.21), which makes the ferron state less favourable, see Fig. 12.14. Thus one should conclude that most probably the distorted spin microregion would not be simply ferromagnetic; only for $J \to 0$ or $U \to \infty$ may they become degenerate. Nevertheless the picture of ferromagnetic 'droplets' accounts for at least a part of the physics involved, and it can be used for qualitative arguments and for crude numerical estimates (mathematically it is much simpler than the picture of confinement of Section 12.4).

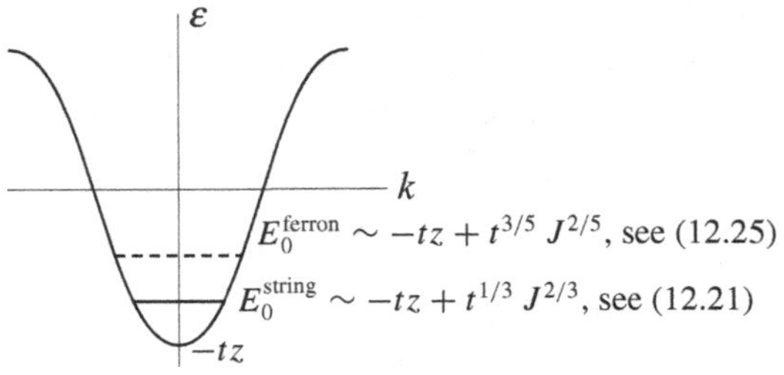


Fig. 12.14

idea of phase separation in the Hubbard model, first put forth in 1974 by Visscher and rediscovered later by many researchers, notably by Emery and Kivelson.

One can indeed give simple arguments that this can be the case in the partially filled Hubbard model. Suppose that phase separation takes place, and let us calculate the energy of such a state. If we have the average doping concentration $\delta = N_\mathrm{h}/V = (N - N_\mathrm{el})/V$ ($N_\mathrm{h}$ is the number of holes), and out of the total volume of the system $V$ the part $V_\mathrm{f}$ is ferromagnetic, with all $N_\mathrm{h}$ holes (or extra electrons) inside it, their actual concentration in this region will be $\delta_\mathrm{f} = N_\mathrm{h}/V_\mathrm{f} = (V/V_\mathrm{f})\delta$. The remaining part of the sample $V_\mathrm{a} = V - V_\mathrm{f}$ is antiferromagnetic. The energy of this antiferromagnetic region is

$$E_\mathrm{af} \sim -JV_\mathrm{a} = -JV\left(1 - \frac{V_\mathrm{f}}{V}\right) . \tag{12.26}$$

The energy of the ferromagnetic region consists of two parts: the magnetic energy $+JV_\mathrm{f}$ and the energy of holes or extra electrons in this region, which is $\sim -tz\delta_\mathrm{f}V_\mathrm{f} + tV_\mathrm{f}\delta_\mathrm{f}^{5/3}$: holes move freely on the ferromagnetic background and occupy the states from the bottom of the hole band, $-tz$, up to the Fermi energy, $-tz + tp_\mathrm{F}^2$, with $p_\mathrm{F}^3 \simeq \delta_\mathrm{f}$. Then the energy of electrons occupying the states up to $p_\mathrm{F}$ is $\sim \int_0^{p_\mathrm{F}} \frac{p^2}{m} d^3\boldsymbol{p} \sim p_\mathrm{F}^5/m$, which, with $m \sim t^{-1}$ and $p_\mathrm{F} \sim \delta_\mathrm{f}^{1/3}$, gives the energy presented above. Thus the total energy of the ferromagnetic region is

$$E_\mathrm{f} = +JV_\mathrm{f} - tz\,V_\mathrm{f}\delta_\mathrm{f} + tV_\mathrm{f}\delta_\mathrm{f}^{5/3} . \tag{12.27}$$

Denoting $V_\mathrm{f}/V$ by $y$ and minimizing the total energy $E_\mathrm{af} + E_\mathrm{f}$ in $y$, we obtain finally (we recall that $\delta_\mathrm{f} = (V/V_\mathrm{f})\delta$):

$$y_0 = \left(\frac{V_\mathrm{f}}{V}\right)_0 = \left(\frac{U}{t}\right)^{3/5} \delta , \tag{12.28}$$

and the total energy of this phase-separated state is

$$E_0/V = -\frac{t^2}{U} - tz\delta + t\delta\left(\frac{t}{U}\right)^{2/5} \tag{12.29}$$

(cf. equation (12.25)). Thus we see from (12.28) that formally the doped system described by the Hubbard model (with only short-range interactions!) would phase separate for all doping $\delta = |1 - n| \lesssim (t/U)^{3/5}$, until the ferromagnetic fraction $y = V_f/V$ occupies the whole sample.

One should compare this state with other possible states. The possible homogeneous states are, e.g. the antiferromagnetic one, the ferromagnetic one and the state with canted sublattices. The antiferromagnetic state is evidently not good: in the first approximation electrons in it are immobile, see, e.g. (12.19), (12.20). As we have seen above, the ferromagnetic state may become favourable for the concentration of extra charge carriers $\delta$ exceeding a certain critical value, which for the homogeneous case is given by (12.23). The energy of the homogeneous canted phase may be easily calculated using the scheme similar to the one used above, see (12.27), but with the effective hopping matrix element $t \to t_{\text{eff}} \simeq t \frac{1/4 + \langle \boldsymbol{S}_0 \cdot \boldsymbol{S}_1 \rangle}{\sqrt{1/4 - S^2}}$, see (12.19). For the canting angle $\alpha$ we find $\langle \boldsymbol{S}_0 \cdot \boldsymbol{S}_1 \rangle = -\frac{1}{4}(1 - 2\sin^2\alpha)$, $S = \langle S^z \rangle = \frac{1}{2}\cos\alpha$, and $t_{\text{eff}} = t \sin\alpha$. Putting these expressions into the total energy

$$E^{\text{canted}}/V = J\langle \boldsymbol{S}_0 \cdot \boldsymbol{S}_1 \rangle - t_{\text{eff}} z\delta + t_{\text{eff}}\delta^{5/3} \tag{12.30}$$

(cf. (12.27)), and minimizing in $\alpha$, we find for small $\delta$

$$\sin\alpha \sim \frac{\delta}{4}\frac{U}{t}, \tag{12.31}$$

which gives the critical concentration $\delta_c$ for the transition from the canted to the ferromagnetic state ($\alpha \to \pi/2$), a value consistent with (12.23). For $\delta < \delta_c$ the energy of the canted state is

$$\frac{E_0^{\text{canted}}}{V} = -\frac{t^2}{U} - \frac{Uz^2}{8}\delta^2 . \tag{12.32}$$

If one compares (12.32) with (12.29), one sees that the phase-separated state definitely has lower energy than 'the best' homogeneous state – the canted one – for small $\delta$.

From (12.28) we find that with increasing $\delta$ the volume of the ferromagnetic state increases, and this phase will occupy the whole volume, $y = V_f/V \to 1$, when the doping concentration is

$$\delta > \delta_c^{\text{ph. sep.}} \sim \left(\frac{t}{U}\right)^{3/5} . \tag{12.33}$$

Comparing this expression with (12.23), we see that the phase-separated state survives up to higher values of $\delta$ than the homogeneous antiferromagnetic state. The energies of different states discussed have the form shown schematically in Fig. 12.15. Thus we can conclude from these arguments that the phase-separated

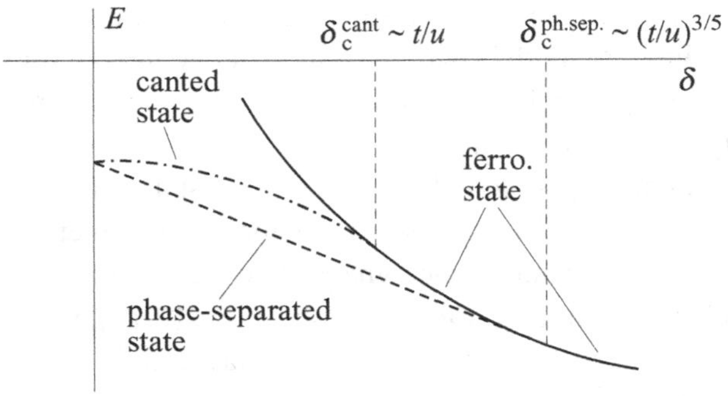


Fig. 12.15

state is indeed quite feasible for the partially filled Hubbard model at least in the strong interaction case $U \gg t$. Thus instead of the phase diagram of Fig. 12.13 we may have not the canted intermediate phase but phase separation in this (and a somewhat wider) region.

The treatment given above is of course not rigorous. We have not taken into account possible local spin distortions close to electrons or holes, and we have not considered quantum fluctuations. We have also compared the energies of only a few possible states. Nevertheless the conclusion that there may occur phase separation in the Hubbard model is very plausible. This is also confirmed by recent numerical calculations (which, however, are mostly carried out in the so-called $t$–$J$ model, which is often used as a substitute for the strongly interacting Hubbard model – see the next section).

In real materials there exists a factor which is missing in the Hubbard model – the long-range Coulomb repulsion between electrons. It is clear that it will strongly counteract the tendency to phase separation discussed above: the requirement of electroneutrality is usually very strong and it would prevent large-scale phase separation. However, such phase separation can still be present if there exist, e.g. mobile ions in the material which can guarantee electroneutrality; this seems to be the case in oxygen-rich high-$T_c$ superconductors, e.g. in $La_2CuO_{4+y}$, where phase separation is observed experimentally. Although it is not yet completely clear whether the mechanism of phase separation in this case is the one described above, this is most probably the case.

Another option could be that the intrinsic tendency to phase separation is still there, but long-range Coulomb forces will prevent the formation of large charged regions, and phase separation will be 'stopped' at a certain length-scale – charged regions forming, e.g. small droplets each containing a finite number (30, or 130, or some other number) of doped electrons and holes ('frustrated phase separation' of Emery and Kivelson). These phase-separated regions can also have different shapes (e.g. linear stripes), they can have their own dynamics, etc.

Note also that the homogeneous canted state described above is actually absolutely unstable. Indeed, according to equation (12.32) its energy is a concave function of the electron density $\delta = N/V$ or of the volume $V$. Correspondingly, this state has negative compressibility, $\kappa^{-1} = V(\partial^2 E/\partial V^2) < 0$, and formally (in the absence of long-range Coulomb forces) it would be absolutely unstable with respect to phase separation.

## 12.8 $t$–$J$ model

When considering strongly interacting systems, one often uses instead of the Hubbard model the so-called $t$–$J$ model. In the Hubbard model each site can have four possible configurations, $|0\rangle$, $|\uparrow\rangle$, $|\downarrow\rangle$ and $|\uparrow\downarrow\rangle$. For strong interactions, $U \gg t$ and $n \leq 1$, the last state – the 'doublet' – has much higher energy than the other three ($\sim U$ instead of $\sim t$). Thus one can project it out and consider only the states with occupation one ($|\uparrow\rangle$, $|\downarrow\rangle$) and zero ($|0\rangle$).[7] Of course, as we have seen above, these doubly occupied states are actually very important: they determine the antiferromagnetic exchange (12.7), and they participate in the spectral weight transfer at doping, Section 12.4.2. However, in the superexchange process they are only virtually occupied; the Hamiltonian (12.7) acts on the subspace of singly occupied states. Thus we can 'get rid' of these states, taking them into account only implicitly, in the exchange interaction (12.7). As a result we have to keep in our model terms of two types: those describing magnetic exchange between neighbouring sites each having one electron, and those describing the motion of the holes, if present. The resulting Hamiltonian is written as

$$\mathcal{H} = \sum_{ij} t_{ij}\, \tilde{c}^{\dagger}_{i\sigma}\, \tilde{c}_{j\sigma} + J \sum_{ij} \boldsymbol{S}_i \cdot \boldsymbol{S}_j \,. \tag{12.34}$$

Here we can treat the parameters $t$ and $J$ as independent parameters, instead of $t$ and $U$ in the Hubbard model. We can even formally consider the situation with $J \gtrsim t$ (although strictly speaking one can obtain the exchange interaction – the second term in (12.34) – from the initial Hubbard model only in perturbation theory in $t/U$, i.e. in reality $J \sim t^2/U$ should be definitely less than $t$).

One should only be careful in dealing with the model (12.34) in that the operators $\tilde{c}^{\dagger}$, $\tilde{c}$ are strictly speaking *not* the usual Fermi operators (although they are often treated as such): one should remember that they describe the motion of electrons, or rather holes, on the background of other electrons. Nevertheless with careful enough treatment the $t$–$J$ model (12.34) may be used to calculate certain properties

[7] Technically one can do this, e.g. by using the so-called Gutzwiller projection, see, e.g. Fazekas (1999).

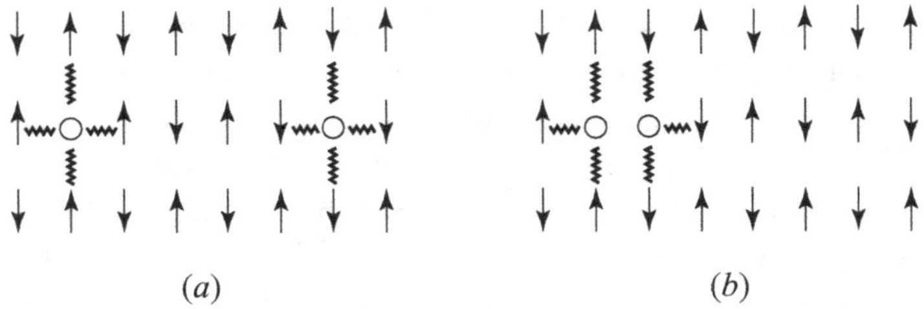

Fig. 12.16

of strongly interacting systems. This is especially useful in numerical calculations, because the reduction of the total number of states in this model as compared to the Hubbard model is very desirable for such calculations.

The possibility of phase separation exists also in the $t$–$J$ model, and it is even much easier to see it here, especially for the somewhat unphysical case of large $J/t$. Indeed, when we have, e.g. two holes far from each other, we lose $z$ antiferromagnetic bonds around each of them, see Fig. 12.16($a$) (in this case four bonds per hole). However, if we put these holes close together, we lose fewer bonds (here only six bonds instead of eight), Fig. 12.16($b$). Thus if $J/t$ is large, we gain energy by putting holes together, i.e. we have a strong tendency to phase separation.

## 12.9 Orbital ordering in the degenerate Hubbard model

The main physical objects for the application of the Hubbard model are transition metal compounds. If one wants to make their description more realistic, one has to take into account the fact that, besides the spins, there exist also orbital degrees of freedom for d electrons. In atoms the d levels ($l = 2$) are five-fold degenerate ($l^z = \pm 2, \pm 1, 0$). This degeneracy is partially lifted in a crystal: e.g. in a cubic crystal field the five-fold degenerate levels are split into a triplet – so-called $t_{2g}$-levels, and a doublet $e_g$. Thus, in an octahedral coordination this splitting is as shown in Fig. 12.17($a$). When one fills these levels by electrons, for instance according to Hund's rule (i.e. having maximal total spin possible, or, in other words, putting one electron after the other with parallel spins until they fill all five levels, and only then starting to fill these levels with electrons with opposite spin), one can still have situations with orbital degeneracy. This is, for example, the case for $Cu^{2+}$ (configuration $d^9$, one hole in a doubly degenerate $e_g$ level) or for $Mn^{3+}$ ($d^4$, three electrons with parallel spins in $t_{2g}$ levels and one electron with the same spin in an $e_g$ level), see Fig. 12.17($b$).

There exists the well-known Jahn–Teller theorem which states that the situation with such degeneracy is unstable (it does not correspond to the energy minimum).

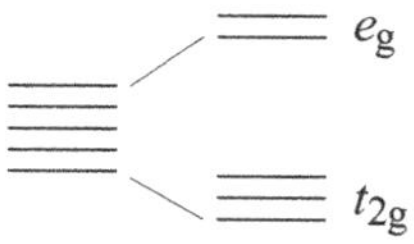


(*a*)

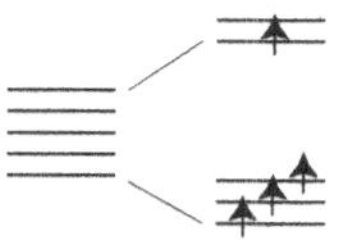

(*b*)

Fig. 12.17

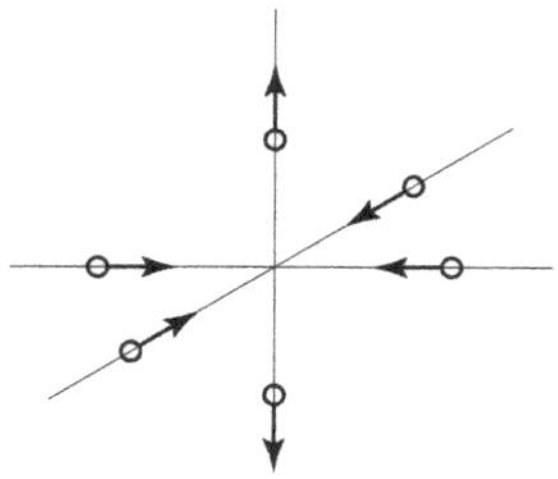

(*a*)

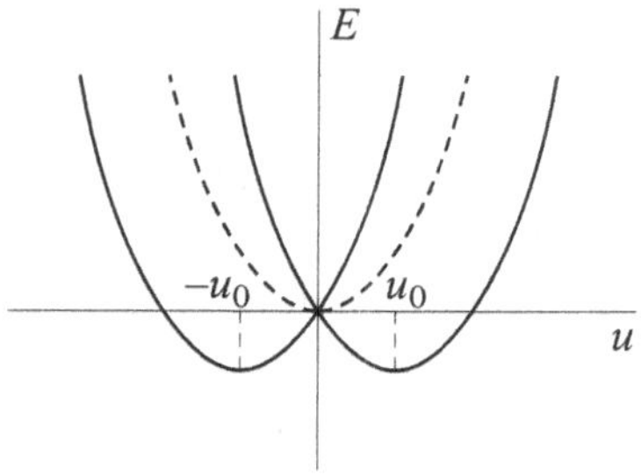


(*b*)

Fig. 12.18

The only degeneracy permitted in the ground state is the Kramers degeneracy, i.e. spin-degeneracy, spins ↑ and ↓.[8] This may be quite simply explained as follows: suppose we make a certain distortion of the atoms surrounding our transition metal ion, for example we make an elongation of the oxygen octahedron surrounding the transition metal ion in an oxide (this is the typical situation in systems such as $LaMnO_3$ etc.), see Fig. 12.18(*a*). Due to this distortion there will appear a certain noncubic potential $\delta V$ which will be a perturbation in our problem. It will be proportional to the distortion $u$, $\delta V \sim \lambda u$.

[8] Interestingly enough, as Teller himself wrote in the 'Historical Note' in the preface to the book of R. Englman *The Jahn–Teller Effect in Molecules and Crystals*, John Wiley, London, 1972, the idea of the Jahn–Teller effect could be attributed to Landau. Teller wrote:

'In the year 1934 both Landau and I were in the Institute of Niels Bohr at Copenhagen. I had many discussions. I told Landau of the work of one of my students, R. Renner, on degenerate electronic states in the linear $CO_2$ molecule. . . . He said that I have to be very careful. In a degenerate electronic state the symmetry on which this degeneracy is based . . . will in general be destroyed. . . .

'I proceeded to discuss the problem with H. A. Jahn who, as I, was a refugee from the German university. We went through all possible symmetries and found that the linear molecules constitute the only exception. In all other cases Landau's suspicion was verified. . . .

'This is the reason why the effect should carry the name of Landau. He suspected the effect, and no one has given a proof that mathematicians would enjoy. Jahn and I merely did a bit of a spade work.'

Of course Teller underestimated the importance of his contribution: this 'spade work' was extremely useful and played a crucial role in starting quite a big field. But it is also striking how often the name of Landau appears in this book: Landau theory of phase transitions; Landau criterion of superfluidity; Ginzburg–Landau equations for superconductors; Landau Fermi-liquid theory. And as we see now, his name is also intrinsically connected with the Jahn–Teller effect, and also with the insulator–metal (Mott) transitions, see Section 12.2.

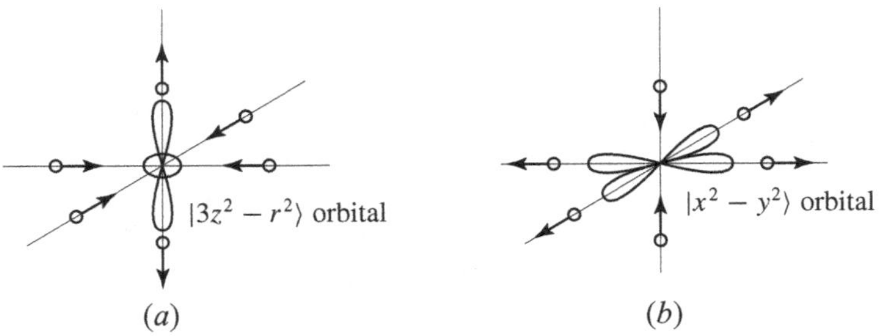


Fig. 12.19

It is known that in the degenerate case the splitting of energy levels is *linear* in the perturbation, $E_{\rm el} = E_0 \pm \lambda u$. At the same time the change of the lattice energy at the deformation is quadratic, $E_{\rm lattice} = \frac{1}{2}Bu^2$ (dashed line in Fig. 12.18(*b*)). The resulting energy levels as a function of distortion look as shown in Fig. 12.18(*b*). As a result there will appear a nonzero distortion at any coupling strength $\lambda$. The instability towards this distortion is called the *Jahn–Teller effect*. Due to coupling between different sites in the concentrated system we have a *cooperative Jahn–Teller effect*, which can lead to a structural phase transition which decreases the symmetry of the crystal and lifts the orbital degeneracy.[9] (Note the similarity of this effect and the Peierls instability discussed in Chapter 11, see (11.6), (11.7).) After this structural transition the degenerate orbital levels at each site are split in a certain way, and electrons occupy the lowest orbitals. Consequently we can call this process *orbital ordering*.

Different orbitals have different wavefunctions and different distributions of electron density. Thus the two degenerate $e_g$ orbitals have wavefunctions $\frac{1}{\sqrt{6}}(3z^2 - r^2) = \frac{1}{\sqrt{6}}(2z^2 - x^2 - y^2)$ and $\frac{1}{\sqrt{2}}(x^2 - y^2)$ (or their linear combinations). Correspondingly, in the first one the electron density is predominantly oriented along the $z$-axis, whereas in the second one it has the form of a flat 'cross' in the basal plane, see Fig. 12.19. It is clear that occupation of the orbital $|3z^2 - r^2\rangle$, Fig. 12.19(*a*), would cause a tetragonal elongation of the $O_6$ octahedra around such an ion, like that shown in Fig. 12.18(*a*): the negative electron density extended along the $z$-direction would push away negatively charged apical oxygens $O^{2-}$ (and to conserve the volume the in-plane oxygens would move in). Vice versa, after such a distortion the Coulomb energy of this orbital would be smaller than that of the $|x^2 - y^2\rangle$ orbital of Fig. 12.19(*b*), i.e. the doubly degenerate $e_g$ levels of Fig. 12.17 would be split by such tetragonal elongation so that the $|3z^2 - r^2\rangle$ level goes down,

[9] For an isolated centre of this type, e.g. for a Jahn–Teller impurity, the situation may be more complicated: because of the degeneracy of several minima, e.g. two minima at $\pm u_0$ in Fig. 12.18(*b*), there may occur quantum tunnelling between them. The resulting situation is known as the *dynamic Jahn–Teller effect*.

and $|x^2 - y^2\rangle$ level goes up. For the orbital $|x^2 - y^2\rangle$ of Fig. 12.19(*b*) the situation would be the opposite: it will be stabilized by the compression of the $O_6$ octahedra along the $z$-axis.[10] (For one hole in $e_g$ orbitals, as, e.g. in $Cu^{2+}$ with nine d electrons, the sign of the charge on the 'hole orbital' and corresponding distortion would be reversed, so that, e.g. the $|x^2 - y^2\rangle$ *hole* orbital on $Cu^{2+}$ would coexist with *elongated* octahedra.)

When an electron occupies one of these orbitals, the ion acquires a quadrupole moment; thus we can also say that such an orbital ordering is simultaneously a *quadrupolar ordering* (this terminology is often used for similar phenomena in rare earth systems). Strictly speaking the order parameter for such a transition, in the sense of Landau expansion, is indeed a second rank tensor – the quadrupolar moment, although in specific cases one often uses different, technically simpler, descriptions, see below.

Let us consider, for example, orbital ordering for the case of double degeneracy for $U \gg t$ and with one electron per site, $n = 1$. In this case each localized electron will be characterized not only by its spin $\sigma$ ($\sigma^z = \pm\frac{1}{2}$), but also by the index of the orbital occupied, $\alpha = 1$ or 2. One can map this extra double degeneracy into an effective pseudospin $\tau = \frac{1}{2}$, so that, e.g. orbital 1 corresponds to $\tau^z = +\frac{1}{2}$, and orbital 2 to $\tau^z = -\frac{1}{2}$. Orbital ordering in this language corresponds to ordering of these pseudospins $\tau$. In different situations it may be, for example, a ferro. orbital ordering (the same orbital is occupied at each site), i.e. local distortions around each transition metal ion are the same – we have then the total ferro. distortion of the whole crystal. Or it may be that different orbitals are occupied at neighbouring sites, and we can then speak about antiferro. orbital ordering.

The mechanisms leading to such orbital ordering may be different. It may be the electron–lattice interaction which was invoked when we discussed the local Jahn–Teller effect, Fig. 12.18: the coupling between local distortions on different sites will finally lead to a cooperative structural transition and to corresponding orbital ordering. Or it may be an exchange interaction, similar to the one discussed for the nondegenerate case in Section 12.3, which would couple orbital and magnetic orderings. The model describing all such situations is the degenerate Hubbard model

$$\mathcal{H} = \sum_{\substack{\langle ij\rangle,\alpha\beta \\ \sigma}} t_{ij}^{\alpha\beta} c_{i\alpha\sigma}^{\dagger} c_{j\beta\sigma} + U \sum_{\substack{i,\alpha\beta \\ \sigma\sigma'}} n_{i\alpha\sigma} n_{i\beta\sigma'} - J_H \sum_i \left(\tfrac{1}{2} + 2\mathbf{S}_{i1} \cdot \mathbf{S}_{i2}\right) . \tag{12.35}$$

[10] Besides the Coulomb, or point-charge contribution, there exists another factor, adding to the coupling between local distortion and splitting of degenerate orbitals with corresponding orbital occupation: the covalency between d orbitals of transition metals and p orbitals of ligands (e.g. oxygens). Usually these factors, Coulomb repulsion and covalency, work together and add up, stabilizing the same orbitals.

| Same orbital – same spin | Same orbital – opposite spins | Different orbitals – same spin | Different orbitals – different spins |
|:---:|:---:|:---:|:---:|

$2$ —  —          —  —          —  ↑          —  ↓

$1$ ↑  ↑          ↑  ↓          ↑  —          ↑  —

$$\delta E = 0 \qquad \delta E = -\frac{2t^2}{U} \qquad \delta E = -\frac{2t^2}{U - J_H} \qquad \delta E = -\frac{2t^2}{U}$$

(a)                        (b)                        (c)                        (d)

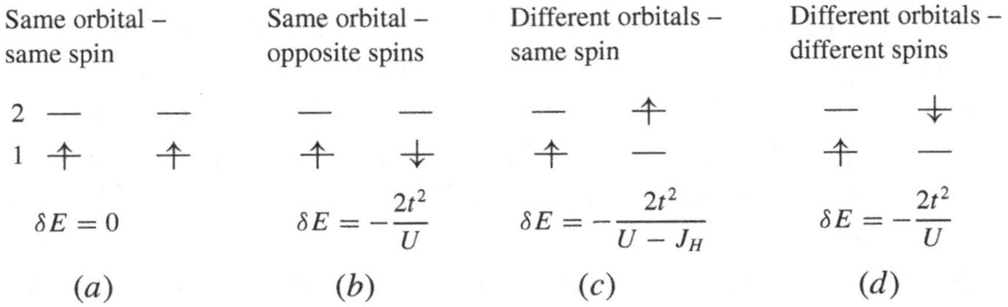

Fig. 12.20

Here $\alpha$, $\beta = 1$, $2$ are the orbital indices. We also included the last term which describes the intra-atomic exchange responsible for Hund's rule (it is absent in the nondegenerate case, in which two electrons at the same site may only have opposite spins).[11]

Suppose that there exists only hopping between the same orbitals, $t^{11} = t^{22} = t$, $t^{12} = 0$. Then we have the following four situations for the pair of doubly degenerate sites, shown in Fig. 12.20. (These four possibilities replace the two configurations which existed in the nondegenerate case, cf. Fig. 12.3.) As in the nondegenerate case, we have here also an extra energy decrease due to virtual hopping of electrons into neighbouring sites.

Whereas in the nondegenerate case these virtual hoppings provided the mechanism for antiferromagnetic ordering, we see that here the same superexchange will lead simultaneously to *both spin and orbital ordering*: in our simple model the third configuration (c) (same spin/different orbitals) has the lowest energy, because in this case in the intermediate state two electrons at the same site have parallel spins, which decreases the energy of this state (the denominator in Fig. 12.20(c)) due to Hund's rule interaction. Thus, one may expect that in a concentrated system in such a situation there will be ordering ferromagnetic in spin and 'antiferro.' orbital ordering. This mechanism of orbital ordering (Kugel and Khomskii, 1972, 1982) acts simultaneously with that due to lattice distortion (the conventional Jahn–Teller mechanism), and there are some indications that in some cases it can even be the dominant mechanism. (Of course, if there occurs orbital ordering due to the exchange interaction, the lattice would 'follow', so that such ordering will always be accompanied by the corresponding structural change.)

Mathematically one can describe this situation writing down the effective exchange Hamiltonian analogous to (12.7). Here, however, each site is

---

[11] Actually the Hund's rule coupling is not really an exchange interaction: it is rather the difference between direct density–density Coulomb repulsion of electrons with parallel and antiparallel spins. Indeed, due to the Pauli principle the electrons with parallel spins 'avoid' each other (they have an antisymmetric coordinate wavefunction), so that on average they are further away from each other and consequently experience weaker Coulomb repulsion, which decreases the energy of such a state. But phenomenologically one can describe this effect by the exchange interaction (the last term in equation (12.35)).

characterized not only by the usual spin, but also by the orbital index which can be mapped onto the pseudospin $\tau = \frac{1}{2}$, as explained above. The resulting Hamiltonian has schematically the form

$$\mathcal{H} = \sum \Big\{ J_S\, \boldsymbol{S}_i \cdot \boldsymbol{S}_j + J_\tau\, \tau_i \cdot \tau_j + J_{S\tau} (\boldsymbol{S}_i \cdot \boldsymbol{S}_j)(\tau_i \cdot \tau_j) \Big\} , \qquad (12.36)$$

i.e. it describes both spin and orbital degrees of freedom which are coupled by the last term in (12.36). (The actual form of this Hamiltonian in real systems may be more complicated, e.g. anisotropic in $\tau$ operators.)

One of the interesting consequences of this treatment is the possibility of having, besides the ordinary spin waves, also a new type of elementary excitations: orbital waves – 'orbitons' (they will, however, be strongly mixed with phonons), and possibly also coupled spin–orbital excitations.

Orbital ordering has important manifestations in many properties of corresponding systems, but especially in their magnetic behaviour. Orbital occupation to a large extent determines the magnitude and even the sign of the exchange interaction and, consequently, the type of magnetic ordering. This constitutes the essence of the so-called Goodenough–Kanamori–Anderson (GKA) rules – see, for example, Khomskii (2001).

The importance of orbitals for magnetic exchange is already clear from the simplified Hamiltonian (12.36). Indeed, if for example $J_{S\tau} > J_S$, then, depending on the type of orbital ordering, characterized by the correlation function $\langle \tau_i \cdot \tau_j \rangle$, the sign of the total magnetic exchange $J_S + J_{S\tau} \langle \tau_i \cdot \tau_j \rangle$ may change.

In general, the GKA rules determine the type of exchange interaction for different local coordinations and for different orbital occupation. Without going into details, one can formulate simplified GKA rules in the following way:

(1) If on neighbouring magnetic sites the half-filled orbitals (having one electron) are directed towards one another, we would have a rather strong antiferromagnetic coupling. (In fact, this case is equivalent to the case of the simple nondegenerate Hubbard model of Section 12.3.)
(2) If, however, on one site we have a half-filled orbital, but on the neighbouring site the corresponding orbital directed towards the first site is empty or full (has two electrons), then the exchange interaction between these two ions will be ferromagnetic, but weaker. This is actually the case schematically shown in Fig. 12.20(*c*), (*d*): the ferromagnetic ordering, case (*c*), is more favourable, but the energy difference between this and the antiparallel spin configuration, case (*d*), is small: instead of the conventional antiferromagnetic exchange $J \sim t^2/U$, see equation (12.7), here the exchange (proportional to the energy difference between parallel and antiparallel spin orientations) is $J \sim (t^2/U)(J_{\mathrm{H}}/U)$ where we have used the expansion in $J_{\mathrm{H}}/U$ in

denominators of the energies in Fig. 12.20(*c*), (*d*) (typical values for transition metal ions are $U \sim 4$–$5\,\text{eV}$ and $J_{\text{H}} \sim 0.8\,\text{eV}$).

Interestingly enough, the second GKA rule gives the possibility to get ferromagnetic ordering in magnetic insulators. This is not a common phenomenon: most often insulators with strongly correlated electrons are antiferromagnetic, and we usually get ferromagnetism in the metallic state. This is also clear from the discussion of Sections 12.5 and 12.6: we saw there that ferromagnetic ordering appears usually when we have partial occupation of corresponding states, $n \neq 1$, which are metallic (see also below, Section 13.5). And in fact the main mechanism of making *ferromagnetic insulators* is that with the appropriate orbital ordering, described above.

In dealing with real magnetic materials one has to take into account many specific details, such as the type of crystal lattice, the detailed character of the 'active' orbitals, etc. Accordingly, the actual GKA rules are much more detailed than the simplest cases described above, see, e.g. Goodenough (1963) and Khomskii (2001). But the general conclusion remains the same; specific types of orbital occupation largely determine the magnetic properties of corresponding systems.

## 12.10 Charge-transfer insulators

Until now, even when discussing systems with strongly correlated electrons, we have only considered the correlated electrons themselves – for example, electrons of partially filled d shells in transition metal compounds. When we want to describe real materials, e.g. transition metal oxides like NiO or $La_2CuO_4$ – the parent compound of high-$T_c$ superconductors – we may need to take into account the other constituent atoms in these compounds, notably oxygen ions. We already partially touched upon this problem in the previous section when we spoke about crystal field splitting of d levels due to interaction of d electrons with the surrounding ions. Since the main objects to which this treatment is usually applied are transition metal oxides, one often speaks about oxygen ions surrounding a transition metal ion, see Fig. 12.18(*a*), although in other cases these may be halogen ions such as $F^-$, $Cl^-$, ..., or $S^{2-}$, $Se^{2-}$, etc. The general term used for such anions surrounding a given metal ion is *ligands*, and the splitting of d levels by the crystal field due to interaction with these anions is also called *ligand field splitting*. However, the role of these ligands does not reduce only to determining the detailed character of d-levels; it may be much more significant.

Consider, as an example, transition metal compounds with the general formula $ABO_3$, of the type of $LaMnO_3$ or $GdFeO_3$ (they are called perovskites). These systems basically consist of a simple cubic lattice of transition metal ions (Mn, Fe, ...)

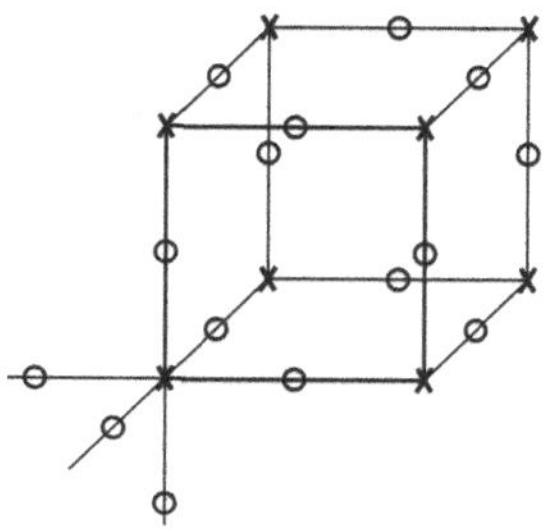

Fig. 12.21

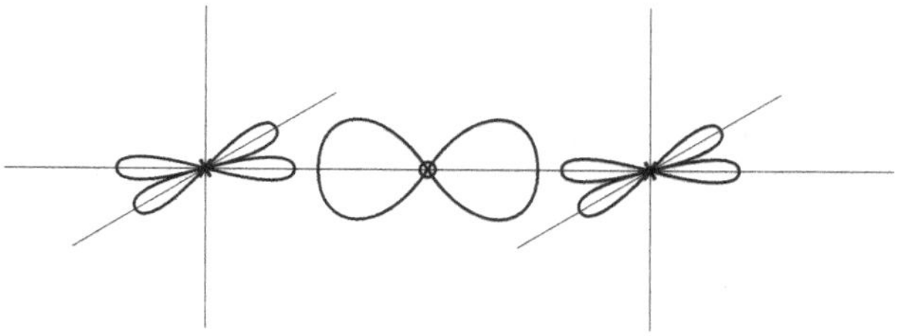

Fig. 12.22

with oxygens located in the middle of the edges, see Fig. 12.21. Here crosses are transition metal ions, and circles are oxygens. $A$-ions (La, Gd) are located at the centre of the cube shown in Fig. 12.21. (In real systems there often appear certain weak distortions of this structure, which, however, are not important for us at the moment.)

We see that here transition metal ions are separated by oxygen ions, which, generally speaking, reduce the dd-hopping $t$ in the Hubbard model, and can make the corresponding system a Mott insulator, with all the consequences thereof. But this also leads to another consequence: in fact in this case the distance between d ions is so large that one can practically neglect direct overlap of the d functions themselves. However, instead we have a relatively large overlap of d orbitals of transition metals with 2p orbitals, the valence orbitals of oxygen. The hopping of d electrons from one site to the other actually occurs via intermediate oxygens, see Fig. 12.22.

In many cases one can exclude these oxygen p orbitals and reduce the problem to that of d electrons themselves, described by the Hubbard model (12.1) or its generalization (12.35). However, this is not always possible: there may exist situations in which one has to take into account oxygen p states in an apparent way.

The general model would then be the one which includes both d electrons of the transition metal ions and p electrons of oxygens, with the hybridization (or hopping) between them. According to our general arguments, for d electrons we definitely have to take into account the electron–electron (Hubbard) repulsion. As the p electrons of oxygen have larger radius, one can in the first approximation neglect the corresponding Coulomb (or Hubbard) repulsion of electrons on oxygens (but generally speaking one has to include it as well).

Thus the general Hamiltonian describing this situation can be written as

$$\mathcal{H} = \varepsilon_d \sum_{i,\sigma} d^\dagger_{i\sigma} d_{i\sigma} + \varepsilon_p \sum_{j,\sigma} p^\dagger_{j\sigma} p_{j\sigma} + \sum_{ij,\sigma\sigma'} (t_{(pd)ij} d^\dagger_{i\sigma} p_{j\sigma} + \text{h.c.})$$
$$+ U_{dd} \sum_i n_{di\uparrow} n_{di\downarrow} + U_{pp} \sum_j n_{pj\uparrow} n_{pj\downarrow} + U_{pd} \sum_{ij,\sigma\sigma'} n_{di\sigma} n_{pj\sigma'} \, . \tag{12.37}$$

Here we have used the self-evident notation $d^\dagger$, $d$ for d electrons and $p^\dagger$, $p$ for oxygen p electrons. We also consider the simplest case, ignoring orbital degeneracy of d electrons, and taking into account only one p orbital – that directed towards neighbouring transition metal ions in Fig. 12.22. In real systems one has to use different p orbitals for different oxygens surrounding a given d ion. We have also included not only the dd interaction, but also pp and pd repulsion, although in most of this section we will only take into account $U_{dd}$, denoting it simply $U$.

The elementary hopping process described by the Hamiltonian (12.37) is the transfer of an electron from the filled 2p shell of $O^{2-}$ to the d ion. This costs the energy

$$\Delta = \varepsilon_d - \varepsilon_p + U_{dd} \, . \tag{12.38}$$

This is called the *charge-transfer energy*. It can also be defined as the energy required to go from the initial configuration $d^n p^6$ to the excited configuration $d^{n+1} p^5$.

Alternatively one could formulate the model in terms of d and p holes. Then the hopping process would be the hopping of a d hole from a transition metal to an oxygen. This is sometimes more convenient, and the corresponding description is usually used, for example, to describe high-$T_c$ cuprates. The corresponding Hamiltonian would look the same as equation (12.37), but one has to interpret $d^\dagger$, $d$, $p^\dagger$, $p$ as the creation and annihilation operators not for d and p electrons, but for d and p holes (and one has to redefine the corresponding energies, see Fig. 12.23).

In the electronic representation the typical energy scheme for the simplest case with one d electron looks as shown in Fig. 12.23(*a*), with the charge-transfer energy marked. One has to remember, however, that this is not a one-electron energy diagram, and the level $\varepsilon_d + U$ is the energy of a state with *two electrons* on the same d ion (thus, in the nondegenerate case, necessarily with opposite spins).

An alternative energy level scheme in the hole representation would look simpler, Fig. 12.23(*b*). Here the arrow denotes the hole, i.e. one missing electron on the d

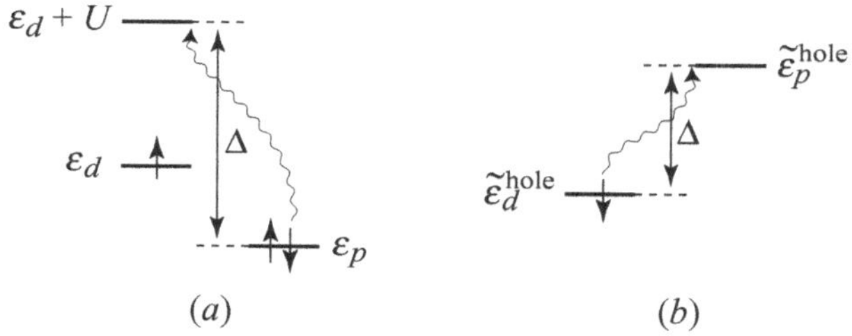


Fig. 12.23

level, and there are initially no holes on the oxygen ($O^{2-}$ with filled $2p^6$ shell). The excitation energy will then be simply the energy difference between oxygen and transition metal hole levels $\tilde{\varepsilon}_d$ and $\tilde{\varepsilon}_p$, and

$$\Delta = \tilde{\varepsilon}_p - \tilde{\varepsilon}_d , \tag{12.39}$$

which makes it much simpler to draw and to deal with. Thus we will mostly use this scheme from now on, using the Hamiltonian (12.37), but having in mind that it describes holes with corresponding hopping and interactions, i.e. $U_{dd}$, $U_{pp}$, etc. would denote the hole–hole interaction. (One has to be aware that the definitions of the hole energies $\tilde{\varepsilon}_d$ and $\tilde{\varepsilon}_p$ in fact should contain the information on the interactions included in the Hamiltonian (12.37), which become especially important when we are dealing with the real situation with possibly many d electrons or d holes at a site.)

Let us consider again the standard situation with one d electron (or, which is the same for a nondegenerate d level, with one d hole) per site. Let us first treat the case when both $U$, $\Delta \gg 1$ (recall that here $U$ is $U_{dd}$ for d holes). In this case the situation would strongly resemble that of the simple Hubbard model of Sections 12.1–12.3 in the case of a strong interaction: in the ground state the d electrons (or holes) would be localized, one at each site, and we would have a Mott insulator. We will again have localized spins, and the magnetic interaction will appear when we take into account virtual hopping of d holes, which we can again treat in perturbation theory.

In contrast to the usual Hubbard model (12.1), however, here virtual hoppings occur via an intermediate oxygen which is taken into account in an apparent way. Thus we have to use the hopping term $t_{pd}p^\dagger d$ in the Hamiltonian (12.37) not twice, but four times, see Fig. 12.24(a). In this figure we show (in the hole representation) two neighbouring d sites with an oxygen ion in between, and by wavy lines we show virtual hoppings of a d hole from one site to the other and back via intermediate p levels (the numbers near the wavy lines show one particular sequence of consecutive hoppings). Once again, this process, if allowed, decreases the total energy,

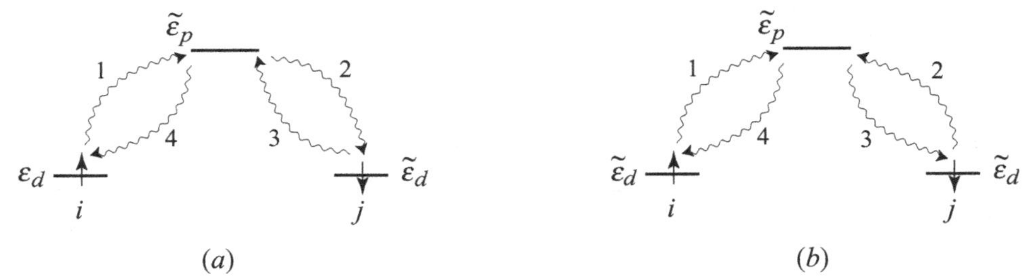


Fig. 12.24

but it is allowed only if the spins of neighbouring d ions are antiparallel. Thus in effect we again obtain the effective antiferromagnetic Heisenberg exchange (12.7), but the exchange integral will be given by the expression

$$J = \frac{2t_{pd}^4}{\Delta^2 U_{dd}} . \tag{12.40}$$

Indeed, each of four hoppings gives a factor $t_{pd}$. After the first hopping the energy of the intermediate state with one hole on oxygen is $\tilde{\varepsilon}_d - \tilde{\varepsilon}_p = \Delta$. After the second hopping this hole moves to the neighbouring d ion, and the energy of this intermediate state is $U_{dd}$. When the hole moves back, we again have the oxygen hole with the energy $\Delta$. These three values of intermediate energies stand in the denominator of equation (12.40).

Note that we can also rewrite the expression (12.40) in exactly the same form as the exchange integral (12.7) in the simple Hubbard model, if we introduce the effective dd-hopping

$$t = t_{dd} = \frac{t_{pd}^2}{\Delta} . \tag{12.41}$$

This is how, it seems, we can reduce the p–d model (12.37) to the conventional Hubbard model (12.1).

However, besides the process described by Fig. 12.24(*a*), in this case there is yet another process which can also contribute to the antiferromagnetic exchange. This process is illustrated in Fig. 12.24(*b*). At first glance it looks identical to that of Fig. 12.24(*a*), but in fact it is different: here after first transferring a d hole to the oxygen from, say, the left d ion, in the next step we transfer to the same oxygen another d hole, from the right. In effect the second intermediate state does not have the energy $U_{dd}$, but instead we have here two holes at the same oxygen. The energy of this state is $2\Delta$, or, if we take into account the repulsion of two holes on the same oxygen, $2\Delta + U_{pp}$. Again, as we can put on the same oxygen p orbital only two holes with opposite spins, this process will also lead to an antiferromagnetic interaction, with the exchange constant

$$J' = \frac{4t_{pd}^4}{\Delta^2(2\Delta + U_{pp})} \tag{12.42}$$

(the factor of 4 instead of 2 in (12.42) comes from the fact that there are twice as many different 'routes' in Fig. 12.24(*b*) than in Fig. 12.24(*a*): one can interchange the sequence of electron hops from sites $i$ and $j$ to the oxygens, and of hops back). The total antiferromagnetic interaction is the sum of both these contributions, $J_{\text{tot}} = J + J'$, i.e. introducing the effective dd-hopping (12.41), we can write it as

$$J_{\text{tot}} = \frac{2t_{pd}^4}{\Delta^2}\left(\frac{1}{U_{dd}} + \frac{2}{2\Delta + U_{pp}}\right) = t_{dd}^2\left(\frac{1}{U_{dd}} + \frac{1}{\Delta + U_{pp/2}}\right) . \tag{12.43}$$

Depending on the ratio of the effective charge-transfer energy $\Delta$ and dd Hubbard repulsion $U_{dd}$, either one or the other term in equation (12.43) will dominate. If $\Delta$ (or $\Delta + U_{pp/2}$) is much bigger than $U_{dd}$, we can keep in (12.43) only the first term, and then our problem will indeed completely reduce to the ordinary Hubbard model. In the opposite limit we should rather keep the processes of the second type, with two oxygen holes, and this regime is indeed somewhat different from the first one.

The difference is not so much in the properties of the ground state: in both these cases, if only $t_{pd} \ll (U_{dd}, \Delta)$, the ground state is a Mott insulator, with localized spins and with antiferromagnetic exchange between them. The real difference is in fact in the nature of the lowest *charge-carrying excitations* (here again we follow our general line: after discussing the type of the ground state we go over to the excitations). In the case of the Hubbard model such excitations correspond to the transfer of an electron from one d site to the other, $d^n(p^6)d^n \to d^{n+1}(p^6)d^{n-1}$, i.e. to the formation of a d hole and a doubly occupied d state. This excitation costs energy $U_{dd}$. The state of the $O^{2-}$ sitting in between remains here the same, ($p^6$). The extra d electron (the state $d^{n+1}$) and the d hole ($d^{n-1}$) thus created can now move through the crystal and carry current.

However, there exists another possibility: if $\Delta < U_{dd}$, then the lowest charge-carrying excitations would be those with charge transfer between a d ion and an oxygen: $d^n p^6 \to d^{n+1}p^5$, which costs energy $\Delta$. Thus this is in a sense a different type of insulator: the ground state is in principle the same as in the Mott, or Mott–Hubbard case, but the lowest charge excitations are quite different; they correspond to the creation of a doubly occupied d state and *an oxygen hole*. Such systems are called *charge-transfer insulators*. This classification was first proposed by J. Zaanen, G. Sawatzky and J. Allen, and it is often called the ZSA scheme. One can illustrate this on the phase diagram of Fig. 12.25 (the ZSA phase diagram). Here we show different possible states of the general d–p model (12.37) as a function of $\Delta$ and $U$, or of corresponding dimensionless quantities $\Delta/t_{pd}$, $U/t_{pd}$ ($U = U_{dd}$).

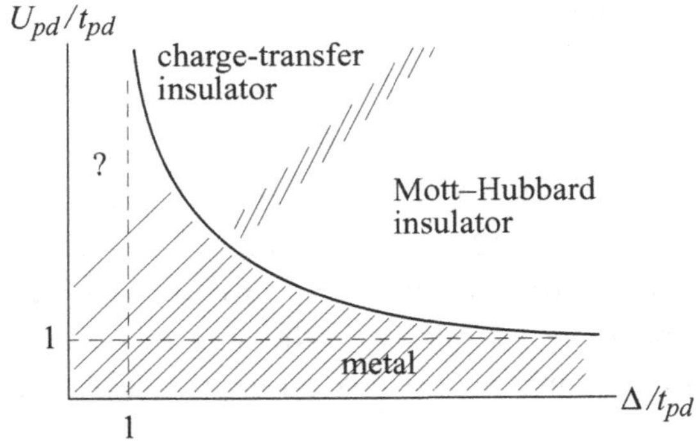


Fig. 12.25

If both $U$, $\Delta \gg t_{pd}$, we have an insulating state with localized electrons, but now this phase is separated into two regions: for $\Delta \gg U$ we can exclude virtual oxygen states and go over to the Hubbard model (12.1), dealing only with d electrons, and this region is a usual Mott–Hubbard insulator. In the opposite limit $U \gg \Delta \gg t_{pd}$ we still have an insulator with localized spins and with antiferromagnetic interaction between them, but the lowest charge excitations are those with the formation of oxygen holes, and the exchange integral will be given by the expression (12.42). This is the region of charge-transfer insulators.

In reality, e.g. among transition metal oxides, the tendency is such that the oxides at the beginning of the 3d series (Ti, V), if insulating, belong to the Mott–Hubbard class, whereas the oxides of heavier transition metals (Co, Ni, Cu) typically are charge-transfer insulators.

For large values of $\Delta$, when we decrease the Hubbard interaction $U$, we expect a Mott transition to a metallic state. The situation in the left-hand part of the ZSA phase diagram of Fig. 12.24 (large $U$, small $\Delta$) is much less clear. Most probably when we reduce $\Delta$, going to the limit of small and possibly negative charge-transfer gap, we would also have here a metallic state, but it would be an 'oxygen metal', with mobile oxygen holes as charge carriers, but with still strongly correlated d electrons, hybridized with oxygen states. This situation resembles that of mixed valence or heavy fermion systems, see the next chapter, and the properties in this regime may be quite nontrivial. There exist also other options in this regime (e.g. different types of insulating states); all these questions remain largely open.

The difference between Mott–Hubbard and charge-transfer insulators becomes especially important if we dope the system by holes. In the Mott–Hubbard regime these holes would go to d sites, and we would have a situation of partially filled Hubbard bands, discussed in Sections 12.4–12.8. In the charge-transfer case, however, it is more favourable to put an extra hole to an oxygen instead of a d level. Of course there is always some mixing, hybridization between d and p states, caused by the third term in the Hamiltonian (12.37), so that the total wavefunction of the hole would always be a superposition $|\psi\rangle = \alpha|d\rangle + \beta|p\rangle$, but if in the Mott–Hubbard regime the main weight in this wavefunction is on the d site, $|\alpha| \gg |\beta|$, in the charge-transfer case the situation would be the opposite, and in the first approximation the hole would be predominantly located on the oxygen. And although even in this case sometimes we can reduce the description to an effective one-band model, treating these hybridized states as new basis functions (this is the picture of Zhang–Rice singlets widely used for high-$T_c$ cuprates), one has to be aware that the real nature of corresponding states, and consequently some of their properties, e.g. such as the distribution of electron and spin density (the magnetic form-factor), details of magnetic interactions, or transport phenomena, may be rather different in doped charge-transfer insulators as compared to doped Mott–Hubbard systems. This one has to keep in mind, especially because many important materials, such as cuprates with high-temperature superconductivity or colossal magnetoresistance manganites, belong to this category.

## 12.11 Insulator–metal transition

Some of the most interesting phenomena occurring in solids are the insulator–metal transitions which exist in certain materials. These transitions can occur due to doping, as discussed above. But they may occur even in stoichiometric systems with a change of temperature, pressure, magnetic field, etc.

There are several possible types of insulator–metal transitions.

1. There may occur insulator–metal transitions which can be explained by the standard band theory, without invoking any special electron–electron interaction. Thus the system having an even number of electrons per unit cell may be an insulator or semiconductor simply because the filled valence band is separated from the empty conduction band by an energy gap, see Fig. 12.26. Such is, e.g. the situation in typical semiconductors like Ge and Si. The structure of the energy spectrum in this case is determined by the type of crystal lattice. If there occurs a structural change, new bands may in principle overlap, so that the energy gap would disappear, and the material would undergo an insulator–metal transition. This indeed happens when Ge and Si melt: the short-range order in liquid Ge and Si is quite different from the tetrahedral coordination in solids (it is actually similar to the structure of liquid Pb or Sn), and as a result liquid Ge and Si are good metals.

Another well-known example is the case of tin (Sn). It exists in two forms: ordinary metallic tin (white tin), and the so-called grey tin which is actually the stable form of Sn at low temperatures (below 13 degrees Celsius) and which is an insulator (strictly speaking it is a zero-gap semiconductor, i.e. it lies on the

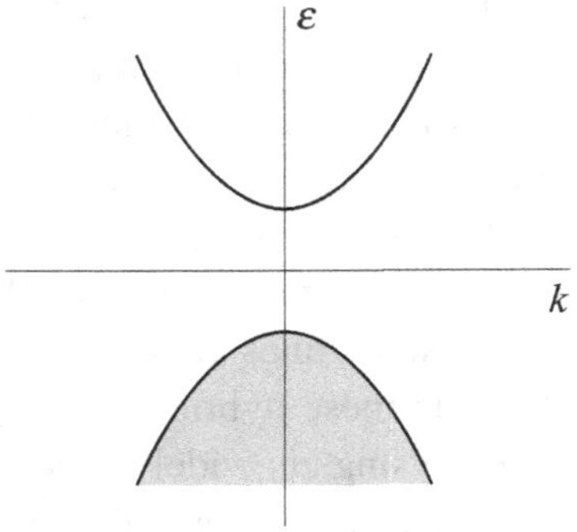


Fig. 12.26

borderline between metals and insulators; but in any case it is not a good metal like white tin). Grey tin has a diamond structure, like that of Ge or Si. The transition of white to grey tin is accompanied by a large change in the lattice and is known as 'tin plague': in the old days sometimes soldiers' tin spoons or tin buttons of their uniforms suddenly 'decayed', becoming grey powder – especially in winter, after being stored for some time in cold weather. In the Middle Ages tin organ pipes also sometimes disintegrated in winter, which was thought to be the work of the Devil. Thus the white–grey tin transition is one of the earliest known examples of metal–nonmetal transitions.[12]

2. In a certain sense the Peierls transition discussed in Chapter 11 is similar to the band structure transition, although we described it completely differently, using all the powerful methods of modern many-body theory. In essence what happens there is a change of the crystal lattice such that the new energy bands in the new crystal potential acquire a gap, this gap appearing at the position of the former Fermi surface. In this case the metal–nonmetal transition is accompanied by the appearance of long-range order – new lattice periodicity.

3. A similar treatment may be given also to insulator–metal transitions accompanying, e.g. spin-density wave formation. An SDW, like the charge-density wave (CDW), does not necessarily lead to an insulating state, e.g. antiferromagnetic Cr, in which SDW exists, remains a metal. However, there are also cases in which the SDW state is really insulating, the SDW energy gap 'covering' the whole

[12] 'Tin plague' could lead not only to unpleasant, but just curious events such as transformation of a box of tin spoons into a box of grey powder. Sometimes it can even have tragic consequences. This is at least one of the explanations of the tragic fate of the famous polar explorer Captain Robert Falcon Scott and his three companions. They perished in the Antarctic in 1912 on the way back after reaching the South Pole, and one of the reasons could have been that in such a cold climate the tin-soldered cans in which they stored their fuel (kerosene) started to leak because of the white–grey tin transition. As a result they lost part of their fuel, which was one of the reasons for the disaster. I am not sure if this is a true explanation, but in any case in his notes, found after his death, Scott himself wrote that loss of fuel played a crucial role in their fate.

Fermi surface. Such is, e.g. the situation in certain quasi-one-dimensional organic compounds.

An important difference here is that in contrast to previous cases the extra potential creating the energy gap is not the external potential due to the lattice, but a self-consistent potential of the electrons themselves. This self-consistent potential is spin-dependent, and it leads to the splitting of the electron subbands. Thus the difference is that here the electron–electron interaction is already essential. The similarity with the first two situations is, however, that the energy gap exists only to the extent to which there exists new long-range order in the system; above the Néel temperature, when magnetic ordering disappears, this energy gap would close and the material would become metallic.

4. And finally there are the situations in which the insulating character of the system is completely determined by the strong electron–electron interactions. This is the case of Mott–Hubbard insulators.[13] In this case the materials are insulating even without any extra order in the system. Thus, e.g. one can in principle describe the insulating nature of the ground state of materials such as NiO and CoO as connected with the antiferromagnetic order which can split the original partially filled (i.e. metallic) bands and create an energy gap. This treatment would then be similar to the treatment of SDW systems. It is indeed possible in some cases to carry out such a programme. Thus the standard band structure calculations, using spin-polarized bands (e.g. LSDA – local spin density approximation) can sometimes give an

[13] There exists yet another, fifth case of insulator–metal transitions – transition to an insulating behaviour in strongly disordered systems. Electron localization in this case (called *Anderson localization*) occurs in systems of electrons interacting with a random potential (e.g. random impurities); electron–electron interactions do not play a crucial role here. The physics of disordered systems is a big special field which I will not discuss here. Suffice it to say that if the disorder is strong enough (for example, if average fluctuations of the random potential $\langle \mathcal{V}^2 \rangle^{1/2}$ exceed the bandwidth), electrons become localized and do not go to infinity as $t \to \infty$, which means the absence of metallic conductivity. The electron density of states in this case remains finite at the Fermi level, i.e. the energy spectrum does not have an energy gap, but the electron mobility is zero – there exists a so-called *mobility gap*.

The proof of the Anderson result (Anderson localization) is rather involved, but one can qualitatively explain what is going on using the following arguments. When disorder in a metal increases, the electron mean free path $l$ decreases. However, it is hardly possible to imagine the situation when, e.g. $l$ becomes less than the interatomic distance $a$; it is clear that the standard description of electron transport (Boltzmann equation, etc.) would fail in this case. One can show that the conductivity in this case ($l \simeq a$) will be

$$\sigma = e^2/3\hbar a \; . \tag{12.44}$$

The limit of the standard description of conductivity related to the condition $l \simeq a$ is known as the Ioffe–Regel limit. If disorder is so strong that formally we would have $l < a$, the conventional description breaks down, and the electron state would become localized – there will occur Anderson localization and the material will become insulating. The borderline between these two regimes (still a metallic one, in which $\sigma(T)$ remains finite as $T \to 0$, albeit with short mean free path and small conductivity, and an insulating one, with $\sigma(T) \to 0$ or $R(T) \to \infty$ for $T \to 0$) corresponds to a value of $\sigma$ similar to the one given by equation (12.44), with a somewhat smaller prefactor,

$$\sigma_{\min} \simeq 0.03\, e^2/\hbar a \tag{12.45}$$

which for $a = 3$ Å corresponds to $\sigma \sim 300\,\Omega^{-1}\,\mathrm{cm}^{-1}$. This value is known as the *Mott minimum metallic conductivity*.

insulating state (although with a quite wrong value of the energy gap). However, this approach predicts that in the disordered state, at $T > T_N$, the gap would close and the system would be metallic. Experimentally this is definitely not the case in these materials, and it is just this fact which actually led to the development of the whole big field of the physics of Mott insulators, strongly correlated electrons, the Hubbard model, etc.

There are several systems among transition metal compounds in which insulator–metal transitions are experimentally observed. These are, for example, many oxides of V and Ti: $VO_2$; $V_2O_3$; many of the so-called Magneli phases $V_nO_{2n-1}$; $Ti_2O_3$; and several systems $Ti_nO_{2n-1}$ (e.g. $Ti_4O_7$). In these materials metal–nonmetal transitions occur as a function of temperature, and can also be induced by pressure, doping, etc. Some of these transitions are accompanied by magnetic ordering, but often such ordering occurs at a temperature not coinciding with $T_{\text{insulator-metal}}$. There is usually a structural change at such transitions, but it is often not clear whether this structural change is the cause and the main driving force of such transition, or it is only a consequence of it: in any case the lattice structure is sensitive to the state of the electronic subsystem, and any change in the latter, especially such a strong change as a metal–insulator transition (e.g. localization of electrons) should be reflected in the structure.

There were a lot of theoretical attempts to describe insulator–metal transitions starting from the Hubbard model. Most of them use certain poorly controlled approximations, and it is difficult to judge how reliable they are. I will only present here some of the results of probably the most successful approach developed by Brinkman and Rice. They used a certain variational scheme (called the Gutzwiller method) and studied, at $T = 0$, the behaviour of the metallic state when the Hubbard interaction $U$ increases. They have shown that in this method the system becomes insulating (electrons become localized) when $U \to U_c = 8|\bar{\varepsilon}_0|$, where $\bar{\varepsilon}_0$ is the average kinetic energy of electrons (of the order of the hopping matrix element $t$). When $U \to U_c$, the electron effective mass diverges as

$$\frac{m^*}{m} = \frac{1}{1 - (U/U_c)^2}\,, \tag{12.46}$$

i.e. the electrons (or rather quasiparticles) become infinitely heavy, and for $U > U_c$ they are localized at particular sites. Using the relations $m^*/m = 1/Z$ where $Z$ is the renormalization factor in the one-electron Green function (8.54), (8.91), one sees that (12.46) corresponds to $Z \to 0$, i.e. the strength of the quasiparticle pole goes to zero at this point. Simultaneously, according to (8.55), the jump of the electron distribution function $n(p_F - 0) - n(p_F + 0)$ goes to zero, i.e. the very Fermi surface disappears, and with it the Fermi liquid. In the insulating state at $U > U_c$ the electron distribution function is smooth, Fig. 12.27($a$), and it tends to a constant value, Fig. 12.27($b$), for $U \to \infty$ (the state localized in space,

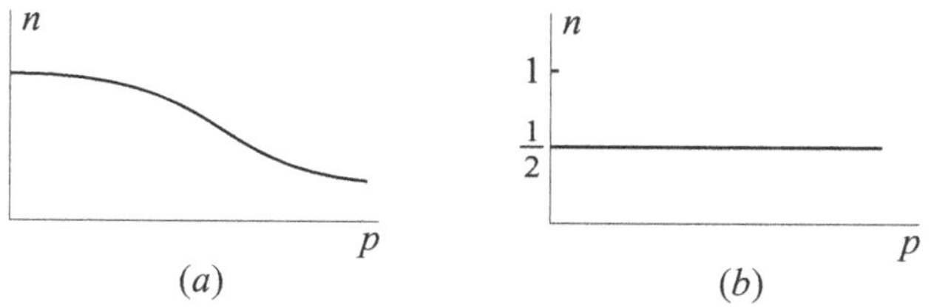


Fig. 12.27

$\Psi(x) \sim \delta(x - x_0)$, gives a constant in momentum space after a Fourier transform, i.e. all momenta are equally represented in the localized state).

One can also show that at least in this approximation the magnetic susceptibility also diverges at $U \to U_c$:

$$\chi = \frac{\mu_B^2 \rho(\varepsilon_F)}{1 - (U/U_c)^2} \sim \frac{m^*}{m}\,, \tag{12.47}$$

i.e. it is enhanced mainly because of the increase of the effective mass. In terms of the Fermi-liquid theory this means that the Landau parameter $F_1^s$ diverges, but $(1 + F_0^a)^{-1}$ remains finite as $U \to U_c$, in contrast to the usual magnetic transitions, cf. (10.16), (10.18). This is consistent with our general idea that a Mott transition is a transition to localized electrons (and consequently localized magnetic moments) but not necessarily a transition to a magnetically ordered state.

Actually of course the localized magnetic moments which appear in a Mott insulator should somehow order at $T = 0$. This effect is missing in the original treatment of Brinkman and Rice. The hope is, however, that this ordering, albeit important, does not play a crucial role in the very phenomenon of electron localization and the corresponding Mott transition, so that the basic physics is accounted for well enough.

Other methods of treating Mott transitions have also been developed recently. One of the most successful of these is the so-called *dynamical mean field theory* (DMFT), see, e.g. Georges *et al.* (1996). In this method, which is somewhat similar to the conventional mean field theory, one reduces the description of the concentrated system to that of one site interacting with the mean field of the surrounding. But, in contrast to the usual treatment, one takes into account dynamic effects both in the singled-out 'impurity' and in the bath. One of the outcomes of this treatment is a rather appealing picture of what happens when a metallic system approaches a Mott transition, e.g. with increasing $U/t$ in the Hubbard model with $n = 1$. This method gives the following 'sequence of events', see Fig. 12.28: far from the Mott transition we have an ordinary metal, with the density of states shown in Fig. 12.28 and with the usual Fermi surface. With increasing electron correlations (increasing $U/t$) there appear 'wings' in $\rho(\varepsilon)$ both below

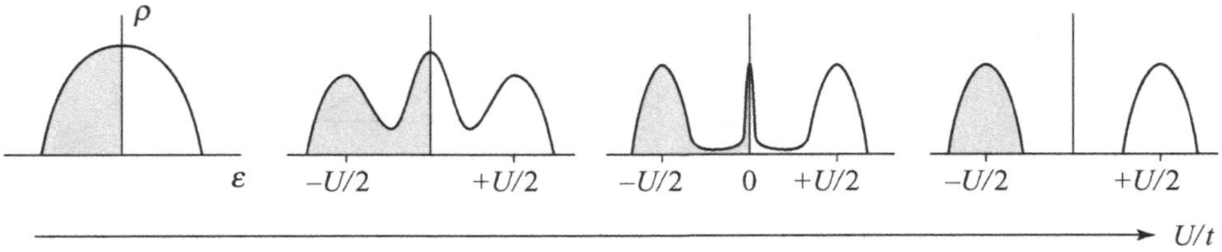


Fig. 12.28

and above the original metallic band (the position of the Fermi energy here is taken as zero energy). These broad peaks in $\rho(\varepsilon)$ appear at $\sim -U/2$ below $\varepsilon_F$ and at $\sim U/2$ above $\varepsilon_F$. When $U/t$ becomes still larger, the main weight is gradually transferred to these side peaks, which actually represent the lower and upper Hubbard bands with the gap $\sim U$ between them. Still, on the metallic side of the transition there remains a quasiparticle peak at the original Fermi level, with the constant density of states at $\varepsilon_F$, but with the width (and correspondingly the total weight $Z$) going to zero as we approach the Mott transition. And at a certain critical value of $U/t$ this quasiparticle peak disappears, and we have only the lower and upper Hubbard bands with the gap between them. This state is a Mott insulator.

Actually a lot of questions in this field still remain open. Can a Mott transition occur 'by itself', without being accompanied by a real ordering of some type (magnetic ordering, structural distortion)? If so, will this transition be first or second order? What is the symmetry (if any!) discriminating between insulating and metallic phases (in the sense of the Landau theory of second-order phase transitions, see Chapter 2)? What would be the corresponding order parameter? It is not even clear which phase, insulating or metallic, should be considered as the ordered and which as the disordered phase. The first impulse is usually to treat the insulating state as the ordered one. However this may not be the case. We have already given earlier the arguments (see the discussion at the end of Section 10.1) that the very existence of the Fermi surface in metals may be a kind of 'ordering'. From this point of view the metal should be treated as a unique ordered phase with zero entropy, and the Mott insulator without magnetic ordering as a state with higher symmetry, or a disordered phase (at least it has spin disorder – cf. the discussion in Section 2.7.3 of the high-temperature metal–insulator phase transition in $V_2O_3$ as driven by magnetic entropy).

All these questions are actually rather deep. It is at present not at all clear whether one can indeed treat Mott transitions on the same footing as the usual second-order phase transitions. Thus, e.g. in the conventional theory we define the order parameter $\eta$ as the average of a certain operator *over the ground state* (e.g.

the magnetization $M = \langle 0|\boldsymbol{M}|0\rangle$, etc.). It is not at all clear whether there exists such a notion for the insulator–metal transitions. Actually the physical definition of the difference between an insulator and a metal is connected with the static conductivity $\sigma(\omega \to 0)$ being zero or finite. But this is not a characteristic of *the ground state*, but rather of *the lowest excited states*: one may define insulators as systems in which the first excited states are separated from the ground state by a finite gap, whereas there is no such gap in a metal.[14] In this sense there may be formally no order parameter in the ordinary sense which would discriminate Mott insulators from metals. If so, the pure Mott transition can only be a first-order transition, if nothing more intricate takes place here.

Concluding this chapter one should say that the whole field of correlated electrons in general and insulator–metal (Mott–Hubbard) transitions in particular is still an active field of research, and a lot remains to be understood.

[14] But be careful with superconductors! In this sense they are more similar to insulators than to metals. Also in strongly disordered systems the electronic states can be localized and consequently the conductivity will be zero, but the energy spectrum may be continuous and have no gap – see the discussion of Anderson localization above. Thus even this definition applies, strictly speaking, only to nonsuperconducting systems without strong disorder.

# 13

# Magnetic impurities in metals, Kondo effect, heavy fermions and mixed valence

In the previous chapter we have considered the properties of strongly correlated electrons. The systems we mostly had in mind were the compounds of transition metal and maybe rare earth elements with partially filled inner d or f shells. We have discussed only the correlated electrons themselves, the prototype model being the Hubbard model (12.1).

When turning to real materials, several extra factors missing in the model (12.1) are important. One of them is the possible influence of orbital degrees of freedom, especially in cases with orbital degeneracy, treated in Section 12.9.

In many situations there is yet another very important factor. There may exist in a system, besides correlated electrons, also electrons of other bands, e.g. electrons in wide conduction bands, responsible for ordinary metallic conductivity. Such is for instance the situation for magnetic impurities in metals, or in the concentrated systems like rare earth metals and compounds in which localized f electrons coexist with the metallic electrons in broad spd bands. The interplay between localized, or, better, strongly correlated electrons and itinerant electrons of the wide bands can lead to a number of very interesting consequences; these will be discussed in this chapter.

## 13.1 Localized magnetic moments in metals

When we put transition metal impurities in ordinary metals (e.g. Mn or Fe in Cu, Au), the result may be two-fold. In certain cases the impurities retain their magnetic moment, but in others they lose it. Qualitatively this second possibility is connected with the following process: when the localized electron level, e.g. the d level of the impurity $\varepsilon_d$, overlaps with the continuous spectrum, electrons on this level may have a finite lifetime, i.e. the d electron with spin $\sigma$ can escape into the conduction band, and in its place another electron from the conduction band may be transferred, possibly with the opposite spin $-\sigma$. As a consequence of this

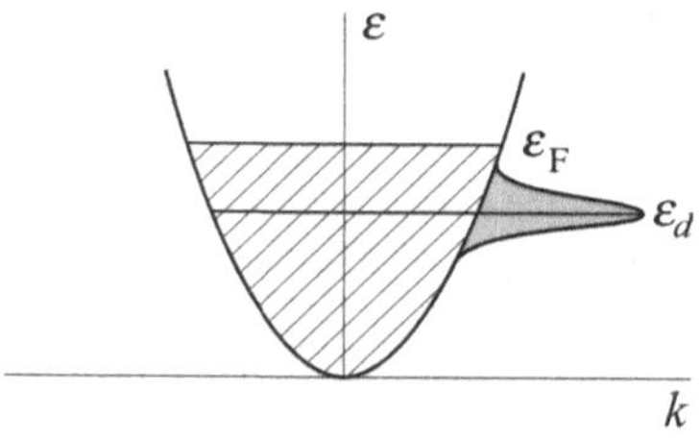


Fig. 13.1

process the localized level acquires a finite width, Fig. 13.1, and there appears a tendency to reduce or completely quench the moment.

The situation, however, is not so simple. When the d level lies deep below the Fermi level, all states in the conduction band with energies close to $\varepsilon_d$ are occupied, and this process is suppressed. The large Coulomb interaction of localized electrons $U_{dd}$ also acts to preserve the localized magnetic moment.

The problem of the appearance of localized magnetic moments at impurities in metals was treated in detail by P. W. Anderson. He used the model qualitatively discussed above, with the Hamiltonian

$$\mathcal{H} = \sum_{\boldsymbol{k},\sigma} \varepsilon_{\boldsymbol{k}}\, c^{\dagger}_{\boldsymbol{k},\sigma}\, c_{\boldsymbol{k},\sigma} + \varepsilon_d \sum_{\sigma} d^{\dagger}_{\sigma}\, d_{\sigma} + U n_{d\uparrow}\, n_{d\downarrow} + \sum_{\boldsymbol{k},\sigma} \left( V_{\boldsymbol{k}}\, c^{\dagger}_{\boldsymbol{k},\sigma}\, d_{\sigma} + \text{h.c.} \right) . \tag{13.1}$$

Here $d^{\dagger}_{\sigma}$, $d_{\sigma}$ are creation and annihilation operators for electrons on the d level of the impurity, $c^{\dagger}_{\boldsymbol{k},\sigma}$, $c_{\boldsymbol{k},\sigma}$ are those of conduction electrons, and $V_{\boldsymbol{k}}$ is the matrix element of d–c hybridization which describes the process of mixing of d and c electrons. (Note that here and below $V$ is the d–c hybridization, not the volume!) This model is known as the *Anderson model* of magnetic impurities.

The approximation used in the first paper by Anderson was the unrestricted (in other words, spin-dependent) Hartree–Fock approximation, in which the following decoupling was made:

$$U\, n_{d\uparrow}\, n_{d\downarrow} \longrightarrow U \langle n_{d\uparrow} \rangle\, d_{d\downarrow} + U\, n_{d\uparrow} \langle n_{d\downarrow} \rangle - U \langle n_{d\uparrow} \rangle \langle d_{d\downarrow} \rangle . \tag{13.2}$$

With this decoupling one can easily find the energy spectrum by diagonalizing the resulting quadratic Hamiltonian. A convenient way to do this, used by Anderson, is to write down the equations of motion for the operators $c_{\boldsymbol{k},\sigma}$, $d_{\sigma}$, or for the corresponding Green functions, which have the form

$$\begin{aligned} \omega\, d_{\sigma} &= \varepsilon_d\, d_{\sigma} + U \langle n_{d,-\sigma} \rangle\, d_{\sigma} + \sum_{\boldsymbol{k}} V^{*}_{\boldsymbol{k}}\, c_{\boldsymbol{k},\sigma} , \\ \omega\, c_{\boldsymbol{k},\sigma} &= \varepsilon_{\boldsymbol{k}}\, c_{\boldsymbol{k},\sigma} + V_{\boldsymbol{k}}\, d_{\sigma} . \end{aligned} \tag{13.3}$$

From these equations, or from the equivalent equations for the corresponding Green functions, we find for the Green function of the impurity electrons the expression

$$G_{dd}^{\sigma} = \frac{1}{\omega - \varepsilon_d - U\,\langle n_{d,-\sigma}\rangle - \sum_{\boldsymbol{k}} \frac{|V_{\boldsymbol{k}}|^2}{\omega - \varepsilon_{\boldsymbol{k}}}}\,, \tag{13.4}$$

or, in other words, $(G_{dd}^{\sigma})^{-1} = \omega - \varepsilon_d - \Sigma_d^{\sigma}(\omega)$, where the self-energy is

$$\Sigma_d^{\sigma}(\omega) = U\,\langle n_{d,-\sigma}\rangle + \sum_{\boldsymbol{k}} \frac{|V_{\boldsymbol{k}}|^2}{\omega - \varepsilon_{\boldsymbol{k}}}\,. \tag{13.5}$$

When the renormalized d level $\varepsilon_d^{\sigma} = \varepsilon_d + U\langle n_{d,-\sigma}\rangle$ lies within the energy band $\varepsilon_{\boldsymbol{k}}$, $\Sigma_d^{\sigma}$ has both real and imaginary parts, which determine the width of the d level. Thus one can write down the density of states of d electrons as

$$\rho_d^{\sigma}(\omega) = \frac{1}{\pi}\frac{\Gamma}{\left(\omega - \varepsilon_d - U\,\langle n_{d,-\sigma}\rangle\right)^2 + \Gamma^2}\,, \tag{13.6}$$

where the width $\Gamma$ is given by

$$\Gamma = \pi \sum_{\boldsymbol{k}} |V_{\boldsymbol{k}}|^2 \delta(\omega - \varepsilon_{\boldsymbol{k}}) = \pi\,|V_{\boldsymbol{k}}|^2 \rho(\varepsilon_{\mathrm{F}})\,. \tag{13.7}$$

The average occupation of the d level with spin projection $\sigma$ is then given by

$$\langle n_{d,\sigma}\rangle = \int_{-\infty}^{\varepsilon_{\mathrm{F}}} \rho_d^{\sigma}(\omega)\,d\omega = \frac{1}{\pi}\arctan\left(\frac{\varepsilon_d + U\,\langle n_{d,-\sigma}\rangle - \varepsilon_{\mathrm{F}}}{\Gamma}\right). \tag{13.8}$$

One can also obtain this result directly using the formula (8.37) expressing the electron density through the Green function (however, one should not integrate in (8.37) over $d^3\boldsymbol{k}/(2\pi)^3$ because we are now studying an isolated impurity).

The equation (13.8) and a similar equation for $\langle n_{d,-\sigma}\rangle$ constitute two self-consistent equations which should be solved together. There exist in general two types of solutions depending on the values of the parameters $\varepsilon_{\mathrm{F}} - \varepsilon_d$, $U$, $\Gamma$. There is always a nonmagnetic solution $\langle n_{d\uparrow}\rangle = \langle n_{d\downarrow}\rangle = \frac{1}{2}\langle n_d\rangle$. However for large $U$ and small $\Gamma$ this solution corresponds not to a minimum, but to a maximum of the energy, and there exists another, magnetic solution with $\langle n_{d\uparrow}\rangle \neq \langle n_{d\downarrow}\rangle$. This solution is doubly degenerate ($\langle n_{d\uparrow}\rangle > \langle n_{d\downarrow}\rangle$ or vice versa). For the symmetric case $\varepsilon_{\mathrm{F}} - \varepsilon_d = U/2$ (i.e. when the levels $\varepsilon_d$ and $\varepsilon_d + U$ are situated symmetrically below and above the Fermi level) the condition for the appearance of localized moments is

$$U\rho_d^{\sigma}(\varepsilon_{\mathrm{F}}) > 1\,, \tag{13.9}$$

where $\rho_d^{\sigma}$ is given by (13.6), i.e. it coincides with the Stoner criterion (11.62) for ferromagnetic instability of itinerant electrons. The total density of states then has

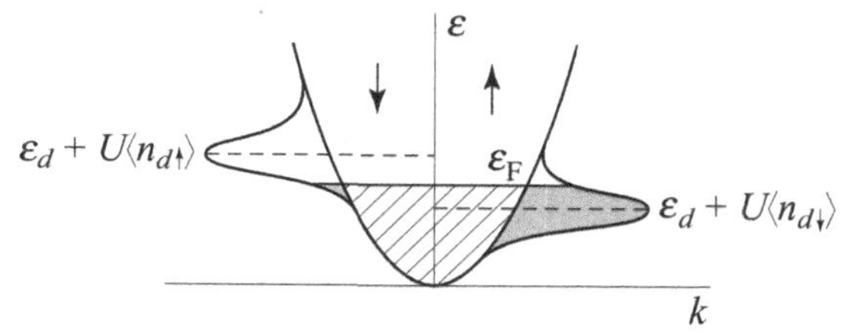


Fig. 13.2

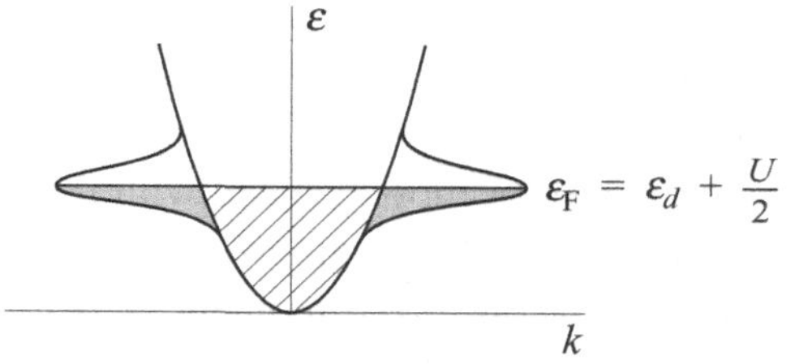


Fig. 13.3

the form shown in Fig. 13.2, i.e. we have two virtual levels with finite width (13.7) and with the Lorentzian density of states (13.6), $\langle n_{d\uparrow}\rangle$ being given by the (grey) area of the peak on the right-hand side of the figure, with energies up to $\varepsilon_F$, and $\langle n_{d\downarrow}\rangle$ being given by the corresponding grey area ($\varepsilon < \varepsilon_F$) of the left peak. The nonmagnetic solution would correspond here to the situation shown in Fig. 13.3, i.e. to equal occupations of the d levels with up and down spins. (We stress once again that this situation is realized for the symmetric Anderson model $\varepsilon_F = \varepsilon_d + U/2$; the conditions for the existence of localized magnetic moments for asymmetric situations are more stringent.)

For the symmetric case the criterion (13.9) may also be rewritten in the form

$$\frac{U}{\pi\Gamma} > 1\,. \qquad (13.10)$$

(Here we approach the instability from the nonmagnetic side, $\langle n_\uparrow\rangle = \langle n_\downarrow\rangle = \frac{1}{2}$, and we have used (13.6) and the condition $\varepsilon_F = \varepsilon_d + U/2$, which gives $\rho_d(\varepsilon_F) = 1/\pi\Gamma$.) The region in the $(\varepsilon_d, U/\Gamma)$ plane where there exists a magnetic solution is shown schematically in Fig. 13.4. We see that the existence of localized magnetic moments of the impurity is facilitated in cases of large Coulomb interaction, small d–f mixing $V$, small density of states of the conduction band $\rho(\varepsilon_F)$ (as we see from (13.7), $\Gamma = \pi\rho(\varepsilon_F)V^2$) and symmetric position of the localized level.

Thus the qualitative picture of the appearance (or of preservation) of the localized magnetic moment in the Anderson model is the following: if the state at energy

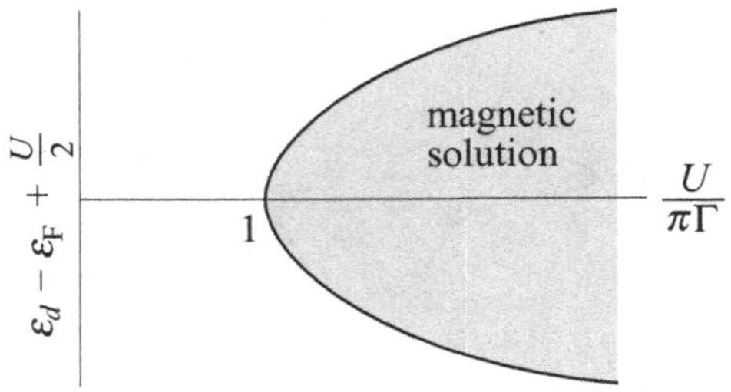


Fig. 13.4

$\varepsilon_d$ with spin $\uparrow$ is below the Fermi level and is occupied, then to put another electron with spin $\downarrow$ at the same d level would cost energy $U$, i.e. one may say that the second electron will occupy the state with the energy $\varepsilon_d + U$, which for large enough $U$ will lie above the Fermi level and consequently should be empty. In this case the impurity has localized moment. When, due to d–c hybridization the number of spin-up electrons decreases, so does the energy of the spin-down state located at $\varepsilon_d + Un_{d\uparrow}$. In effect these two levels, with spins $\uparrow$ and $\downarrow$, would move towards each other and approach $\varepsilon_F$, and if this process is strong enough (if the hybridization $V$ is large and $U$ small), one may end up in the nonmagnetic state of Fig. 13.3. However, if the conditions (13.9), (13.10) are satisfied, the solution with localized magnetic moments, illustrated in Fig. 13.2, would be stable.

## 13.2 Kondo effect

The treatment of the Anderson model given above is still a mean field one. When one goes beyond the mean field approximation, new effects appear which make the situation much more difficult but also much more interesting. Notably, it turns out that the remaining interactions between the impurity electron (or spin) and conduction electrons, which were not taken into account in the Hartree–Fock treatment of the previous section, lead to effective screening and eventual 'disappearance' of localized magnetic moments as $T \to 0$. Thus the very moment which we 'created' with such difficulties only two pages ago would disappear! This is the famous Kondo effect.

Actually Kondo considered not the Anderson model (13.1), but the so-called s–d exchange model with localized spin $\boldsymbol{S}$ interacting with the conduction electrons via an exchange interaction

$$\mathcal{H}_{sd} = \sum_{\boldsymbol{k},\boldsymbol{k}'} J_{\boldsymbol{k}\boldsymbol{k}'} c^{\dagger}_{\boldsymbol{k},\sigma} \boldsymbol{\sigma}_{\sigma\sigma'} c_{\boldsymbol{k}',\sigma'} \cdot \boldsymbol{S} \simeq J\boldsymbol{s}(\boldsymbol{0}) \cdot \boldsymbol{S} \tag{13.11}$$

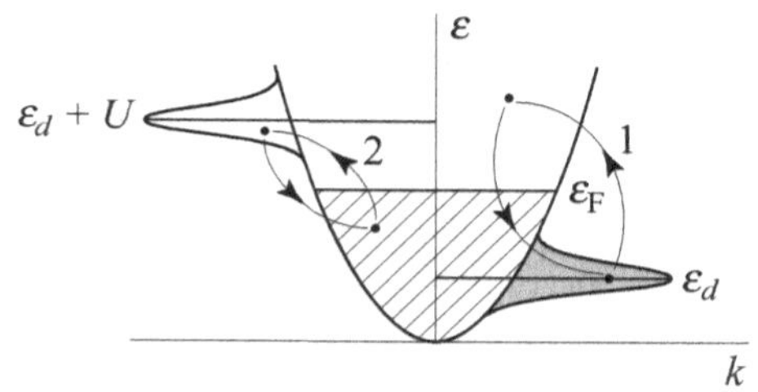


Fig. 13.5

where $\boldsymbol{\sigma}$ are the $\sigma$-matrices and $\boldsymbol{s}(\mathbf{0})$ is the spin density of conduction electrons at the position of the impurity. Actually this interaction can be obtained from the Anderson model (13.1) in the limit in which there exist localized magnetic moments, see the previous section. In this case, if the d level is deep enough and $\Gamma$ is small, the total occupation of the d level is close to 1, e.g. $\langle n_{d\uparrow}\rangle \simeq 1$, $\langle n_{d\downarrow}\rangle \simeq 0$. However, as we have already mentioned, there is another degenerate solution with the same energy, corresponding to $\langle n_{d\downarrow}\rangle \simeq 1$, $\langle n_{d\uparrow}\rangle \simeq 0$. Thus, the average occupation of the localized levels in this regime is close to integral (in our case 1), but there remains spin degeneracy, i.e. spin degrees of freedom. Consequently we may project out charge degrees of freedom and keep only the spin ones, obtaining the effective exchange interaction (13.11). This procedure is similar in spirit, and actually in mathematics, to the one which led us from the Hubbard model (12.1) to the superexchange Hamiltonian (12.7); here it is known as the *Schrieffer–Wolf transformation*. In the simplest form one can obtain the s–d exchange interaction (13.11) starting from (13.1) and treating the hybridization term in perturbation theory up to second order in $V$. The processes which contribute to the exchange are virtual transitions of the d electron to the conduction band and back (process 1 in Fig. 13.5), and the transition from the conduction band to the already occupied d state and back (process 2 in Fig. 13.5); to put the second electron at the d level costs extra energy $U$. As in our case $n_\uparrow \sim 1$, $n_\downarrow \sim 0$, we have $\varepsilon_{d\uparrow} \simeq \varepsilon_d$, $\varepsilon_{d\downarrow} \sim \varepsilon_d + U$. Again, similar to the Hubbard model, the resulting exchange interaction is antiferromagnetic, with the exchange constant

$$J = 2V_{k_F}^2\left(\frac{1}{\varepsilon_F - \varepsilon_d} + \frac{1}{\varepsilon_d + U - \varepsilon_F}\right) = 2V^2\frac{U}{(\varepsilon_F - \varepsilon_d)(\varepsilon_d + U - \varepsilon_F)}\,. \tag{13.12}$$

Let us consider now the consequences of the exchange interaction (13.11). First we do it more formally. One of the processes is the scattering of electrons on the localized spin. There are two possible processes contributing to this scattering: ordinary potential scattering and exchange, or spin-flip scattering, in which the electron exchanges its spin with the impurity.

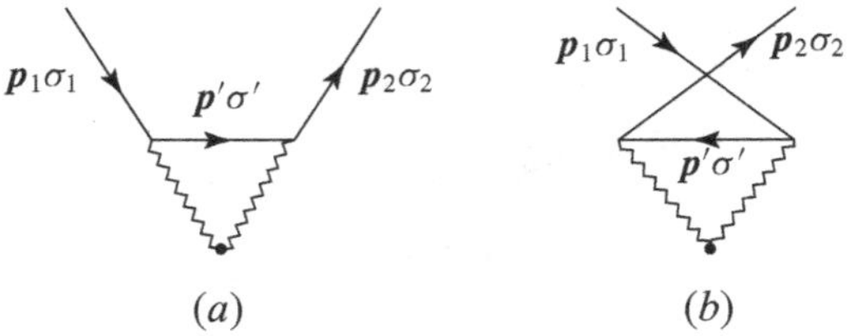


Fig. 13.6

In the lowest approximation (the first Born approximation) nothing special happens. The first-order contribution to the scattering amplitude $A^{(1)}$ will be proportional to $J\boldsymbol{\sigma}\cdot\boldsymbol{S}$, and after summation over all spin orientations we obtain for the scattering rate $W \sim |A^{(1)}|^2 \sim J^2 S(S+1)$. However, in second order the spin-flip scattering gives a term in the scattering amplitude which is logarithmically divergent. The corresponding processes may be represented by the two diagrams shown in Fig. 13.6. The diagram 13.6($a$) describes the process in which an incoming electron with momentum $\boldsymbol{p}_1$ and spin $\sigma_1$ is scattered into an intermediate state $\boldsymbol{p}'\sigma'$ and then goes to a state $\boldsymbol{p}_2\sigma_2$. The summation over intermediate states should be carried out over unoccupied states above the Fermi energy, i.e. it should contain a factor $(1 - f(\boldsymbol{p}'))$ where $f(\boldsymbol{p})$ is the Fermi factor. This term thus has the form

$$J^2 \sum_{\sigma'} \int \frac{d^3\boldsymbol{p}'}{(2\pi)^3} \frac{\left(1 - f(\boldsymbol{p}')\right)\left(\boldsymbol{\sigma}_{\sigma_1\sigma'}\cdot\boldsymbol{S}\right)\left(\boldsymbol{\sigma}_{\sigma'\sigma_2}\cdot\boldsymbol{S}\right)}{\varepsilon(\boldsymbol{p}_1) - \varepsilon(\boldsymbol{p}')} . \tag{13.13}$$

The diagram 13.6($b$) describes the process in which initially an electron–hole pair is created, the electron being in the state $\boldsymbol{p}_2\sigma_2$ and the hole in $\boldsymbol{p}'\sigma'$ (left vertex in Fig. 13.6($b$)), and then the initial electron is scattered from the state $\boldsymbol{p}_1\sigma_1$ into the state $\boldsymbol{p}'\sigma'$ which is now empty. Correspondingly, now $|\boldsymbol{p}'| < p_{\mathrm{F}}$, which is taken care of by the function $f(\boldsymbol{p}')$, which gives the term

$$-J^2 \sum_{\sigma'} \int \frac{d^3\boldsymbol{p}'}{(2\pi)^3} \frac{f(\boldsymbol{p}')\left(\boldsymbol{\sigma}_{\sigma'\sigma_1}\cdot\boldsymbol{S}\right)\left(\boldsymbol{\sigma}_{\sigma_2\sigma'}\cdot\boldsymbol{S}\right)}{\varepsilon(\boldsymbol{p}') - \varepsilon(\boldsymbol{p}_1)} . \tag{13.14}$$

It is important that the second contribution has the opposite sign which is a consequence of the antisymmetric character of the electron wavefunction.

After summation over spin indices we obtain finally the contribution to the scattering amplitude

$$A^{(2)}_{\boldsymbol{p}_1\boldsymbol{p}_2} = J^2 \int \frac{d^3\boldsymbol{p}'}{(2\pi)^3} \frac{2f(\boldsymbol{p}') - 1}{\varepsilon(\boldsymbol{p}_1) - \varepsilon(\boldsymbol{p}')} \left(\boldsymbol{\sigma}_{\sigma_2\sigma_1}\cdot\boldsymbol{S}\right) . \tag{13.15}$$

The integral in (12.18) is logarithmically divergent for $|\boldsymbol{p}_1| \to p_{\mathrm{F}}$ or $\varepsilon(\boldsymbol{p}_1) \to \varepsilon_{\mathrm{F}}$ (note that the scattering is elastic, i.e. $|\boldsymbol{p}_2| = |\boldsymbol{p}_1|$ and $\varepsilon(\boldsymbol{p}_2) = \varepsilon(\boldsymbol{p}_1)$). Thus the

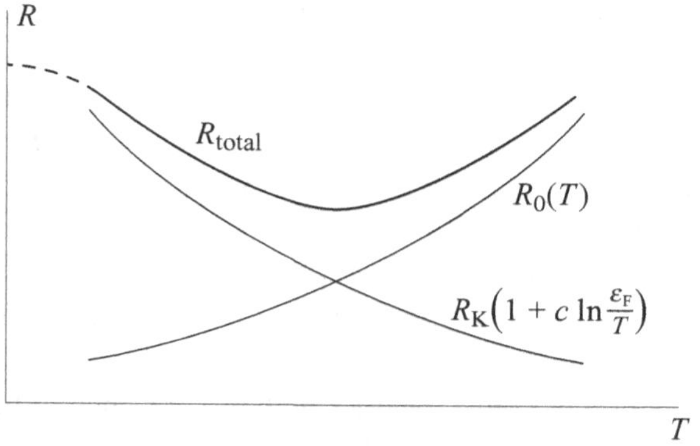


Fig. 13.7

second-order term in the scattering amplitude is proportional to $\ln \frac{\varepsilon_F}{\max\{|\varepsilon(\boldsymbol{p}_1)-\varepsilon_F|, T\}}$, and the total scattering amplitude up to second order in $J$ is

$$A_{\boldsymbol{p}_1\boldsymbol{p}_2} = J(\boldsymbol{\sigma}\cdot\boldsymbol{S})_{\sigma_2\sigma_1}\left[1 + J\,\rho(\varepsilon_F)\ln\frac{\varepsilon_F}{\max\{|\varepsilon(\boldsymbol{p}_1)-\varepsilon_F|, T\}}\right]. \tag{13.16}$$

This scattering amplitude determines the electron lifetime $\tau^{-1} \sim |A|^2$, which enters the resistivity through the usual formula $R^{-1} = \sigma = ne^2\tau/m$. As a result the resistivity behaves as

$$R(T) = R_0(T) + R_K\left(1 + 2J\,\rho(\varepsilon_F)\ln\frac{\varepsilon_F}{T}\right), \tag{13.17}$$

i.e. it contains a term logarithmically increasing with decreasing temperature. This term, together with the usual scattering $R_0(T)$, gives the famous minimum of the resistivity, schematically shown in Fig. 13.7. When the second order in perturbation theory diverges, as in our case, it means that we have to sum many terms of higher order, which will diverge as $(\ln \varepsilon_F/T)^2$, $(\ln \varepsilon_F/T)^3$, etc. Summation of the simple geometric series of this type gives an expression of the type

$$\frac{J\rho}{1 - J\rho\ln(\varepsilon_F/T)}. \tag{13.18}$$

This expression formally diverges for antiferromagnetic interactions $J > 0$ at a certain 'critical', or rather characteristic temperature

$$T_K = \varepsilon_F e^{-1/J\rho(\varepsilon_F)}, \tag{13.19}$$

which is called the Kondo temperature – cf. the case of the BCS theory of superconductivity, equations (11.47), (11.48). Of course, in contrast to superconductivity in the case of a single impurity in a metal there can be no real phase transition.

Rather $T_K$ (13.19) gives the *temperature scale* at which there occurs a change of the behaviour of our system: above $T_K$ the perturbation theory described above is valid, and below $T_K$ it is no longer applicable. Thus, e.g. above $T_K$ there is a logarithmic term in the resistivity; below $T_K$ the behaviour of $R(T)$ changes, and actually resistivity saturates. Similarly, there is a cross-over in the behaviour of the impurity susceptibility: it also increases with decreasing temperature at $T > T_K$, but saturates at the value $\sim 1/T_K$ at $T < T_K$.

The divergent contribution we have obtained can be traced back to the fact that the spin operators which enter into the spin-flip processes do not commute. Thus the quantum nature of spins, together with the presence of the Fermi surface which make scattering processes essentially two-dimensional, are responsible for the Kondo effect (as $T \to 0$ the electrons scatter only at the Fermi surface, i.e. their momenta are confined to a two-dimensional manifold, to the Fermi surface). Actually it is this two-dimensionality which gives the logarithm in (13.16); in that sense the situation is analogous to the formation of the well-known Cooper pairs in superconductivity – the 'bound state' of two electrons which attract each other but can scatter only above the Fermi energy.

This analogy can actually be made even closer. One can interpret the results obtained above as a tendency to form a singlet 'bound state' of the conduction electrons with the localized spin. Indeed we can obtain such a 'bound state' with the 'binding energy'

$$E_\text{b} \simeq \varepsilon_\text{F}\, e^{-1/J\rho} \tag{13.20}$$

if we consider the scattering of an electron outside the Fermi surface on a localized spin. Mathematically the corresponding treatment is analogous to the general treatment of the formation of impurity states given in Section 6.5. The difference is that here we have to sum in the equation of type (6.133) over unoccupied states $q > p_\text{F}$, which makes the problem essentially two-dimensional. Of course, this 'bound state' is not an actual bound state: the energy $E_\text{b}$ (13.20) would correspond to a decrease of energy relative to $\varepsilon_\text{F}$, and for $E_\text{b} \ll \varepsilon_\text{F}$ this level would still overlap with the continuum and would have a certain width. It should rather be treated as a *resonance* which for $T_K \sim E_\text{b} \ll \varepsilon_\text{F}$ lies actually at the Fermi surface or very close to it. This resonance is called Kondo resonance, or sometimes Abrikosov–Suhl resonance. Actually it describes the *screening* of localized spins: due to the antiferromagnetic interaction (13.11) the electrons with spins opposite to $\boldsymbol{S}$ have a tendency to come closer to the impurity, forming a screening cloud with the typical size $\xi = \hbar v_\text{F}/T_K$. At high temperatures such screening is not efficient, and the system behaves as a real localized magnetic impurity, but with decreasing temperature the screening becomes more and more efficient until it looks as though the localized magnetic moment completely disappears at $T \ll T_K$ (when we look

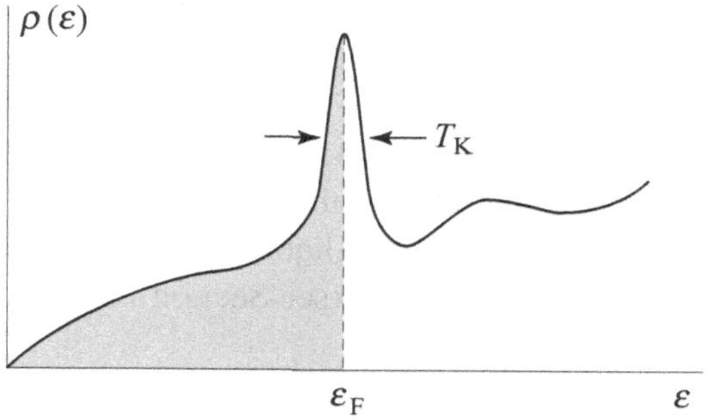


Fig. 13.8

from outside). Thus qualitatively we may say that the essence of the Kondo effect is the screening of the localized magnetic moment of an impurity by conduction electrons, occurring at low temperatures $T \lesssim T_K$.

What then is the state of the system as $T \to 0$? It is now established that in normal cases (e.g. for an impurity spin $S = \frac{1}{2}$ interacting with a nondegenerate band) the limiting behaviour is again that of the Fermi liquid, but with strongly renormalized parameters. The Kondo temperature and the corresponding energy scale $\sim T_K$ actually play the role of a new effective 'Fermi energy'. As a result all the usual expressions for thermodynamic properties presented at the beginning of Chapter 7 remain valid, but with $\varepsilon_F \to T_K$, $\rho(\varepsilon_F) \to 1/T_K$, etc. Thus, e.g. the specific heat due to such Kondo impurities tends to $c(T) \sim T/T_K$, and the susceptibility $\chi|_{T<T_K} \simeq \mu_B^2/T_K$, so that the Wilson ratio $R_W = \pi^2\chi/(3\mu_B^2 c/T)$ remains $\sim 1$. (Of course all these contributions are proportional to the concentration of Kondo impurities.)

Qualitative interpretations of this behaviour may be given using the picture of Kondo resonance: the low-temperature behaviour of this system is that of a Fermi system with a narrow peak in the density of states (Kondo resonance) very close to the Fermi energy, see Fig. 13.8 (of course the intensity, the area, or the number of states in this peak is proportional to the concentration of such impurities). The width of this peak is $\sim T_K$. At $T \ll T_K$ we feel a large density of states, which is reflected, e.g. in the large $\gamma$-value ($\gamma = c(T)/T \sim 1/T_K$). However, when the temperature becomes larger than $T_K$, the region $\sim T$ around $\varepsilon_F$ contributes to the thermodynamic properties such as specific heat, etc., i.e. we have to average the density of states shown in Fig. 13.8 over the interval $\sim T$ around $\varepsilon_F$, and the average density of states decreases with increasing temperature (actually the Kondo peak itself starts to disappear).

The theoretical problem of how to describe the cross-over from the magnetic behaviour at high temperatures to a nonmagnetic Fermi liquid as $T \to 0$ turned out to be very difficult, and a lot of effort was required to reach a full understanding.

And even now the Kondo effect still brings about some surprises. In particular, relatively recently it was realized that there exist certain special situations (e.g. the so-called multichannel Kondo effect – the situation when the total number of degrees of freedom of conduction electrons is larger than the spin of the impurity) in which the behaviour is very different from the conventional one. This multichannel Kondo effect can lead, e.g. to non-Fermi-liquid behaviour as $T \to 0$, a problem which now attracts considerable attention (see Section 10.2).

## 13.3 Heavy fermion and mixed-valence systems

In the previous section we have considered the case of an isolated magnetic impurity. We have discussed predominantly the situation when the impurity level lies deep below the Fermi energy, so that the total occupation of this level is close to integer (e.g. $\sim$1), and only spin degrees of freedom remain. It is very interesting to lift these restrictions and to study concentrated systems in which the d or f levels lie relatively close to $\varepsilon_F$ or even cross it. It turns out that in this situation we meet many very interesting phenomena, not all of which are now understood well enough. These are the phenomena of mixed valence and heavy fermions. There exist a lot of materials belonging to this class, especially rare earth (4f) and actinide (5f) compounds. Typical examples of such systems are, e.g. $CeCu_6$, $CeAl_3$, $UBe_{13}$, $U_2Zn_{17}$, and many others. As the heavy fermion and mixed-valence states are predominantly found in rare earth (4f) and in actinide (5f) compounds, we will speak below about f electrons, using also the corresponding notation.

The properties of these materials are very rich. There are among them normal metals with very large effective mass, $m^* \sim 10^2$–$10^3$ electron mass, and with huge linear specific heat $c = \gamma T$, $\gamma$ $(\sim m^*) \simeq 10^3\,\text{mJ/mole} \cdot \text{K}^2$ (in ordinary metals like Al or Cu $\gamma \sim 1\,\text{mJ/mole} \cdot \text{K}^2$); examples are $CeCu_6$, $CeAl_3$. In other materials heavy fermion behaviour coexists with magnetic ordering, but often with very small magnetic moment ($CeAl_2$, $U_2Zn_{17}$). There also exist heavy fermion superconductors ($CeCu_2Si_2$, $UPt_3$, $UBe_{13}$) with rather unusual properties, most probably having unconventional ($d$-wave or $p$-wave) pairing. (Actually the study of superconductors with unconventional pairing, which is now so popular for example in high-$T_c$ superconductors, was started before the discovery of HTSC in cuprates, in connection with superconductivity in heavy fermion systems.) There exists among these systems also a relatively small number of insulating materials with small energy gap – the so-called Kondo insulators ($SmB_6$, $YbB_{12}$, CeNiSn). Some of these substances display very interesting phase transitions as a function of temperature, pressure, etc. at which the electronic structure and other properties change drastically (the valence-change transitions), e.g. the $\gamma$–$\alpha$ transition in Ce, 'black–gold' transition in SmS, etc. And finally, many of these systems display quantum critical points and non-Fermi-liquid behaviour.

All specific properties of these materials are connected with the interplay of strongly correlated electrons and the electrons of wide conduction bands. Thus we are dealing here with the generalization of the Anderson or Kondo models (13.1) and (13.11) of magnetic impurities to the case of concentrated systems. We thus obtain the so-called Anderson lattice

$$\mathcal{H} = \sum_{\boldsymbol{k},\sigma} \varepsilon_{\boldsymbol{k}} c^{\dagger}_{\boldsymbol{k},\sigma} c_{\boldsymbol{k},\sigma} + \varepsilon_f \sum_{i,\sigma} f^{\dagger}_{i\sigma} f_{i\sigma} + U \sum_{i} f^{\dagger}_{i\uparrow} f_{i\uparrow} f^{\dagger}_{i\downarrow} f_{i\downarrow}$$

$$+ \sum_{i,\boldsymbol{k},\sigma} \left( V_{i\boldsymbol{k}} c^{\dagger}_{\boldsymbol{k},\sigma} f_{i\sigma} + \text{h.c.} \right) , \qquad (13.21)$$

or Kondo lattice

$$\mathcal{H} = \sum_{\boldsymbol{k},\sigma} \varepsilon_{\boldsymbol{k}} c^{\dagger}_{\boldsymbol{k},\sigma} c_{\boldsymbol{k},\sigma} + \sum J_{i\boldsymbol{k}\boldsymbol{k}'} c^{\dagger}_{\boldsymbol{k},\sigma} \boldsymbol{\sigma}_{\sigma\sigma'} c_{\boldsymbol{k}',\sigma'} \cdot \boldsymbol{S}_i , \qquad (13.22)$$

where $i$ is the lattice index, and where we have used for correlated electrons the notation $f$, $f^{\dagger}$ to stress that we predominantly have in mind f electron systems.

We see that here, besides all the difficult problems we have met for the usual Anderson or Kondo impurities, we also have to take into account the effects connected with the presence of *many* such f sites. As a result the properties of these systems are much richer than in the impurity case.

We cannot cover here all aspects of the theory of these systems; they can be found in special books and reviews, e.g. Hewson (1993). We will give here only the basic scheme which permits one to systematize the behaviour of these systems in different regimes.

The f–f repulsion $U$ is usually very large, and we can take it as the biggest parameter in our model, as well as the bandwidth $W$ ($\sim\varepsilon_{\mathrm{F}}$) of conduction electrons. Consider first the case of a deep f level, $\varepsilon_f \ll \varepsilon_{\mathrm{F}}$. In this case the f level is occupied by one electron, and we can go over from the Anderson lattice model (13.21) to the Kondo lattice (13.22).

There are two physical effects (and, respectively, two possible regimes) in this case. One of them is the RKKY exchange interaction between f electrons mediated by conduction electrons, see (9.29). This interaction would lead to a certain magnetic ordering at the critical temperature

$$T_{\mathrm{c}} \sim J_{\mathrm{RKKY}} \sim \rho(\varepsilon_{\mathrm{F}})\, J^2 \sim J^2/\varepsilon_{\mathrm{F}} , \qquad (13.23)$$

where $J$ is the s–f exchange in (13.22). However, there also exists the opposing tendency – the tendency to screen out the magnetic moment and to create a nonmagnetic ground state due to the Kondo effect at each site. The scale of this effect is given by the Kondo temperature

$$T_{\mathrm{K}} \sim \varepsilon_{\mathrm{F}}\, e^{-1/J\rho(\varepsilon_{\mathrm{F}})}\,, \tag{13.24}$$

see (13.19).

When $T_{\mathrm{c}} > T_{\mathrm{K}}$, the RKKY interaction 'wins', and the ground state will be magnetically ordered. Such are most of the rare earth metals (Gd, Er, etc.) and their compounds. However, if $T_{\mathrm{K}} > T_{\mathrm{c}}$, then the magnetic moments 'disappear' with decreasing temperature at $T \sim T_{\mathrm{K}}$ – before they have a chance to order. In this case we would end up in a nonmagnetic state. In analogy with the impurity Kondo effect the energy scale will be given here by $T_{\mathrm{K}}$, so that the thermodynamic properties ($c(T)$, $\chi(T)$) will be similar to those of the Kondo impurities, with the important difference that we now have such 'impurities' at each site, i.e. we have not 1, but $10^{22}$ of them. Consequently all anomalous properties will be strongly enhanced (in dilute systems they are of course proportional to the impurity concentration).

Indeed, as we have discussed in the previous section, cf. Fig. 13.8, now these systems behave at low temperatures as systems with extremely high density of states at the Fermi energy. Consequently these systems behave at low temperatures as Fermi liquids, but with a huge enhancement of the effective mass, reaching $\sim 10^3\, m_0$, and all this in systems with $10^{22}$ such states. This is the *heavy fermion* state. All the anomalies in this regime are really giant, and, as one used to say, these systems present 'a paradise for experimentalists' – but simultaneously 'a nightmare for theoreticians'.

An important difference with respect to the impurity case is seen in the behaviour of the resistivity and in other transport properties. Whereas for the Kondo impurity the resistivity behaves as shown in Fig. 13.7, i.e. with decreasing temperature it increases and saturates at $T \ll T_{\mathrm{K}}$, in regular systems such as the Kondo lattice there should be no residual resistivity (if there are no extra impurities). Thus, $R(T)$ should decrease and formally should go to zero for $T \to 0$. Experimentally this is indeed the case: the typical behaviour of the resistivity in heavy fermion systems looks as shown in Fig. 13.9.

Indeed at high temperatures $T \gg T_{\mathrm{K}}$ each f site acts independently, and the whole system may be visualized as a collection of independent impurities, with the Kondo-type behaviour of resistivity of Fig. 13.7. However, at low enough temperatures a new coherent regime will be formed, and as $T \to 0$ the system behaves as a Fermi liquid with very low degeneracy temperature $T^*$ and high effective mass $m^* \gg m$. The resistivity at $T < T^*$ is mainly determined by the electron–electron scattering, and according to the general treatment given in Chapter 10, it behaves as $R(T) = (T/T^*)^2 = AT^2$, see (10.7). The coefficient $A \sim (1/T^*)^2$ scales with the coefficient $\gamma$ in the linear specific heat, $\gamma \sim 1/T^*$. An important problem, yet unsolved, is whether the coherence energy scale $T^*$ coincides with

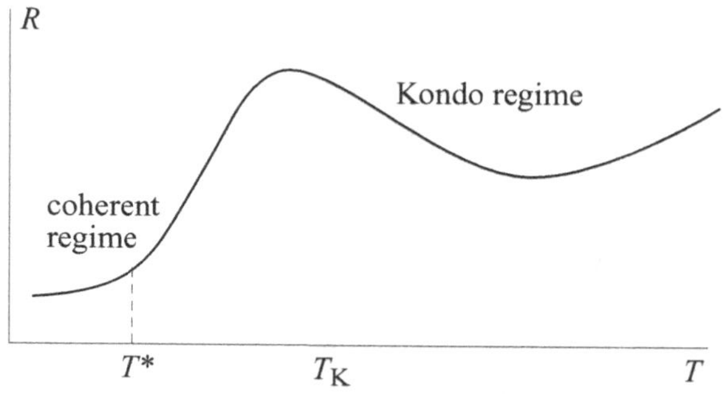


Fig. 13.9

the Kondo temperature for an isolated impurity $T_K$, or whether they are different. Experimentally, in most cases, the $T^2$-law of resistivity is observed at temperatures much lower than the single impurity $T_K$, but whether it is because of the existence of two different energy scales with $T^* < T_K$, or is simply due to a large cross-over region between the well-developed Fermi-liquid behaviour with $R \sim T^2$ and the Kondo behaviour at $T > T^*$, is not really clear.

We have stated above that the Kondo-like, or heavy fermion behaviour can be observed if $T_K > T_c \sim J_{RKKY}$. As $J_{RKKY} \sim J^2$ and $T_K \sim \exp(-1/\rho J)$, this is possible for large enough exchange integral $J$; at small $J$ the magnetic regime always wins. If we now remember that according to our previous treatment, see (13.12), for large $U$ (e.g. $U \to \infty$) we have

$$J = \frac{2V^2}{\varepsilon_F - \varepsilon_f}\,, \tag{13.25}$$

we see that $J$ increases when the f level approaches $\varepsilon_F$. Of course, we cannot use the expression (13.25) for $\varepsilon_F - \varepsilon_f \to 0$; in this case the corresponding derivation is not valid, but the qualitative tendency given by equation (13.25) is still correct. Thus we can draw a 'phase diagram' of our system in the plane $(\varepsilon_f, T)$ (the Doniach phase diagram), see Fig. 13.10. In our previous treatment we have considered the situation with the (almost) integral occupation of f levels, and we have used the concepts of localized moments, Kondo effect, etc. This picture is valid if the f levels lie still relatively deep below $\varepsilon_F$. What would happen if we increase $\varepsilon_f$? This is illustrated in the Doniach phase diagram.

We start with a very deep f level, $\varepsilon_f \ll \varepsilon_F$. As we have discussed above, in this case $J$ (13.25) is very small, and $T_c \sim J_{RKKY} \sim \rho(\varepsilon_F)J^2 \sim J^2/\varepsilon_F$ (13.23) is much larger than the Kondo temperature $T_K \sim \varepsilon_F e^{-1/J\rho(\varepsilon_F)}$ (13.19). In this regime the system is magnetically ordered at low temperatures. Such are the majority of rare earth metals and compounds.

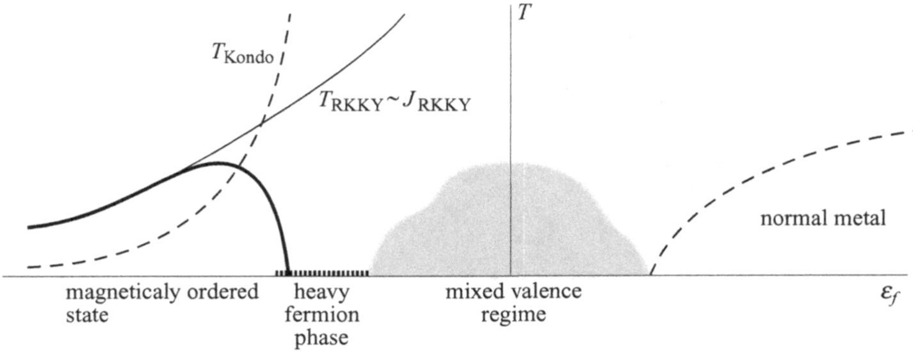


Fig. 13.10

When we move the f level closer to the Fermi level, the Kondo energy scale starts to gain, and it may become comparable to or even exceed $T_c$. One can get this regime for a still relatively deep f level, with occupation $n_f$ close to one. This is the heavy fermion regime. The critical temperature of magnetic ordering by then starts to decrease because the magnetic moments themselves are reduced due to Kondo screening, and we see that we can naturally get here the situation with $T_c \to 0$, which is the typical situation with quantum critical points. That is why many heavy fermion systems show quantum critical behaviour.

When we move the f level still further up and reach the situation when the f level $\varepsilon_f$ itself approaches the Fermi level and $\varepsilon_F - \varepsilon_f$ becomes comparable with the f level width $\Gamma = \pi \rho V^2$ (13.7), we cannot use the Kondo lattice model any longer, and have to go back to the Anderson lattice Hamiltonian (13.21). The occupation of the f level in this case will be less than 1 (or in general will be noninteger) – we enter the regime of *mixed valence*. (By mixed, or *intermediate valence* we mean here just that – noninteger occupation of the usually deep f level with a small radius of f orbitals, which can normally be treated as belonging to the ionic core and which do not participate in chemical bonding.) Here we have the situation intermediate between two different occupations of the f shell. For example, in compounds like $CePd_3$ the average f shell occupation is neither $4f^1$, corresponding to valence $Ce^{3+}$, nor $4f^0$, which would correspond to $Ce^{4+}$, but has an intermediate value.

Intermediate valence means that there are quantum fluctuations between different configurations, so that the ground state is a superposition of states with different distributions of electrons between the f orbitals and the conduction band, e.g. $|\Psi\rangle = \alpha|f^1c^n\rangle + \beta|f^0c^{n+1}\rangle$. The average occupation $\langle n_f\rangle$ then is the weight $|\alpha|^2$ with which this configuration is represented in the total wavefunction. The mixing is due to the hybridization term $Vc^\dagger f$ in (13.21). However, the difficult theoretical problem is how to hybridize correlated bands. The f and c electrons have completely different properties. The f electrons are strongly correlated: because of strong

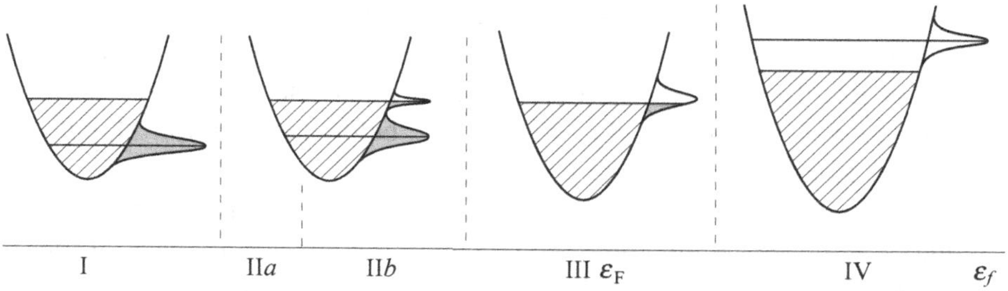


Fig. 13.11

Hubbard f–f repulsion we can put at each f level not more than one electron. At the same time, conduction electrons are ordinary fermions, and each level may be occupied by *two* electrons, with spins up and down. Thus if we have to hybridize two such different bands, with different 'capacities' and actually with different statistics (ordinary Fermi statistics for c electrons and 'atomic' statistics for f electrons), then we cannot say in advance, for example, how many electrons we can put in the hybridized band: is it one electron? or two? or 1.5? This is of course a very crude way to formulate the emerging problems, but it essentially reflects the difficulties encountered in the theory of mixed valence and heavy fermion compounds. This field, as well as other topics in the physics of correlated electrons, is now at the forefront of condensed matter theory.

The properties of the mixed-valence phase resemble those of the heavy fermion (dense Kondo) phase. Notably, the magnetic susceptibility is Curie-like at $T > T^* \sim \Gamma$ (13.7), and saturates at the value $\sim \mu_B^2/\Gamma$ for $T \to 0$. The difference is that usually, according to equations (13.12), (13.19), the Kondo scale $T_K$ increases with decreasing $\varepsilon_F - \varepsilon_f$, but remains small. In the mixed-valence regime the appropriate scale is larger: if for heavy fermions usually $T_K \sim 1$–$10\,K$, for mixed-valence compounds typically $T^* \sim 10^3\,K$.

And finally, if we continue to raise the f level, sooner or later it will be above the Fermi energy $\varepsilon_F$; all f electrons would then 'spill out' into the conduction band, and we will end up with a normal nonmagnetic metal with empty f levels. Thus, the phase diagram displaying the sequence of events as the f level moves upwards may be finally drawn as in Fig. 13.10, and the evolution of the energy structure with $\varepsilon_f$ would look as shown in Fig. 13.11:

Region I: Magnetic metal with localized magnetic moments. The f level lies deep below the Fermi level. Usually this state is magnetically ordered at low temperatures.

Region II: Dense Kondo system. The f level with width $\Gamma = \pi\rho V^2$ is closer to $\varepsilon_F$. There exists a collective Kondo peak in the density of states close to $\varepsilon_F$, with width $\sim T_K \simeq \varepsilon_F e^{-1/\rho J}$. This region is divided into two subregions, II*a* and II$b$. In region II$a$, Kondo-like behaviour may coexist with magnetic ordering of the remaining (often strongly reduced) magnetic moments. In region II$b$, the nonmagnetic (heavy Fermi liquid) state becomes stabilized down to $T = 0$. The f level $\varepsilon_f$ itself is still below the Fermi level, but there appears Kondo resonance at $\varepsilon_F$.

Region III: Mixed-valence region. The ground state is the nonmagnetic Fermi liquid, with moderately enhanced susceptibility and specific heat. It may be visualized as a fluctuating valence state (quantum fluctuations). This regime is reached when the f level itself is close to the Fermi level.

Region IV: Normal nonmagnetic metal with an empty f shell (f level above the Fermi level).

This is a qualitative scheme which describes the general behaviour of concentrated systems with coexistent correlated and ordinary itinerant (metallic) electrons. We cannot discuss here the many very interesting phenomena occurring in the heavy fermion and mixed-valence compounds. Suffice it to say that due to the very large effective mass and low characteristic temperature $T_K$ or $T^* \sim 1$–$100\,$K (which plays the role of the Fermi temperature $T_F \sim \varepsilon_F$ for normal electrons) we have here a unique possibility to study experimentally both the regimes with $T \ll T^*$ (degenerate electrons) and $T \gg T^*$ (nondegenerate regime) in one system, which for other systems would require extreme conditions (extremely high pressures and temperatures, magnetic fields, etc.). Thus, besides their own interest, the heavy electrons in these systems give us a unique opportunity to model in experimentally accessible conditions the behaviour of general condensed matter in extreme regimes.

## 13.4 Kondo insulators

Despite the large variety of properties, all different possible states of the heavy fermion and mixed-valence systems described above, see, e.g. Fig. 13.11, were metals – maybe with very strongly renormalized parameters, very large effective mass, with rather unusual properties, but still metallic. There exists, however, a small group of these compounds which have very small gaps in their energy spectrum, i.e. which are strictly speaking insulators. Several examples were already mentioned above. They are usually called Kondo insulators, although some of them, e.g. 'gold' SmS, should be better called mixed-valence insulators. What is so specific in these materials, which makes them fundamentally different from the other mixed-valence and heavy fermion systems?

When analysing their electronic structure, we can notice one common feature: these are the materials which, if the f level had integral occupation, would have been insulators. Thus, in ordinary conditions SmS ('black' SmS), containing $Sm^{2+}$

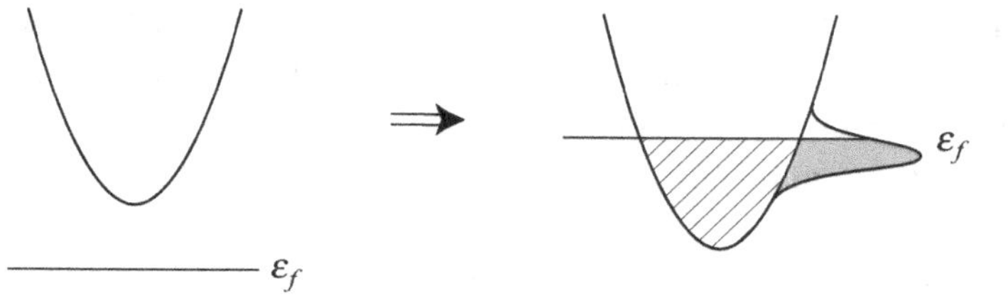


Fig. 13.12

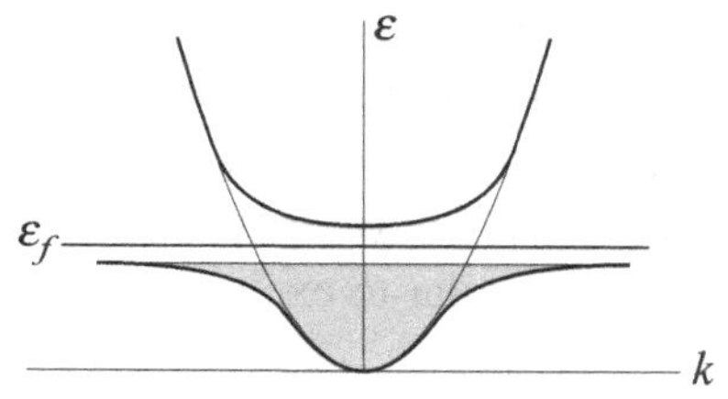


Fig. 13.13

with the configuration $4f^6$, is a semiconductor with the energy gap $E_g \simeq 0.2$–$0.3\,$eV. Under pressure it undergoes a phase transition into a 'gold' modification which looks like, and for a long time was treated as, a metal; the corresponding transition was treated as an insulator–metal transition. However, more detailed studies have shown that the high-pressure 'gold' phase is in fact a small-gap semiconductor. The standard explanation of the transition of SmS is that under pressure the f level crosses the bottom of the initially empty conduction band, see Fig. 13.12, and there appear electrons in the conduction band. Note, however, that the situation here is rather specific: all the electrons in the conduction band are actually 'former f electrons', so that their number is always equal to the number of 'empty places', holes in the f level, $n_{\text{cond.}} = n_{f\,\text{holes}}$. This is just the situation which is most favourable for the creation of a gap in the spectrum.

There are several factors which can lead to the opening of the gap at the Fermi surface in this situation. If we simply use the f–c hybridization (the term $Vc^\dagger f$ in (13.21)) we may think that the bands would simply hybridize and 'repel' each other, as shown in Fig. 13.13. And in the case considered (the total number of electrons is integer) the lower hybridized band may be completely filled and separated by a gap from the upper empty band. That is why the gap in these insulators is sometimes called the hybridization gap.

However, this explanation has several drawbacks. First of all we cannot treat different electrons on an equal footing and consider the usual band hybridization as though f electrons are uncorrelated. As we have already mentioned above, the problem of 'how to hybridize correlated electrons' is one of the main problems in the entire field of heavy fermion and mixed-valence materials. The second difficulty is that even if we could obtain the hybridized spectrum of this kind, usually due to different dispersion in different directions the actual gap would not form.

Thus we have to look for an alternative explanation of the insulating nature of these materials. The fact that the number of conduction electrons here is equal to the number of the remaining f holes is still very important, but it can play a different role. We notice that the situation here is rather similar to the one in excitonic insulators, considered in Section 11.4. As in that case, we can expect here the formation of 'excitons' due to the attraction of s electrons to f holes. As we have here an equal number of them, each of the conduction electrons can find a 'mate', and the formation of such a state would produce an insulator. In this picture Kondo insulators would be very similar to excitonic insulators, and the gap would have a collective nature.

There is yet another picture which can in principle lead to an insulating ground state. When we considered above the Kondo lattice, we did not specify what was the relative concentration of the localized moments $n_{\text{spin}}$ and of the conduction electrons $n_{\text{c}}$; or rather we always assumed that $n_{\text{c}} > n_{\text{spin}}$. In that case, even if the Kondo effect dominates, we would get a (nonmagnetic) metallic ground state, albeit with strongly renormalized parameters. However, if $n_{\text{c}} = n_{\text{spin}}$, again we would have a similar situation: each conduction electron would bind to one spin, and in effect we may get an insulator. (This is actually what gave the name 'Kondo insulator' to these systems.) This would definitely be the case if the f–s exchange interaction $J$ and the binding energy of the Kondo singlet $\sim T_{\text{K}}$ were very large, larger than the bandwidth of conduction electrons. Usually, however, we are in the opposite limit, and the result is not so evident. Theoretically this problem is still far from being solved: however, qualitatively it seems that the condition discussed above (equal number of conduction electrons and localized spins of f holes) is a necessary (and sufficient?) condition for the formation of Kondo insulators, and that the physical mechanisms discussed above are indeed responsible for the creation of this state.

## 13.5 Ferromagnetic Kondo lattice and double exchange mechanism of ferromagnetism

As we have seen in the previous sections, there exists in the Kondo lattice a competition between the nonmagnetic (e.g. heavy fermion) and magnetically ordered ground state. Magnetic ordering in this case is realized due to the RKKY exchange interaction (9.29) of localized spins via conduction electrons

$$\mathcal{H} \sim J_{\text{RKKY}} \boldsymbol{S}_0 \cdot \boldsymbol{S}_r \ , \qquad J_{\text{RKKY}}(r) \sim \frac{J^2}{\varepsilon_{\text{F}}} \frac{\cos(2k_{\text{F}}r + \varphi)}{r^3} \ . \tag{13.26}$$

What if we have a system with a small number of conduction electrons, $k_{\text{F}} \to 0$? It is clear from (13.26) that the interaction between spins will be nonoscillating (or rather oscillating with a very large wavelength), and at least for the ferromagnetic sign of $J_{\text{RKKY}}$ we can expect the formation of net ferromagnetic ordering.

Actually in this case we cannot use the standard treatment leading to the RKKY interaction: this treatment was based on perturbation theory in $J/\varepsilon_{\text{F}}$, and for small electron density (small $\varepsilon_{\text{F}}$) this approach breaks down. One can nevertheless treat this case proceeding 'from the opposite end' – treating the on-site exchange between conduction electrons and localized spins (often called sd-exchange) as strong and taking into account the electron motion (the kinetic energy of localized electrons) later. Typically one considers in this case not an antiferromagnetic, but a ferromagnetic on-site exchange

$$\mathcal{H}_{\text{H}} \sim -J_{\text{H}} \boldsymbol{S}_i \cdot \boldsymbol{\sigma}_i \ ; \tag{13.27}$$

physically this corresponds to the Hund's rule exchange which tends to make the spins at the same atoms parallel.[1] In this case we have a system which can be called a ferromagnetic Kondo lattice; it is described by the model (13.22), but with the opposite (ferromagnetic) sign of the exchange coupling. In application to ferromagnetic metals, this model is known as the *double exchange* model.

Suppose that in this case the interaction $J_{\text{H}}$ is larger than the conduction electron bandwidth. It is convenient in this case to go back to the coordinate representation and rewrite the Hamiltonian (13.22) (in the tight binding approximation) as

$$\mathcal{H} = -t \sum c^{\dagger}_{i\sigma} c_{j\sigma} - J_{\text{H}} \sum c^{\dagger}_{i\sigma} \boldsymbol{\sigma} c_{j\sigma} \cdot \boldsymbol{S}_i \ . \tag{13.28}$$

One can also have an interaction between localized spins on different sites, due to some other mechanism not connected with the conduction electrons; it often has the antiferromagnetic sign. Then the total Hamiltonian will have the form

$$\mathcal{H} = -t \sum c^{\dagger}_{i\sigma} c_{j\sigma} - J_{\text{H}} \sum c^{\dagger}_{i\sigma} \boldsymbol{\sigma} c_{j\sigma} \cdot \boldsymbol{S}_i + J \sum \boldsymbol{S}_i \cdot \boldsymbol{S}_j \ . \tag{13.29}$$

If there are no conduction electrons, the material would be antiferromagnetic due to the exchange interaction $J > 0$. However, one can easily see that the inclusion

[1] This approach is definitely valid for systems in which both localized and itinerant electrons belong to the same partially filled d shell (to different d subbands thereof). This is the situation we are dealing with when considering, e.g. ferromagnetism in metals like Fe or Ni and in compounds of the type $La_{1-x}Ca_xMnO_3$. When, however, localized and itinerant electrons belong to different ions or have a completely different nature, e.g. f electrons of Ce and conduction electrons in materials like $CeAl_3$, the exchange is of Schrieffer–Wolf type (13.12) and is antiferromagnetic.

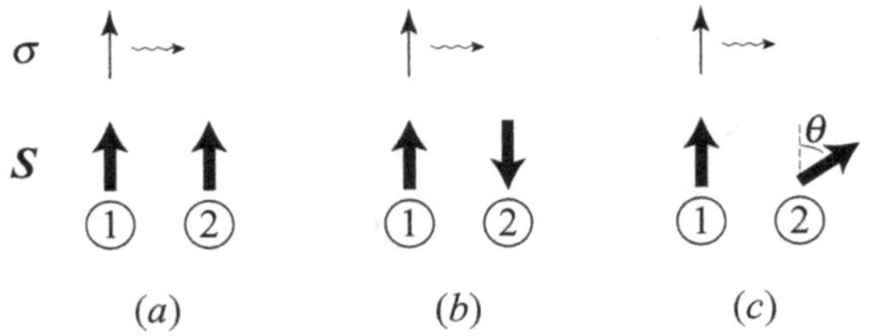


Fig. 13.14

of conduction electrons (e.g. due to doping) would lead to an opposite tendency, trying to make the system ferromagnetic.

The easiest way to see this is to consider the situation with $t \ll J_{\rm H}$, treating the spins quasiclassically. Due to the strong Hund's rule exchange (13.27) the spins of conduction electrons tend to be parallel to the localized spins. In this case we can have two situations; either the neighbouring spin is parallel, or antiparallel to the first one, see Fig. 13.14. In the first case (Fig. 13.14($a$)) the conduction electron can easily hop to the nearest neighbour (this hopping is shown by the wavy line in Fig. 13.14($a$)). Thus in this case the conduction electron can delocalize and decrease its kinetic energy. However in the second case this process is forbidden: the electron at site 2 would have spin *opposite* to the local spin there, which for $J_{\rm H} \gg t$ would prevent such hopping (the energy loss $\sim J_{\rm H}$ would exceed the energy gain $\sim t$). To decrease its kinetic energy, the conduction electron should make all localized spins parallel, i.e. it will induce ferromagnetic ordering.[2]

We can also use slightly different language and explain the tendency to ferromagnetism due to electron motion using Fig. 13.14($c$): the electron from site 1 hops to the neighbouring site 2 and there, due to strong Hund's rule interaction, 'pulls up' the spin $\mathbf{S}_2$, orienting it parallel to itself, i.e. in the result again making all its neighbours and finally the total sample ferromagnetic (of course we need a finite concentration of conduction electrons to accomplish this). This mechanism of ferromagnetism in metals is called *double exchange*, it was first suggested in the

[2] Strictly speaking, these arguments are not completely correct: if one takes into account the quantum nature of spins, one can see that the on-site exchange (13.27) does not really require that the spins of the conduction and localized electrons are *parallel*, but it tells us that the lowest state should be the one with the *maximum total spin* (e.g. for $S = \frac{1}{2}$, a triplet state with $S_{\rm total} = 1$). This state, however, besides the 'classical' possibilities $\sigma\!\uparrow\! S\!\uparrow$ and $\sigma\!\downarrow\! S\!\downarrow$, can have the $z$-projection of the total spin $S^z_{\rm total} = 0$; the corresponding wavefunction is $\frac{1}{\sqrt{2}}(\sigma\!\uparrow\! S\!\downarrow + \sigma\!\downarrow\! S\!\uparrow)$. We see that there is a part of this wavefunction which corresponds to the state $(\sigma\!\uparrow\! S\!\downarrow)$ obtained in the situation of Fig. 13.14($b$) after electron transfer. Thus the electron hopping will not be completely forbidden in this case, but will be reduced by the factor $1/\sqrt{2}$ (or $1/\sqrt{2S+1}$ in the general case) due to the normalization factor above. This can lead to certain observable consequences, though the gross picture of ferromagnetism caused by kinetic energy gain of conduction electrons remains the same.

Fig. 13.15

1940s by Zener, and then studied in detail by Anderson, de Gennes, Nagaev, and others.

One can notice that the physics of ferromagnetism in this case is rather similar to the one we saw in the single-band model (Hubbard model) in Chapter 12: in both cases the motion of electrons is hindered by the antiferromagnetic ordering of the background (due to the same electrons in Chapter 12, but due to other, localized electrons here), and the tendency to decrease the electron kinetic energy leads to the suppression of antiferromagnetic ordering and eventually to the establishment of ferromagnetism (or possibly some more complicated magnetic structure in the case of the Hubbard model, e.g. RVB-like states).

When there exists direct **S**–**S** exchange (the last term in (13.29)), these two mechanisms of magnetic ordering compete with one another. As a result a 'compromise' may be reached, e.g. in the form of a canted state: a two-sublattice structure, but with the spins of the sublattices not exactly antiparallel, but canted at a certain angle $\theta \neq \pi$, see Fig. 13.14(*c*) and Fig. 13.15. As we have seen, the electron hopping is strongly influenced by the background magnetic structure. Treating spins classically, one can get the following expression for the effective hopping matrix element:

$$t_{\text{eff}} = t \cos \frac{\theta}{2} \, . \tag{13.30}$$

Thus for the ferromagnetic ordering ($\theta = 0$) we have the full hopping $t_{\text{eff}} = t$, and correspondingly the full bandwidth of conduction electrons, and for the antiferromagnetic case ($\theta = \pi$) we have $t_{\text{eff}} = 0$ (this approximation is definitely valid for large spin $S$, but, strictly speaking, should be modified for smaller spins due to quantum effects – see the footnote above; we shall nevertheless ignore these corrections and use below the expression (13.30)).

One can easily write down the energy of the system with electron concentration $x$ ($\ll 1$) and with canted spin structure. From (13.29), (13.30) one gets (taking $J_{\text{H}} \gg t,\, J$) the energy per site

$$\frac{E}{N} = J S^2 z \cos\theta - zxt \cos \frac{\theta}{2} \, . \tag{13.31}$$

Here $z$ is the number of nearest neighbours. The first term in (13.31) is the energy of localized spins forming two sublattices at an angle $\theta$, Fig. 13.15. The second term describes the kinetic energy of doped electrons with concentration $x$, moving in a tight binding band with the hopping matrix element $t \cos\frac{\theta}{2}$ (13.30) (for $x \ll 1$ we can put all electrons at the bottom of the band and ignore the filling of higher-lying states in the band, giving the energy $\sim x^{5/3}$).

Minimizing the expression (13.31) in $\theta$, we obtain

$$\cos\frac{\theta}{2} = \frac{tx}{4JS^2} \,. \tag{13.32}$$

Thus in this approximation already at small doping $x$ the antiferromagnetic sublattices start to cant (de Gennes). At $x = x_c = 4JS^2/t$, the value of $\cos\frac{\theta}{2}$ reaches 1, i.e. the angle between the sublattices becomes zero, $\theta = 0$, and the system goes over to a ferromagnetic state. This mechanism of ferromagnetism is essentially due to the same tendency which we already saw before in Sections 12.5 and 12.6: the ferromagnetic state is here stabilized by the decrease of the kinetic energy of electrons which move much more easily in the ferromagnetic background than in the antiferromagnetic background.

The double exchange mechanism of ferromagnetism described above may well be the main mechanism of ferromagnetism in metallic ferromagnets such as iron or nickel. It is also widely used to describe the properties of metallic ferromagnetic oxides, e.g. systems with 'colossal' magnetoresistance $La_{1-x}M_xMnO_3$ ($M$ = Ca, Sr). And although many details of the behaviour of such systems are still not clear[3] the tendency which we saw above – that the ferromagnetic state usually goes hand in hand with metallic conductivity, and the antiferromagnetic state is more typical for insulators – is definitely observed in most of the magnetic materials (although of course there are also many exceptions to this general tendency).

---

Concluding this book, I want to say once again that I have tried to give here at least an impression of the problems physicists are now working on in condensed matter physics, the language used and the methods employed. Of course I could not

[3] Besides the possible importance of quantum effects mentioned above, there exists in the double exchange model (13.29) the tendency to an instability of the homogeneous canted state towards phase separation into metallic ferromagnetic regions containing all the doped electrons, and antiferromagnetic insulating regions. This tendency is again similar to the one we saw in the single-band Hubbard model, Section 12.7: one easily sees that the total energy of the homogeneous canted state which we would obtain by putting the value $\cos\frac{\theta}{2}$ (13.32) into the expression (13.31), would be such that $d^2E/dx^2 < 0$. But this is nothing else but the inverse compressibility of the system. A negative value of the compressibility is forbidden by the general rules of thermodynamics; it means absolute instability of the corresponding homogeneous state. There are, however, other contributions to the total energy of the system besides the one considered above, notably the long-range Coulomb forces which favour electroneutrality and thus oppose the tendency to phase separation (which nevertheless seems to persist, albeit in a modified form) – see the discussion in Section 12.7.

cover all the interesting and exciting problems in this big field, as well as introduce all the theoretical methods; for instance I have left out the big and very important field of transport phenomena. I hope, however, that I have been able to present some of the unifying concepts underlying modern condensed matter physics, especially the concepts of order and elementary excitations. I have also tried, and I hope was able to succeed to some extent, to show, that this relatively old and 'classical' part of physics is still full of surprises, 'alive and kicking'.

# Bibliography


A. A. Abrikosov, L. P. Gor'kov and I. E. Dzyaloshinsky, *Methods of Quantum Field Theory in Statistical Physics*, Dover, New York, 1975.

N. W. Ashcroft and N. D. Mermin, *Solid State Physics*, Harcourt College, New York, 1976.

G. Baym and Ch. Pethick, *Landau Fermi-liquid Theory: Concepts and Applications*, Wiley-Interscience, New York, 1991.

P. M. Chaikin and T. C. Lubensky, *Principles of Condensed Matter Physics*, Cambridge University Press, Cambridge, 2000.

S. Doniach and E. H. Sondheimer, *Green's Functions for Solid State Physicists*, Benjamin/Cummings, New York, 1974.

P. Fazekas, *Lecture Notes on Electron Correlations and Magnetism*, World Scientific, Singapore, 1999.

A. L. Fetter and J. D. Walecka, *Quantum Theory of Many-Particle Systems*, Dover, New York, 2003.

A. Georges, G. Kotliar, W. Krauth and M. Rozenberg, *Rev. Mod. Phys.*, **68**, 13 (1996).

J. B. Goodenough, *Magnetism and the Chemical Bond*, Interscience, New York, 1963.

A. C. Hewson, *The Kondo Problem for Heavy Fermions*, Cambridge University Press, Cambridge, 1993.

D. I. Khomskii, in *Spin Electronics*, ed. M. Ziese and M. J. Thornton, Springer, Berlin, 2001, p. 89.

C. Kittel, *Quantum Theory of Solids*, John Wiley, New York, 1987.

C. Kittel, *Introduction to Solid State Physics*, John Wiley, New York, 2004.

K. I. Kugel and D. I. Khomskii, *Sov. Phys. Uspekhi*, **25**, 231 (1982).

L. D. Landau and I. M. Lifshits, *Statistical Physics*, Addison-Wesley, Reading, MA, 1969; 3rd edn Part 1, Butterworth–Heinemann, London, 1980.

L. D. Landau and I. M. Lifshits, *Quantum Mechanics: Non-Relativistic Theory*, 3rd edn, Butterworth–Heinemann, London, 1977.

L. D. Landau and I. M. Lifshits, *Fluid Mechanics*, Butterworth–Heinemann, 1987.

L. D. Landau and Ya. B. Zeldovich, ZhETF, **32**, 1944 (1943a).

L. D. Landau and Ya. B. Zeldovich, Acta Phys. Chem. USSR, **18**, 194 (1943b).

H. von Löhneisen, A. Rosch, M. Vojta and P. Wölfle, *Rev. Mod. Phys.*, **79**, 1015 (2007).

G. D. Mahan, *Many-Particle Physics*, Springer, Berlin, 2000.

M. P. Marder, *Condensed Matter Physics*, Wiley-Interscience, New York, 2000.

R. D. Mattuck, *A Guide to Feynman Diagrams in the Many-Body Problem*, Dover, New York, 1992.

R. E. Peierls, *Quantum Theory of Solids*, Oxford University Press, Oxford, 2001.
R. E. Peierls, *More Surprises in Theoretical Physics*, Princeton University Press, Princeton, NJ, 1991.
S. Sachdev, *Quantum Phase Transitions*, Cambridge University Press, Cambridge, 1999.
J. R. Schrieffer, *Theory of Superconductivity*, Perseus Books, New York, 1999.
J. S. Smart, *Effective Field Theories of Magnetism*, Saunders, New York, 1966.
H. E. Stanley, *Introduction to Phase Transitions and Critical Phenomena*, Oxford University Press, New York, 1987.
R. M. White, *Quantum Theory of Magnetism*, Springer, New York, 2006.
K. Yosida, *Theory of Magnetism*, Springer, New York, 1996.
J. M. Ziman, *Principles of the Theory of Solids*, Cambridge University Press, Cambridge 1979.
D. N. Zubarev, *Usp. Fiz. Nauk*, **71**, 71 (1960).

# Index

Printed in the United States
By Bookmasters